S0-BSZ-971

ROWAN UNIVERSITY
CAMPBELL LIBRARY
201 MULLICA HILL RD.
GLASSBORO, NJ 08028-1701

# Designs on Nature

# Designs on Nature

Science and Democracy in Europe and the United States

Sheila Jasanoff

PRINCETON UNIVERSITY PRESS  •  PRINCETON AND OXFORD

Copyright © 2005 by Princeton University Press
Published by Princeton University Press, 41 William Street, Princeton, New Jersey 08540
In the United Kingdom: Princeton University Press, 3 Market Place, Woodstock,
Oxfordshire OX20 1SY

All Rights Reserved

Library of Congress Cataloging-in-Publication Data

Jasanoff, Sheila.
    Designs on nature : science and democracy in Europe and United States / Sheila Jasanoff.
        p. cm.
    Includes bibliographical references and index.
    ISBN 0-691-11811-6 (cloth : acid-free paper)
    1. Democracy and science—Europe. 2. Democracy and science—United States. I. Title.
Q127.E8J37 2005
338.9'26—dc22
                                                                    2004055296

British Library Cataloging-in-Publication Data is available

This book has been composed in Goudy

Printed on acid-free paper. ∞

pup.princeton.edu

Printed in the United States of America

10 9 8 7 6 5 4 3 2 1

3 3001 00901 3789

# Contents

CONTENTS

# Figures and Tables

## Figures

## Tables

# Acknowledgments

This book is the product of much travel and many transitions, and the debts I owe are correspondingly various. I would like first to acknowledge two U.S. government organizations that encouraged me to look at biotechnology as a subject of political analysis long before the hot winds of politics began blowing across this field in the 1990s. The now dissolved Office of Technology Assessment (OTA) was an early supporter. The case studies I wrote for OTA's 1984 study of international developments in biotechnology and for its 1987 report on the bicentennial of the U.S. Constitution laid the foundation for an enduring interest in the politics of the life sciences. I am also extremely grateful for a grant from the National Science Foundation ("The 'New' Politics of Biotechnology: A Comparative Study," grant no. 8911157) that allowed me to convert my diffuse interests into a systematic comparative project.

Several universities and research centers provided crucial intellectual and logistical resources at key moments in the project's development. Cornell University offered a splendidly collegial home for many years of border-crossing work that included the early stages of this study. Yale Law School and Wolfson College generously supported my research during leaves in New Haven and Oxford. Two remarkable institutions—the Rockefeller Foundation's heavenly Bellagio Study Center and the incomparable Wissenschaftskolleg in Berlin— made room for imagination to flower and thought to deepen while facilitating research and writing in every possible way. The John F. Kennedy School of Government at Harvard, where I have been since 1998, offered many valued opportunities for interaction with students and colleagues during the final phases of the project.

The book has benefited greatly from the opportunities I have had to present some of its arguments at several universities. I would like to thank in particular Brown University, Iowa State University, the University of Minnesota, the University of Iceland, and the University of California at San Francisco for their hospitality and engaged responses to lectures based on this book.

A study of this scope could not have been completed without a great deal of help from many sources, and I regret that it is not possible to acknowledge all of them by name. I would like to thank Rebecca Efroymson, Peter Mostow, Xandra Rarden, and Tania Simoncelli for their superb research assistance at various stages in the book's evolution. Gesine Bottomley and her library staff at the Wissenschaftskolleg worked wonders with every request for materials, even the most off-beat and trivial, and Beata Panagopoulos helped me locate economic data on the biotech industry at the Kennedy School. I am grateful to the many scientists, government officials, and public interest representatives who generously gave their time, in some cases through repeat interviews, during my research trips in all three countries that this book compares. Many of my sources are acknowledged in references throughout the text, but I would like to single out Sue Davies and Sue Mayer for their information, advice, and friendship over the years; and Mark Cantley, whose hospitality and provocative, critical engagement were indispensable to my understanding of the European politics of biotechnology. I would also like to thank two extraordinary assistants, Deborah van Galder at Cornell and Constance Kowtna at Harvard, for their invaluable support at the beginning and end of the project.

Debts to readers are impossible to capture through a mere listing of names, especially when, as in this case, several of them are also close friends who have helped me refine my thinking about science, politics, and the meaning of scholarship over many years of sustained and sustaining interaction. However inadequately, I would like especially to thank John Carson, Robin Grove-White, Rob Hagendijk, Stephen Hilgartner, Frank Laird, Angela Liberatore, and Brian Wynne for commenting on some or all of the manuscript and, with their unfailingly accurate insights, helping me to make the book's arguments both clearer and more cogent. Martin Lengwiler, Dagmar Simon, and other colleagues at the Wissenschaftszentrum in Berlin provided much appreciated reactions that enabled me to refine my comparative approach and to sharpen some key aspects. The book has also gained immensely from two exceptionally thoughtful reader reports to Princeton University Press, to which my editor Chuck Myers added his own judicious and helpful suggestions. Needless to say, none of these readers and critics is responsible for any errors or failings in the book.

Lastly there are the debts that are too deep for words: to my late friend Sheila McKechnie, whose faith in all I did, but most particularly in this book, was a comfort through most of a working life; to Stefan Sperling, with whom the completion of the book has been an extended conversation; to my mother, who has looked forward to this book's publication almost as earnestly as I have myself; and to my family—Jay, Alan, Maya, and now Luba—whose love and not inconsiderable powers of discernment are the most constant influences in my life and work.

# Abbreviations and Acronyms

AAAS  American Association for the Advancement of Science
ACGM  Advisory Committee on Genetic Modification
*AcNPV*  *Autographa californica*
ACRE  Advisory Committee on Releases to the Environment
AEBC  Agriculture and Environment Biotechnology Commission
AGS  Advanced Genetic Sciences
ANT  actor-network theory
BÄK  Bundesärztekammer
BBI  Berlin-Brandenburg Institute for German-French Cooperation
BBSRC  Biotechnology and Biological Sciences Research Council
BIO  Biotechnology Industry Organization
BfR  Bundesinstitut für Risikobewertung (Federal Institute for Risk Assessment)
BMBF  Bundesministerium für Bildung und Forschung (Federal Ministry for Education and Research)
BMFT  Bundesministerium für Forschung und Technologie (Federal Ministry for Research and Technology)
BMGS  Bundesmininsterium für Gesundheit und Soziale Sicherung (Federal Ministry of Health)
BMVEL  Bundesministerium für Verbraucherschutz  Ernährung und Landwirtschaft (Federal Ministry of Consumer Protection, Food and Agriculture)
BSAC  Biotechnology Science Advisory Committee
BSC  Biotechnology Steering Committee
BSCC  Biotechnology Science Coordinating Committee
BSE  bovine spongiform encephalopathy
CA  Consumers' Association

CDU/CSU  Christlich-Demokratische Union/Christlich Soziale Union
(Christian Democratic Party)
CGS      Center for Genetics and Society
CNR      cell nuclear replacement
COPUS    Committee on Public Understanding of Science
CRG      Council for Responsible Genetics
CUBE     Concertation Unit for Biotechnology in Europe
DEFRA    Department of the Environment, Food and Rural Affairs
DFG      Deutsche Forschungsgemeinschaft (German Research
Foundation)
DHSS     Department of Health and Social Security
DNA      deoxyribonucleic acid
DOE      Department of the Environment
ECSC     European Coal and Steel Community
EEC      European Economic Community
EGE      European Group on Ethics
ELSI     Ethical, Legal, and Social Implications
EP       European Parliament
EPA      Environmental Protection Agency
EPC      European Patent Convention
ES cells embryonic stem cells
ESchG    Embryonenschutzgesetz (Embryo Protection Law)
EST      expressed sequence tag
EU       European Union
FAST     Forecasting and Assessment in Science and Technology
FAZ      *Frankfurter Allgemeine Zeitung*
FDA      Food and Drug Administration
FDP      Freie Demokratische Partei (Free Democratic Party)
FET      Foundation on Economic Trends
FP6      Sixth Research Framework Programme
FSA      Food Standards Agency
GAEIB    Group of Advisers on the Ethical Implications of Biotechnology
GeN      Gen-ethisches Netzwerk (Gen-Ethics Network)
GenTG    Gentechnikgesetz (Genetic Engineering Law)
GIVF     Genetics and IVF Institute
GM       genetically modified
GMAG     Genetic Manipulation Advisory Group
GMO      genetically modified organism
HFEA     Human Fertilisation and Embryology Authority
HGP      Human Genome Project
HSE      Health and Safety Executive
IVEM     Institute of Virology and Environmental Microbiology
IVF      in vitro fertilization

| | |
|---|---|
| MAFF | Ministry of Agriculture, Fisheries and Food |
| MRC | Medical Research Council |
| NAS | National Academy of Sciences |
| NBAC | National Bioethics Advisory Commission |
| NCHGR | National Center for Human Genome Research |
| NEPA | National Environmental Policy Act |
| NERC | National Environment Research Council |
| NFR | Novel Foods Regulation |
| NGO | nongovernmental organization |
| NIH | National Institutes of Health |
| NRDC | National Research Development Corporation |
| NSF | National Science Foundation |
| OECD | Organization for Economic Cooperation and Development |
| OPUS | Office on Public Understanding of Science |
| OSTP | Office of Science and Technology Policy |
| PBC | People's Business Commission |
| PCBE | President's Council on Bioethics |
| PTO | Patent and Trademark Office |
| PUS | public understanding of science |
| PUSH | public understanding of sciences and humanities |
| PUST | public understanding of science and technology |
| RAC | Recombinant DNA Advisory Committee |
| rBGH | recombinant bovine growth hormone |
| rBST | recombinant bovine somatrotopin |
| RCEP | Royal Commission on Environmental Pollution |
| rDNA | recombinant DNA |
| RKI | Robert Koch Institute |
| SET | Science, Engineering and Technology |
| SNP | single nucleotide polymorphism |
| SPD | Sozialdemokratische Partei Deutschlands (Social Democratic Party) |
| SPUC | Society for Protection of Unborn Children |
| S&TS | science and technology studies |
| USDA | U.S. Department of Agriculture |
| WTO | World Trade Organization |
| ZERP | Zentrum für Europäische Rechtspolitik (Center for European Law Politics) |
| ZKBS | Zentrale Kommission für die Biologische Sicherheit (Central Commission for Biological Safety) |

Designs on Nature

# Prologue

## Science in High Places

On a somber fall weekend in mid-November 2001, Europe was forming in the oddest of places. The scene was Genshagen, a nondescript small town with an ancient pedigree[1] just south of Berlin in the former East German state of Brandenburg. Site of the largest Daimler-Benz aircraft engine plant in wartime Germany,[2] Genshagen is now home to the Berlin-Brandenburg Institute (BBI) for German-French Cooperation, a privately supported organization dedicated to furthering cross-national exchanges in the fields of economics, politics, science, and culture. Schloss Genshagen, the institute's headquarters, provides an elegant if modest venue for consolidating the new Europe. Built in 1878 as the manor house of Baron Leberecht von Eberstein, the imposing, four-story building and its surrounding seven hectares of parkland are among the region's protected monuments. Inside, the DM 4 million (€ 2 million) renovations undertaken since BBI's founding in 1993 have restored the main salon to something like its former dignity. Whitewashed walls, tall windows, and a painted and gilded coffered ceiling form a quiet backdrop for reflecting on matters of state. German Chancellor Gerhard Schröder and French Prime Minister Lionel Jospin have attended events here, as have scores of French and German ministers, academics, and intellectuals.

Outside on this particular weekend, Genshagen was not at its most inviting: gray sky and bare branches, a faded façade, a drizzle of snow on crumbling steps and rutted drive created an atmosphere of gentle melancholy. The subject, too, was suitably grave, though not at first glance political. Officially it was a conference on basic European values in bioethics (*europäische Grundwerte in der Bioethik*), but politics lurked palpably beneath the surface. Conceived in two parts, with a follow-up scheduled for January

2002, the BBI meeting was a precursor to the German Parliament's (Bundestag's) debate at the end of January 2002 on whether embryonic stem (ES) cells could be imported into Germany. Members of the ethics advisory councils of both participating nations were meeting with leading scientists, lawyers, clerics, and politicians to consider the issue. Their task was to discuss in French and German (this was definitely a *continental* European gathering) whether Europe has a common basis for deciding if and when research in the life sciences violates fundamental human values. Is an embryo entitled to full protection under constitutional guarantees of human dignity? Do ES cells merit similar consideration? Do German and French experts agree on these points? And what consequences might there be for "Europe" if the continent's two most powerful legal cultures hold different views from Britain, their skeptical partner and ally across the English Channel?

For Germany's ruling "red-green" coalition of Social Democrats (SPD) and Greens in 2001, these questions were not only metaphysical and moral. With national elections less than a year away, the government's future was on the line, hostage to a stagnant economy, rising unemployment, an aging population with insufficient high-tech skills, and the continuing fiscal burdens of reunification. Like other Western states, Germany too was increasingly looking to technological breakthroughs to boost the economy. A seemingly arcane debate about the embryo's moral status thinly concealed programmatic concerns about the relationship of science to the state and of innovation to economic recovery. Britain's relatively unproblematic embrace of embryo research posed a particular challenge. Not for nothing did the German sociologist Wolfgang van den Daele, a member of Chancellor Schröder's National Ethics Council (Nationaler Ethikrat), take pains to defend British policy. The important point, van den Daele argued, was that Britain's decision had been reached by democratic means in an open society; it was the process, not the particular outcome, that conferred legitimacy.

But politics ran deeper at the Genshagen meeting than the Schröder government's immediate electoral concerns. At stake as well was the meaning of citizenship in the emerging politics of Europe. The question on the table was about *European* values, and by no means trivially so. If there was a transcendent European identity, would it not be defined around the very kinds of issues under consideration in that airy, dignified room? The problem of bioethics centered, after all, on finding agreed upon moral spaces for the new entities brought into the world through developments in genetics, molecular biology, and industrial biotechnology. Embryos and stem cells had to be located within a discourse of rights and duties once reserved for fully developed human beings. Are these new biological constructs continuous with our existing selves, and hence entitled to protection under already elaborated notions of individuality and personhood, or are there principled reasons to divide the conscious, reasoning, human self from these products of human invention? As the list of conference participants bore witness, it was not enough simply to produce answers. Who spoke on the issues was just as important. Religious viewpoints had to be represented, for instance, along with legal and cultural expertise, and the secular, progressive imagination of science.[3] At issue as well was the degree of internal dissension that Europe could tolerate on fundamental values and still remain, meaningfully, a single Europe. In secluded Genshagen, with its complex memories of war and reconciliation, under the improbable rubric of bioethics, participants were deliberating on the constitution of Europe for the twenty-first century.

.   .   .

*Andere Länder, andere Sitten,* the Germans like to say, or "other lands, other customs." With regard to the life sciences, however, this adage holds only partly true. Though doubts persist about the appropriate limits of biological research and development, Western nations are united in thinking that the issues deserve attention at the highest political levels. On January 22, 2001, for example, Britain's House of Lords voted to permit the cloning of human stem cells by a 120-vote majority.[4] Just a year later, on February 13, 2002, a Select Committee of the Lords issued a report endorsing research with stem cells.[5] The report's conclusions set U.K. policy apart from Germany's on some key points, as we will see later. More interesting for now is that this issue was deemed suitable at all for consideration by Britain's unelected upper legislative chamber. Simon Jenkins, conservative commentator for the *Times of London,* noted the irony of the Lords' 2001 vote, but also its oddly constitutional significance.[6] "On Monday night," Jenkins sourly observed, "British stem cell research was left in the hands of a group of people with no democratic, professional or territorial legitimacy."[7] Moral authority to speak about stem cells should not be separated, he implied, from the political authority to speak for the nation on a question of such gravity. The unelected peers, in Jenkins' view, did not possess the necessary standing.

Across the Atlantic in the United States, stem cells were also on the national political agenda, although here the issue was entangled with presidential politics. On August 9, 2001, a month before the terrorist attacks that transformed his presidency, George W. Bush delivered his first televised address to the nation. His topic was not national security, tax policy, or education, all of which had figured prominently in his lackluster 2000 election campaign; rather, it was federal funding for stem cell research. Steering between the Christian fundamentalists to his right and the business and scientific interests of the center-left, Bush announced an uneasy compromise. Federal funds would not be spent on creating new embryonic stem cells, but they could be used to fund research on cell lines that already existed. The news made headlines in the quiet days before September 11, 2001, but it was not the first time that presidential pronouncements about biological research had drawn so much media attention. Only the previous year, on March 14, 2000, President Bill Clinton and U.K. Prime Minister Tony Blair had issued a joint statement calling for "fundamental information" about the human genome to be made freely available to all researchers.[8] Inaccurate reports of their press briefing caused an immediate, precipitous drop in the price of biotechnology stocks, wiping out some $10 billion or 25 percent of their value in one day.[9] In that respect, it was quite unlike the presidential accord of March 1987, when America's Ronald Reagan and France's Jacques Chirac publicly resolved the priority dispute over which country's researchers had first identified the AIDS virus. By agreeing to share the credit, the two heads of state tacitly acknowledged the discovery's huge economic potential—and signaled their unwillingness to compromise those gains through continued uncertainty over the allocation of credit.

These vignettes dramatize the central role of science and technology in contemporary economic and social development and support sociologists' claims that we are in transition from the old industrial societies of the nineteenth and twentieth centuries to a new form of global social organization called "knowledge societies."[10] In this emerging formation, knowledge has become the primary wealth of nations, displacing natural resources, and knowledgeable individuals constitute possibly the most important form of capital. State policies, correspondingly, are geared more and more toward nurturing and exploiting knowledge, with scientific knowledge and technical expertise commanding the highest premiums. These far-reaching alterations in the nature and distribution of resources, and in the roles of science, industry, and the state, could hardly occur without wrenching political upheavals. Within the biotechnology sector alone, disagreements about the moral status of stem cells are only one of a series of flash points that also include controversies about the risks of transgenic crops, transatlantic battles over the acceptability of genetically modified (GM) foods, discord about the international management of biosafety and biodiversity, and rising worldwide concern about the limits of human genetic manipulation. The salience of these debates underlines the deeply contested character of the transition to the tightly interdependent, knowledge-dominated, high-tech economies of the twenty-first century. They also spotlight the life sciences as the sector in which these restructurings are preeminently taking place.

Conflicts about the management of biotechnology within and among nations point to wider uncertainties about the relationship of science and democracy at the threshold of the third millennium. What consequences will the shift from industrial to knowledge societies have for organized power, social stratification, and individual liberty? What will happen to core democratic values such as citizen participation and governmental accountability in such a transformation, and who will be the winners and losers? How will rapid developments in science and technology affect and transform more stable elements of national politics and culture? What will it mean for existing institutions of governance if science and technology, far from operating as objective legitimators of policy, themselves appear as catalysts of domestic and international political turmoil? And are there criteria by which we can evaluate responses across countries, in order to judge whether some are handling change more effectively, ethically, or democratically than others?

These are the questions I set out to explore in this book through a comparative study of the politics of biotechnology in Britain, Germany, the United States, and the European Union (EU). The debates of the 1970s concerning the safety of recombinant DNA (rDNA) experiments supply the background for my story, but the book's primary focus is on events from 1980 to the present. There are three main reasons for choosing this period. First, there is an extensive and multifaceted literature on the historic Asilomar conference of 1975 and its impact on the development of guidelines for rDNA research by the U.S. National Institutes of Health (NIH).[11] Later developments have received less attention from historians and sociologists of science, and still less from students of democratic theory. Second, whereas U.S. biotechnology policy led Europe's in the 1970s, distinctively European forms of politics and policymaking took shape in the 1980s, inviting systematic comparative analysis. Europe's growing economic and political integration propelled the European Commission to new activism in both sponsoring and regulating biotechnological research and development. At the same time, EU member states emerged as independent players, with their own stakes in the future of biotechnology. Third, as researchers' dreams moved closer to industrial production, the ethical, social, and environmental ramifications of biotechnology began to attract more serious attention. Monolithic positions in support of or opposition to genetic engineering dissolved into more nuanced conversations about the appropriate objectives of research and development in the life sciences. In Europe as well as the United States, new forums, actors, instruments, and discourses arose to grapple with a significantly broader and more diversified political agenda. The implications of all these transitions for democratic politics and governance are only now becoming apparent and merit careful study.

## Questions for Democracy

The political reception of biotechnology serves as a window for looking into a number of large contradictions confronting democratic governments in the twenty-first century. Science and technology have been regarded for centuries as instruments of social progress and personal liberation. Yet, as scientific knowledge becomes more closely aligned with economic and political

5

power, producing new expert elites, the distance between the governors and the governed can be expected to grow—a dismal prospect in societies where low levels of electoral participation and citizen engagement are already causes for concern. Science, moreover, has historically maintained its legitimacy by cultivating a careful distance from politics.[12] As state-science relations become more openly instrumental, we can reasonably wonder whether science will lose its ability to serve either state or society as a source of impartial critical authority. New questions about access and equality can be expected to arise as biotechnology becomes more global, as they already have in connection with existing techniques such as in vitro fertilization and promised ones such as "genetic enhancement." Will continued advances in the life sciences produce a new genetic underclass, and will they simultaneously increase the state's already immense power to define, classify, and regulate life itself?[13] These are some of the considerations that prompt a detailed investigation of the politics of the life sciences. As I hope to show, there are particular gains to be had from making this inquiry comparative.

The stories told in this book are partly about invention, both scientific and social. They relate how public and private actors in three Western nations, and to some extent the European Union, assisted in the production of new phenomena through their support for biotechnology, and how they reassured themselves and others about the safety of the resulting changes—or failed to do so. Inventiveness in the life sciences and technologies went hand in hand with institutional and procedural inventiveness in the political realm, as national actors developed new capacities for assessing and regulating the processes and products of genetic engineering. Just as importantly, though, the stories in this book are about reinvention. They show how and with what degrees of success attempts to master the concerns generated by biotechnology drew on, reproduced, or reinforced old ways of coping with hazards. In this respect, the politics of biotechnology serves as a theater for observing democratic politics in motion.

The comparative accounts in this book develop and expand upon three major arguments that have featured in my earlier work, though perhaps never with quite the centrality they are accorded here. The first is that democratic theory cannot be articulated in satisfactory terms today without looking in detail at the politics of science and technology. That contemporary societies are constituted as *knowledge* societies is, of course, an important part of the reason. It follows that important aspects of political behavior and action cluster around the ways in which knowledge is generated, disputed, and used to underwrite collective decisions. It is no longer possible to deal with such staple concepts of democratic theory as citizenship or deliberation or accountability without delving into their interaction with the dynamics of knowledge creation and use. More specifically, biological sciences and their applications have brought about ontological changes and reclassifications in

6

the world, producing new entities and new ways of understanding old ones. Such changes entail a fundamental rethinking of the identity of the human self and its place in larger natural, social, and political orders. We will see throughout the book that unexpected innovations in administrative and judicial practices, forms of citizen participation, and discourses of public persuasion happened around genetics and related areas of science and technology. Together, these developments suggest that some of the liveliness of contemporary democracy is to be found away from the polling booths, where one often looks for it in vain, in the less examined machinery of science and technology policy—that is, in technical advisory committees, court proceedings, regulatory assessments, scientific controversies, and even the ephemeral web pages of environmental groups and multinational corporations.

The book's second major argument is that, in all three countries, policies concerning the life sciences have become embroiled to varying degrees in more or less self-conscious projects of nation-building or, more accurately, projects of reimagining nationhood at a critical juncture in world history.[14] The case is clearest in Germany, where deliberation on what is at stake in biotechnology policy has been tied to two recurrent narratives of nationhood: the still unfinished project of reconstituting German identity after two world wars and the Holocaust, and more recent questions about how that identity should be articulated in the aftermath of reunification; competing and increasingly intense discussions of Europeanization only make more urgent the need to work out the meanings of German nationhood. In Britain, too, questions of national identity have been woven into the conflicts around biotechnology, although understandably in a lower key than in Germany. British debates on the life sciences were caught up with two larger sets of *fin de siècle* concerns: the reinvention of the Labour Party in the post-Thatcher years, and Britain's ongoing struggle to modernize and democratize institutions seen to be out of touch with the economic and social realities of the twenty-first century. How to regain a technological edge, and what social compromises to make or not make in that process, figured as subtexts in virtually every major structural reform initiative, from the devolution of political power to Scotland and Wales to creating an independent high court and imposing fees on university students.[15]

In Brussels, no less than in Berlin or London, dilemmas about political identity and institutional legitimacy became wrapped up with those related to biotechnology policy. In the early millennial years, the EU wrestled with, and often seemed stymied by, the problem of enlarging its borders without increasing already troubling levels of electoral apathy and distrust. "When a reality TV show attracts more votes than an election," a British journalist lugubriously observed about European electoral politics, "democracy is in trouble."[16] How much diversity could the Union tolerate and still remain a viable union? When should national values and political traditions trump policies put forward at the European level? The answers to these questions

7

were worked out in part in discussions of what to do about biotechnology—both within the EU framework and in relation to the ever-present competitive challenge from the United States.

On its face, nation-building is hardly a term one would think to apply to the United States at the turn of the millennium. Secure in its borders, and victorious in its military interventions in Afghanistan and Iraq, the United States in the early twenty-first century seemed untroubled by wrenching doubts about its territorial integrity, identity, or purpose. Yet the end of the cold war and the beginning of the "war on terror" brought a necessary reevaluation of the U.S. position in the world and a tacit renegotiation of what American democracy means in relation to a host of issues on the national and transnational political agendas.[17] Triumphalism about the market, with attendant reassertion of domestic ideologies of technological leadership and deregulation, profoundly shaped the U.S. environment for the life sciences. A conceded world leader in research and development, the United States encountered unexpected opposition in finding global markets for the early fruits of its inventiveness. Resistance to biotechnology became almost a surrogate for resisting America's imperial power. In tracing connections between the macropolitical dynamics of nationhood and the micropolitics of biotechnology in the United States or Europe, I will not argue for any simplistic notions of causality, but I will show at many points how the framing of particular debates at once fed into and was influenced by deeper concerns about national identity at a time of significant geopolitical ferment.

The book's third argument, not unrelated to the second, is that political culture matters to contemporary democratic politics: however slippery this concept may seem to analysts, students of politics in a globalizing world must try to come to grips with it. In much relevant literature on politics, there has been a tendency to relegate political culture to "other" places and times—much in the way that nineteenth- and early twentieth-century cultural anthropologists found culture only in alien, primitive, or marginal societies, assuming that their own social beliefs were founded on the universals of science and reason. Accordingly, political culture has been invoked primarily in studies of non-Western political systems or of older, premodern polities.

Comparative analysis of the sort undertaken here reveals disconcerting problems with the understanding of political culture as exotic or foreign. To begin with, even economically and socially integrated Western nations are seen to differ importantly in their reception of science and technology. These differences cannot be explained in terms of discrepant ideologies, national interests, policy priorities, or states of technological development. They occur despite the leveling effects of protechnology state policies, global movements of knowledge and capital, and the role of transnational actors such as scientists, social movements, and industry. There are persistent differences in national ways of meeting common economic and social challenges, and the fact that these are hard to pin down and account for, and are contested even as they are reproduced, only makes the task the more intellectually engaging. An important locus of difference is in the systematic practices by which a nation's citizens come to know things in common and to apply their knowledge to the conduct of politics. I term these culturally specific ways of knowing "civic epistemologies" and discuss them in detail in chapter 10. How democratic polities acquire communal knowledge for purposes of collective action emerges in my telling as a particularly significant feature of political culture.

A renewed appreciation of political culture allows us to make sense of particular puzzles in each of the three country studies. In the United States, we address why a once robust debate on environmental issues such as nuclear power and chemical pollution has given way to a relatively complacent acceptance of the risks and benefits of genetic engineering. In Britain, the problem is almost the opposite, for here a nation historically tolerant of pollution and technological risk and relatively resistant to institutional innovation has emerged, in some ways, as the most active experiment station for the politics of biotechnology. By contrast, in Germany, an extremely sophisticated public debate conducted in expert committees, academia, and the elite mass media has failed to produce comparable innovation in the institutions of public policy.

There are also some comparative puzzles that I hope to elucidate. Some of these focus on the divergent political pathways traced by the same event or issue across the three countries. Why, for example, have agricultural biotechnology and GM food not become openly controversial in the United States or Germany but did turn into matters of intense concern in Britain? How, to the contrary, did Britain succeed in carving out a relatively uncontested space for embryo research, while American politics on this issue remained deeply divided, and Germany refused to allow the most difficult choices to rise to political salience in the first place? Why is patenting life forms seen as an ethical issue in Europe but not in the United States? And what accounts for the fact that bioethics, simultaneously and energetically embraced as a policy discourse in the EU and in three sovereign nations, nevertheless is understood in vastly different ways in each of its contexts of development?

The range and specificity of these cross-boundary differences militate against the easy generalizations about Europe and the United States offered by Robert Kagan in his light and lively essay on Western power at the millennium.[18] Kagan wishes us to "stop pretending that Europeans and Americans share a common view of the world, or even that they occupy the same world." I, too, challenge the notion of a globally shared "common view of the world," but my arguments begin and end in different places. In the context of biotechnology, I show, to start with, that terms like "European" and "American" are far more fluid and contested than is presumed by monolithic accounts of culture such as Kagan's. Clashes are endemic both within and between these cultures, particularly in relation to scientific and technological change, and the analyst's task is to probe how cultural identities are dynamically reasserted or transformed in these processes. Europe in particular is a multiply imagined community in the minds of the many actors who are struggling to institutionalize their particular visions of Europe, and how far national specificities should become submerged in a single European nationhood—economically, politically, or ethically—remains far from settled.[19] Moreover, if Europe and the United States do not occupy the same world, it is because the nature of that world is itself a thing that remains uncertain and contested. The world occupied by nation states never was a single place, but always a work in progress, represented and fought for according to different normative conceptions of the appropriate kinds of economic, political, social, and technological integration to be attained. Globalization has not resolved the tensions; it has if anything made the problems of coexistence more self-evident. Whose vision of the world should be naturalized or made "real" under these circumstances is of the utmost political and epistemological consequence. The politics of biotechnology, I suggest, is a remarkably productive site in which to observe competing ways of worldmaking being contentiously, often forcefully, negotiated, though not by military means.

Throughout the book, I use the methods of the interpretive social sciences to make sense of complex social and political phenomena, including most especially resources from the field of science and technology studies. Through a combination of historical reflection, close textual reading, personal interviews, observation of key institutions, and qualitative analysis of legal and political developments, I try to characterize how three wealthy, technologically advanced, deliberative democracies have tried to come to terms with one of the most far-reaching advances in the human ability to intervene in nature. In the process, I take issue with or reject as incomplete some commonly held views about cross-national divergences in the politics of biotechnology. One is the notion that U.S.–European differences on such matters as genetically modified crops and foods are simply the result of European protectionism, and hence are bound to persist as a form of international political gridlock.[20] Another is the countervailing proposition that convergence across countries is bound to happen, and is in fact happening under the prod of scientific and economic rationality. A third is the asymmetric invocation of "history" as an explanation for German opposition to some forms of genetic engineering, but not for British or American acceptance of the same developments—which by extension are seen as natural and inevitable. A fourth is the equally asymmetric attribution of the rejection of GM products in each country to public hysteria, media hype, or the public misunderstanding of science—without invoking comparable social explanations for the acceptance of the same technologies.[21]

As a comparatist of many years standing, I am aware of course that some readers will approach these arguments skeptically. To those inclined to view the world as embarked upon a course of increasing economic and social convergence, any attempt to characterize policy outcomes in terms of national political cultures may seem backward-looking, overdoing the differences between nations at the expense of flows that are increasingly drawing us all closer together. Some will charge that cross-*national* comparison, in particular, is full of intellectual dangers: it reifies national boundaries, overlooks heterogeneity and change, and perhaps reinforces parochial stereotypes of national identity. In answering similar criticisms of her seminal work on Dutch art fo the seventeenth century, the noted art historian Svetlana Alpers had this to say: "To those who will protest that . . . I exaggerate differences within European art by slighting the continuous interplay between the art of different countries, I would reply that they are mistaking my purpose. I do not want either to multiply chauvinisms or to erect and maintain new boundaries, but rather to bring into focus the heterogeneous nature of art."[22] What Alpers wished to do for northern European art, I aim to do here for Western democracy, that is, to bring into sharper relief its own heterogeneity, especially as displayed in its multifaceted, culturally differentiated encounters with science and technology.

The comparative accounts I offer are not designed to make it easier to predict where and when the next crisis over biotechnology will erupt or what procedures will then be best suited to restoring trust in science and government (although readers of this book may find it easier to appreciate which kinds of scientific and technological issues are most likely to become sensitive in each national setting). Far more, I want to display the separate logics that have driven three closely similar political traditions toward disparate ends in managing fateful encounters with biology and biotechnology. My purpose is to enhance our capacity for political and cultural appreciation of these developments—or, in terms elaborated by MaxWeber and other german political philosphers, to aim for *Verstehen* (understanding) rather than *Erklärung* (causal explanation).[23] It is to set aside reductionist, linear accounts of some of the most significant sociopolitical transformations of late modernity in favor of a kind of story-telling that does justice to the ambiguity of these experiences, and to their richness.

# 1

## Why Compare?

Biotechnology politics and policy are situated at the intersection of two profoundly destabilizing changes in the way we view the world: one cognitive, the other political. This unique position makes the project of using the life sciences to improve the human condition anything but straightforward. It also makes biotechnology a particularly apposite lens through which to compare the triumphs and tribulations of late capitalistic technological democracies.

On the cognitive front, the shift is from a realist to a constructivist view of knowledge. Years of work on the social construction of science and technology, and the contingency of similarity and difference judgments,[1] have taught us to be skeptical of absolutist claims concerning objectivity and progress. Scientific knowledge, it is now widely accepted, does not simply accumulate, nor does technology invariably advance benign human interests. Changes in both happen within social parameters that have already been laid down, often long in advance.[2] In the field of environmental regulation, for example, concepts of risk and safety, methods of compiling and validating data, ideas of causation and blame, and (crucially for biotechnology) even the boundary between "nature" and "culture" have all been shown to reflect deep-seated social assumptions that rob them of universal validity.[3] The methods with which policymakers carry on their business similarly cannot be taken as neutral, but must be seen as the result of political compromise and careful boundary maintenance, favoring some voices and viewpoints at others' expense.[4] The criteria by which one measures policy success or failure are likewise products of negotiation; in applying them, one implicitly adopts contingent, locally specific standards of reliability and validity. The special authority of scientific claims is in competition with other representations of reality diffused through the global media, and scientific expertise is subject to appropriation by multiple, diffracted social identities and interests.[5] Any attempt to compare the performance of national policy systems today must take these complexities into account.

On the political front, the shift is toward a fracturing of the authority of nation-states, with consequent pressures to rethink the forms of democratic governance. State sovereignty is eroding under the onslaught of environmental change, financial and labor mobility, increased communication, the global transfer of technical skills and scientific knowledge, and the rise of transnational organizations, multinational corporations, and social movements.[6] Supranational concerns, such as the demand for free trade or globally sustainable development, are gaining political salience,[7] but they are at the same time encountering resistance from tendencies toward greater local autonomy based on particularities of culture and place.[8] As a result, the "old" politics of modernity—with its core values of rationality, objectivity, universalism, centralization, and efficiency—is confronting, and possibly yielding to, a "new" politics of pluralism, localism, irreducible ambiguity, and aestheticism in matters of lifestyle and taste.

These flows and movements have attenuated the connections between states and citizens, calling into question the capacity of national governments to discern and meet their citizens' needs. Yet we live in a time when knowledgeable citizens are more than ever demanding meaningful control over the technological changes that affect their welfare and prosperity. Many therefore see this epoch as a proving ground for new political orders whose success will depend, in part, on our learning to live wisely with our growing capacity to manipulate living things and our equally growing uncertainty about the consequences of doing so.

There is little question that genetic engineering, along with the cognitive, social, and material adjustments made to accommodate it, will form an essential part of the politics of the twenty-first century, just as it did of the political history of the preceding three decades. Attempts to deploy biotechnology for the public good, and to ensure democratic control over it, touch the political and cultural nerve centers of industrial nations in the global economy. These efforts are *political* in the sense that they centrally concern the production and distribution of societal benefits and risks; they are *cultural* in that, by intervening in nature, biotechnology forcefully impinges on social meanings, identities, and forms of life. Comparison among national and regional debates surrounding biotechnology should therefore help us identify and make sense of the wider political realignments that are taking place around us at this moment. Comparison may even help us decide which courses of action we wish to follow, as individuals or as political communities.

But how should such a project be organized? What should we compare, using what methods, and with what ultimate hopes of illumination?

Comparison, particularly in the policy field, has historically been driven by a faith in the possibility of melioration through imitation.[9] Analysts assumed that they could objectively evaluate which agency, nation, or political system was "doing better" at implementing particular policy goals; such findings then were supposed to assist policymakers elsewhere in deciding which course of action to follow. While one should not denigrate this practical ambition, one should likewise not take its feasibility for granted. With growing awareness of the culturally embedded character of both knowledge and policy, there are reasons to be skeptical of unproblematic learning from others' experiences. The insights gained from comparative analysis suggest, indeed, that neglecting cultural specificities in policymaking may be an invitation to failure within any political community's own terms of reference. Comparative studies of science and technology policy today need a different justification than simply the propagation of improved managerial techniques. Rather than prescribing decontextualized best practices for an imagined global administrative elite, comparison should be seen as a means of investigating the interactions between science and politics, with far-reaching implications for governance in advanced industrial democracies.

But if deeper social and political understanding is our goal, what conceptual tools should we bring to the task of comparison, and how should these differ from past approaches? This chapter lays out the case for a new kind of comparative analysis—one that retains nation states as units of comparison but is organized around the dynamic concept of political culture, rather than the more static categories of political actors, interests, or institutions. My aim is to explore the links among knowledge, technology, and power within contemporary industrial democracies and to display these links from the standpoints of those situated within particular cultures of action and decision. This approach illuminates how political culture plays out in technological debates and decisions—most particularly how it affects the production of public knowledge, constituting what I call the civic epistemologies of modern nation states. The methods I adopt for this purpose owe as much to the history and sociology of knowledge and the anthropology of technological cultures as they do to comparative politics, policy studies, or law. Interpretive methods, I hope to show, are especially well suited to investigating the complex reception of novel science and technology into a nation's political life.

I begin the chapter with the theoretical considerations that will guide my comparison of biotechnology debates in Britain, Germany, and the United States. I then discuss the organization of the study, including the reasons for selecting these three countries as cases for comparison and biotechnology as the lens through which to compare them. I conclude with a brief outline of the remaining chapters.

15

## Beyond State and Structure: Theoretical Considerations

Comparative analysis is a relative newcomer to the study of social engagements with science and technology. As little as twenty years ago, the comparison of national policies significantly implicating technical questions—on issues such as public health, pharmaceutical drug regulation, industrial and occupational safety, and environmental protection—was still in its infancy. Up to that point, cross-national research on the politics of science and technology was constrained by a number of unspoken assumptions that cast doubt on the utility of comparison.

Reasons for the initial neglect included, to begin with, a firm belief in the universality of science. Political systems might differ, but science was held to be everywhere the same. The influential American sociologist Robert K. Merton spoke for this viewpoint when he represented "universalism," or the invariability of knowledge across political and cultural domains, as one of the core norms of science.[10] Also militating against expectations of cross-national variation was the widely accepted thesis of technological determinism, which holds that technology's inner logic, founded on its material characteristics, bends human institutions to suit its development trajectories.[11] Economic determinism provided an analogous argument from the social sciences, suggesting that, even if national policies initially diverge, competitive pressures in an increasingly interdependent global marketplace will eventually overwhelm such differences.

These ideas resonated in the field of political science, where the dominant school of thought held that technically complex decision making takes its color more from the nature of the issues than from features of national culture or politics. Policymakers everywhere, so the reasoning went, would be compelled by the same scientific, technical, and economic considerations; policies would therefore converge, and little insight would be gained from comparing national approaches over time. These views are still represented in some contemporary political writing, but this book argues that, in its narrow focus on decision outcomes and its failure to problematize the foundations of knowledge, such work misses important differences and regularities among contemporary cultures of democratic politics.[12]

Comparative analysis came into vogue in the 1980s as an instrument for advancing well-recognized and widely appreciated social objectives. In a world increasingly committed to economic and political integration, government and industry (if not always the noneconomic organs of civil society) shared an interest in lowering trade barriers by harmonizing regulations. Comparative research was seen as a useful aid to this project: as a means of highlighting areas where policies and values remain significantly divided, thus paving the way for negotiation and cross-national agreement. The capacity of policy institutions emerged as an important topic of comparison in studies of technically grounded regulatory fields such as environmental protection,[13] where success depended on the will and ability of state authorities to monitor and enforce compliance with complex legal obligations. Comparative study, according to advocates of transnational capacity-building, provided helpful lessons in how to improve the effectiveness of administrative institutions.

In this first wave of comparative analysis, policies were assumed for methodological purposes to be discrete and singular, with ascertainable causes and determinate consequences. A great advantage of this method was that it offered built-in criteria for comparison and evaluation. The policy process could be parsed into separable stages (for example, agenda-setting, legitimation, implementation, evaluation, and revision) that followed each other in linear succession and could be compared from one political context to another.[14] Since impacts were taken as clearly marked and objectively measurable, questions about the relative performance of states in meeting their goals also seemed unproblematic. States and citizens, at least within similar political systems, were presumed to want the same goods: health, safety, jobs, patents, new drugs, higher agricultural productivity, a cleaner environment, and so forth. In this intellectual framework there was nothing awkward about asking which political system produces the most responsive policies, affords the most protective standards, fosters the most innovation, fuels the most economic growth, or most effectively resolves political conflict. Only in the light of empirical research did these presumptions have to be reconsidered and sharply modified.

### The First Wave Breaks: National Styles of Regulation

In the early 1980s several studies of health, safety, and environmental regulation in Western countries put to rest the notion that policy strategies and outcomes are uniquely determined by economic, scientific, or technological considerations. Regulation, it emerged, displayed distinctively national characteristics, leading to observable differences in the timing, priorities, forms, and stringency of interventions.[15] Scientific evidence was shown to carry different weight in different policy environments, its interpretation conditioned by homegrown traditions of legal and political reasoning and habits of deference or skepticism toward expert authority. Cultural influences seeped into the very heart of technical analysis. Confronted by ostensibly the same research results, governmental agencies in one country concluded that a product or activity posed no risks to health or the environment, but in another held that it was unacceptably hazardous and should be banned or strictly regulated.[16] When decisionmakers reached broadly similar policy endpoints, they often did so through different routes of reasoning and public justification.[17] Patterns of interaction between regulators and regulated parties, as well as the

17

reliance on particular policy procedures and discourses, appeared firm enough to warrant the label "national styles of regulation."

Contrasts between U.S. and European approaches to managing risk seemed especially pronounced. Researchers were struck by the open and adversarial processes of rulemaking in the United States, the frequent resort to litigation, and U.S. agencies' significantly greater reliance on formal, quantitative measures of risk, costs, and benefits. Such systematic divergences invited explanations based on differences in the structure of political institutions. Comparative studies, like much other political analysis of the period, initially looked to the state for explanations, and to the relatively fixed "opportunity structures" it provides for political action.[18]

In the U.S. case, it took little prompting to see that the regulatory landscape is molded to an extraordinary degree by institutions that invite public expressions of skepticism and distrust. A constitutionally ordained separation of powers not only facilitates rivalry between Congress and the executive branch but also authorizes the courts to review the basis of administrative rules. Low entry barriers to the courts and an activist judiciary provide generous opportunities for interested parties to challenge decisions contrary to their immediate interests.[19] Citizens' capacity to take issue with, and hence to deconstruct, claims made by the state is strengthened through laws that require open meetings and disclosure of relevant technical information. At the same time, the relative dearth of vertical hierarchies and horizontal networks of cooperation impedes the kinds of informal negotiation and consensus-building that are found in European (and, outside the Western tradition, Japanese) policy formation.[20] All these entrenched attributes of politics heighten the vulnerability of U.S. policymakers, supplying plausible reasons for their distinctive approach to rationalizing policy decisions.

The argument from national political structure was particularly effective in explaining U.S. agencies' hankering for objectivity based on numerical calculations. Operating in a fishbowl of transparency, with significantly less protection from civil service traditions or legal insulation than their European counterparts, American regulators were not free to justify their actions by simply invoking delegated authority or superior expertise; they had to establish through explicit, principled argument that their actions fell within a zone of demonstrable rationality.[21] Numerical assessments of risks, costs, and benefits provided compelling evidence. European regulators, by contrast, seemed generally better able to support their decisions in qualitative, even subjective terms. Expert judgment carried weight in and of itself as a basis for action, the more so when backed by negotiation among relevant parties; there was on the whole less need to refer to an exogenous method, model, or logic to support policy decisions.[22]

These variations, moreover, were not accidental but rooted in longstanding practice, as the historian Theodore Porter documented in his comparative study of social accounting methods.[23] Already by the mid-twentieth century, experts of the U.S. Army Corps of Engineers had begun to insist on the objectivity of the cost-benefit analyses with which they justified flood-control projects. Whereas British actuaries and French railroad engineers admitted that their cost-benefit calculations reflected professional judgments, Corps engineers stoutly maintained that their assessments were not so compromised: *their* numbers were not subjective estimates but reliable representations of reality.

Time, however, has helped expose problems and puzzles that reduce the appeal of explanations based on state structures as basic determinants of national policy choices. The issues are both theoretical and empirical. The theoretical dilemmas reflect most importantly the rise of poststructuralist thought and the attendant difficulty of taking entities such as "science," "state," or "society" for granted as stable units of analysis. All these concepts have to be seen instead as historically situated, contingent, dynamic constructs, whose form and fixity are as much in need of explanation as they are available for explaining other developments. That states remain reasonably constant in their institutional structures, for example, let alone in their modes of action and self-legitimation, cannot simply be assumed as given. The analyst has to show how such continuities are maintained, if indeed they are, and why transformative political opportunities are either embraced or avoided by ruling institutions. For comparative analysts, this means that state structures must be regarded as both dependent and independent variables; similar conclusions hold with respect to science, technology, and society, the other macroformations of importance to this comparative study.

Support for this more fluid way of thinking about "social kinds" such as the state comes particularly from the field of science and technology studies (S&TS).[24] More than most branches of the social sciences, S&TS concerns itself with the nature and power of the categories and objects by which we organize our knowledge of the world.[25] Central to the S&TS enterprise has been to ask how societies produce authoritative knowledge and functioning technological artifacts. Through such investigations, it has been possible to demonstrate that the products of the sciences, both cognitive and material, embody beliefs not only about how the world *is*, but also how it *ought* to be. Natural and social orders, in short, are produced at one and the same time—or, more precisely, coproduced.[26] The apparent firmness of the devices with which we make sense of our existence, then, is maintained through more or less purposive action by identifiable actors. Accordingly, to understand how social entities such as "the state" or natural entities such as "the gene" function in the world, one has to ask how diverse actors use and understand the concept, how it is articulated through formal and informal practices, where and by whom it is contested, and how it reasserts itself in the face of challenges to its integrity or meaning.

Put differently, explanations based on variables such as national styles of regulation run into difficulty by failing to ask why some social structures or processes are seen as more deterministic than others. Why are certain features of the world taken as given or independent, while others are assumed to be shaped and directed by, hence dependent on, those fixed parameters? The structuralist literature on social movements, for example, commonly attributes the strength and effectiveness of activists to differential opportunity structures provided by state institutions; similarly, following John Kingdon, students of political agenda-setting have suggested that new items emerge from the interaction of contingent events and agile social entrepreneurs with relatively unchanging frameworks of politics and policymaking.[27] Yet social theorists from Michel Foucault to the political sociologist Theda Skocpol[28] have shown that social actors can make or remake the opportunities for intervention, often using knowledge as an instrument for modifying existing possibilities.[29] Social structures, in other words, are not immutable; they change in the very process of enabling actors to use them.

Recent work on social movements has engaged with active processes of meaning-creation that frame problems for collective action, build communal identities, and allow actors to mobilize against perceived injustices. To quote one distinguished team of converts:

> We come from a structuralist tradition. But in the course of our work on a wide variety of contentious politics in Europe and North America, we discovered the necessity of taking strategic interaction, consciousness, and historically accumulated culture into account. We treat social interaction, social ties, communication, and conversation not merely as expressions of structure, rationality, consciousness, or culture, but as active sites of creation and change. We have come to think of interpersonal networks, interpersonal communication, and various forms of continuous negotiation—including the negotiation of identities—as figuring centrally in the dynamics of contention.[30]

Political analysis that denies this kind of agency to activists operates with a much reduced, mechanistic model of human behavior. It overlooks the potential for altering the terms and conditions of political debate and overestimates, as a result, the invincibility of the status quo.[31]

Empirical findings from the first wave of comparative policy studies also point to deficiencies in structuralist and state-centered modes of explanation. Over the past thirty years, for example, industrial nations have often converged on which health, safety, and environmental problems merit legislative or regulatory attention, but there is much less uniformity in how the issues are characterized and which solutions are deemed most suitable for resolving perceived problems. Biotechnology is no exception. Although it has called forth roughly similar cycles of regulatory attention in Europe and the United States, biotechnology has given rise to quite different national discourses of risk and safety, naturalness and artificiality, innovation and ownership, constitutional rights, and bioethics.[32] To account for these divergences, we must ask how policy problems were construed in different political cultures. What did biotechnology mean to the actors opposing, advocating, or seeking to manage its further development? Once we turn to this interpretive domain, we quickly discover that the attribution of meaning to new technologies cannot be accounted for in terms of structured variances in national styles of regulation. Binding differences come about without overt action by political leaders or, as noted earlier, by any deterministic guidance from science, technology, bureaucratic organizations, or the invisible hand of the market.

A further empirical challenge has to do with shifts in the articulation of policy objectives, both within nations and across them. Political structure, with the provisos noted above, does predictably well at accounting for continuities such as those captured by the concept of national regulatory styles, but it is dismayingly helpless before questions of change. Why, for example, did the focus of environmental policy change from pollution control in the 1960s to prevention in the 1970s, sustainability in the 1980s, and precaution in the 1990s? Why did U.S. policy in the 1980s display greater concern for chemical risks than policies anywhere in Europe but then shift to greater complacency with regard to biotechnology just a decade later?[33] Where, more generally, do new policy ideas come from, who are the agents of their dissemination, and how are they institutionalized and reembedded into existing political arrangements? Questions such as these call for a dynamic exploration of political discourses and actor coalitions that extends beyond the formal power centers of the state.[34]

### Revisiting Political Culture

All of the foregoing suggests that culture—more particularly political culture—matters in shaping the politics of science and technology. For the purposes of this study, political culture refers to systematic means by which a political community makes binding collective choices. The term encompasses institutionally sanctioned modes of action, such as litigiousness in the United States, but also the myriad unwritten codes and practices with which a polity supplements its formal methods of assuring accountability and legitimacy in political decisionmaking. Political culture in contemporary knowledge societies includes the tacit, but nonetheless powerful, routines by which collective knowledge is produced and validated. It embraces institutionalized approaches to reasoning and deliberation. But equally, as we shall see, political culture includes the moves by which a polity, almost by default, takes some issues or questions out of the domain of politics as usual. An important part of this book's argument is that political authority in the management of science and technology derives not only from the formal and informal rules of political practice, but also from less explicit cultural commitments to forms of legitimation that fill out the routines of what we think of as normal politics.

The analysis of political culture seeks to capture the stabilities in social practices and meaning making while getting below the bland surfaces of formal politics and decision making. Yet the skeptical reader may question whether the transfer of explanatory energy from "state" to "political culture" solves anything or only muddies the analytic waters. After all, while states and their organs are relatively easy to locate, with boundaries demarcated by physical space and formal practices, culture is a notoriously slippery concept, and anthropologists have struggled long and hard against using it in a reified, uncritical, totalizing, even patronizing manner.[35] Can we avoid these difficulties in operationalizing a cultural perspective on politics and, if so, by means of what categories?

We must, to begin with, use the term culture judiciously in this context, not invoking the analyst's labeling prerogative mechanically or asymmetrically (culture as the marker of the other, not of ourselves[36]); we need always to ask how actors within a culture make sense of their own confusions and predicaments. As a "social kind" in its own right, political culture must be seen not only as resilient and resistant to change, but also as constructed, flexible, and subject to renewal. The comparative strategy of this book is geared toward making just such a reading of political culture more tractable and analytically useful.

We should not underestimate the difficulty of this project. Particularly troubling for analysis is the recognition that systems of knowledge and belief about the natural world are not built independently of the social worlds within which they are embedded. Evidence from many quarters points to a subtle and multidimensioned process of coproduction, in which problems of society and problems about nature are simultaneously addressed and resolved.[37] So, at the dawn of the scientific revolution, seventeenth-century English gentleman-scientists like Robert Boyle, a founder of Britain's Royal Society and pioneer in studying the properties of air, devised ways of conducting and publicly reporting their experiments that furthered not only new regimes of fact-finding but also new regimes of governance. As practices of "witnessing" science, such as peer review, gradually took hold, there was an associated move away from absolute monarchical power and the divine right of kings toward a more democratic politics, in which citizens gained standing to evaluate the performance of those in power. The practices of science and of liberal democracy flourished together.[38] In time, new systems of classification, counting, and standardization developed to sustain the modern nation state, with its far-flung economic and military enterprises, its need for centralized administrative control, and its ever-recurring quest for credible and demonstrable successes.[39]

The continual interpenetration of political choices or commitments and the production of reliable knowledge raises obvious difficulties for comparison. One difficulty has to do with the analyst's own standpoint. Lele tribesmen in Africa,[40] sheep farmers in the north of England,[41] and expert U.S. regulatory agencies[42] all determine what is risky in their environments in accordance with specific, situated needs for order and meaning; each chooses to label some things as dangerous and ignore others that visitors from other worlds might find "objectively" more so. Where, then, is the Archimedean point from which we can begin to assess the performance of alternative systems of governance? A second problem concerns the pinpointing of causes for purposes of explanation. The framework of coproduction suggests that the state's instrumental goals, the knowledges and practices adopted for achieving them, and the applicable standards of credibility and legitimacy are all constructed together through a unitary process of ordering the world. How, then, can we presume to explain outcomes in terms of distinct and independent causes? In William B. Yeats's beautifully apposite metaphor, how can we know the dancer from the dance?[43]

Clearly, there is a need for new methods and new conceptual approaches. The field of comparison has to be more creatively mapped and explored than in earlier studies that looked primarily at clusters of similar policy actions, with identical, context-free life-cycles, and tried to identify their alleged causes and consequences. To understand how policy domains are carved out from the political sphere and rendered both comprehensible and manageable, we must employ analytic categories different from those of decision makers operating within the policy process. We need a conceptual language that can grapple with both continuity and change, while rejecting some of the rigidities of structure.

### Framing

A good way to begin is by asking how issues are framed for public action in democratic societies. Erving Goffman's path-breaking sociological work showed that there is nothing intrinsic or externally determined about how people organize their experiences or how they choose to imagine the causes and effects of particular phenomena.[44] Goffman's approach has been augmented by a rising interest across the social sciences in the narrative, discursive, and textual dimensions of human behavior, which in turn has opened up a rich seam for the interpretation of political action.[45] From this perspective, the regulation of science and technology, whether to further innovation or control risk, can fruitfully be seen as a kind of story-telling by communities situated in particular times and places who are attempting to deal with unsettling or disruptive changes in their environments.

Stories told in the policy arena attempt to order and make sense of complex experiences; they enable people to take meaningful action and so reduce their feelings of helplessness and alienation. The intersubjective, or communally held, cognitive frames constructed in this process, often embedded in material objects and routinized social practices, impose discipline on unruly events by creating understandable causal relationships, identifying agents of harmful behavior, and finding solutions that convey a sense of security and moral order. Of course, "real" disruptive events happen all the time, such as the birth of the cloned sheep Dolly in Scotland or the destruction of the World Trade Center in New York, and theories of framing do not deny those realities. Rather, framing allows us to see that events do not in and of themselves dictate the pathways along which public responses will move—nor even necessarily provoke any political action.[46] Events first have to be set within an interpretive context that allows them to function as a starting point for deliberation or concerted action: so Dolly's birth announcement became a challenge for "bioethics," and the September 11, 2001, attacks were cast as grounds for a "war on terror."[47]

Framing policy problems is an intensely social activity, as Goffman and other analysts have argued;[48] yet as frames embed themselves in social behavior and material culture, they fundamentally alter people's perceptions of what is real in the world around them. The sociologist Joseph Gusfield offered an instructive, policy-relevant example.[49] Years of "random" car accidents that killed mostly people in their teens and early twenties were at one point reinscribed on the U.S. national consciousness as the "problem" of "drunk driving." New coalitions of distraught mothers and other accident victims began to push for legislation to curb what, until that moment, had been dismissed as random tragedies, permitting no mandatory social correctives.

The ensuing reframing was not preordained to happen, nor did it have to happen in exactly the way it did. To illustrate some of the contingencies of framing, we can revisit Gusfield's account of drunk driving with insights gained from a theory of sociotechnical systems, *actor-network theory*, which sees technology as a heterogeneous network of human actors and nonhuman actants.[50] As the frame of social awareness shifted from random accidents to drunk driving, the automobile emerged, if only for a moment, from its casing of enameled steel as a thing of many parts, tied to various hard and soft components—objects, actors, rules, practices—in a complex (and hazardous) network of road transportation. As if endowing its users with x-ray vision, the frame of drunk driving permitted society's movers and shakers to detect all kinds of once invisible nodes in the network where intervention now seemed possible in the interest of saving lives: raising the drinking age; penalizing innkeepers and even private party-givers who allowed drinkers to go on the road; mandating seatbelt use; reducing speed limits; and requiring cars themselves to be engineered with new safety features such as airbags and antilock brakes. These heterogenous elements were pulled together through a messy period of social experimentation, producing a new regime of automobile safety regulation.

In contrast to the notion of political agenda-setting, which takes for granted the shape of political issues, framing implicitly makes room for the contingency of social responses and the partiality of the imaginative space that is carved out for political action in any society. Not every culture that looks at a set of events necessarily frames them in the same way. Things within the frame for one group of actors (such as innkeepers, party-givers, and speed limits in Gusfield's case[51]) may fall outside the frame for others, even if all have accepted drunk driving as a policy priority. Selectivity is inevitable in the construction of frames, but so too, we will see, is cultural conditioning. Framing in this way usefully occupies a middle ground between the contingent and the determined.

Sociological in its origins, the idea of framing has moved into wider worlds of political analysis, some with specific resonance for this study. A raft of disciplinary specialties, from social movements theory to cognitive psychology, have taken on board that representations matter as much as whatever we may choose to call reality in shaping social behavior. Political inquiry correspondingly has expanded to include both the making of powerful representations and their effects on public attitudes and actions. In one working out of these ideas, the comparatist Juan Díez Medrano has argued that national differences in attitudes toward European integration reflect how citizens in various member states frame the idea of Europe.[52] Medrano conceives of framing largely as a cognitive process and is concerned primarily with how European nationals speak about integration. He uses the concept of *cultural preoccupation* "to encompass very general beliefs, symbols, and images as well as more concrete topics of discussion in a particular society."[53] While interviews and survey data can be used to identify some of these recurrent discursive and cognitive elements, the problem for S&TS scholars interested in framing is more complex. The durability of frames, too, has to be accounted for. To explain the cognitive and political staying power of frames, we need to know about the diversity of materials with which they are constructed, how they achieve taken-for-granted status, and what happens to make frames change.

Frames offer rich resources for interpretive analysis, but they are also potentially treacherous instruments. Frames may overlap, for instance, with the "same" object, action, actor, or relationship occupying a position in more than one but carrying quite different meanings and entailing different normative obligations in each. Thus, the human embryo may be represented as a person-in-the-making in the informal practices of an in vitro fertilization (IVF) clinic or an "adoption agency" that "places" it for implantation.[54] At the same time, it may be denied personhood under laws governing abortion or compensation for personal injury. Such slippages, if widely perceived as contradictions, may produce incentives for reorganizing collective understandings of kinds and categories into new, more coherent and encompassing frames. Framing, in short, provides an effective way of accommodating the solidity as well as the interpretive flexibility of the worlds in which policy gets made—a feature that will emerge with great clarity in the comparisons that lie ahead.

## Boundaries

Genetic engineering threatens or calls into question many of the categories that have been accepted as foundational in the ordering of societies, both ancient and modern. These include the fundamental divisions between nature and culture, moral and immoral, safe and risky, god-given and human-made. The molecularization of the life sciences, which enables us both to "read" and to manipulate characteristics beneath the visible surfaces of living entities, poses particular challenges to principles of governance based on older orderings and classifications. The grand project of mapping and sequencing the human genome, hailed as the holy grail of genetic science, has revealed that the lowly mustard weed has almost as many genes as we do.[55] We can import genes from spinach into pigs, from jellyfish into rabbits, and from fish into tomatoes; the technique of xenotransplantation allows cells from genetically altered pigs or chimpanzees to be inserted into biologically compatible humans. We can contemplate altering the human genome so as to produce enhanced human beings, with characteristics that today would be regarded as out of the ordinary, even superhuman. What, then, is nature, and what is being human?

The process of rebuilding order from confusion on points like these requires that problematic entities or behaviors be fitted if possible on one side or the other of conceptual divides that societies take to be foundational. The divisions that matter to a culture must not be allowed to break down. They are essential to maintaining cultural integrity, as anthropological studies incessantly remind us. Sociologists use the term "boundary work" to describe the creation and maintenance of essential social demarcations.[56] Boundaries are everywhere at play in the world, exercising enormous influence on thought and action, although they are produced in many cases through processes that are all but invisible even to the most energetic participants. Lawyers, for instance, make and remake the boundaries between acceptable and unacceptable behaviors while claiming to "find" these demarcations within the law. A major function of policymaking for the life sciences is to create and maintain boundaries that correspond to people's preexisting ethical and social sensibilities concerning the products of biotechnology.

How risky is it to admit into society biotechnological products that ambiguously straddle cognitive and social boundaries? Bruno Latour has argued that the urge to "purify" the "hybrid" networks of our high-tech world—for instance, the genetically engineered mouse or the ozone hole—into metaphysically pristine categories of the natural and the social is essential to maintaining order. He sees the nature/culture divide as so basic to modernity as we understand it that he terms it "constitutional."[57] The sociologist Zygmunt Bauman, too, attributes to modernity's "gardening instinct" a relentless desire to root out ambivalence in all its forms;[58] nature/culture hybrids that cross the conventional boundaries of moral thought represent but one challenge to this thirst for purity and order. Approaching similar problems as a cultural anthropologist, Mary Douglas argued that judgments about purity and danger are linked to the need for stability in social structures, whether against foreign invasion or within the hierarchy of groups. Work in the sociology of science, however, has also called attention to products of science and technology that come to occupy a valued social or moral position precisely because they resist being disambiguated: as "boundary objects," they serve as repositories of multiple meanings.[59] Legal terms often perform this function, as for example in international law, where a concept like "sustainable development" gains adherents through its very ability to accommodate diverse viewpoints and interpretations.

Indeed, perhaps the most influential form of boundary work in contemporary societies is done by legal institutions as they try to sort the infinite variety of human actions and their consequences into finite and pragmatic conceptual categories.[60] Should animals with altered genes be regarded as inventions for purposes of patent law? Should embryos be seen as persons, property, or some sort of hybrid, partaking of the properties of both? Should people be allowed to "own" their tissues and cells once scientists have extracted them for research purposes and given them independent existence as "immortal" cell lines? But politically significant boundary work also takes place in a multitude of more specialized forums that are less transparently in the business of boundary maintenance than legislatures or courts, such as expert advisory committees, parliamentary commissions, ethics review boards, and nongovernmental organizations.[61] The functions, processes, and methods, as well as the achievements, of boundary work will form important dimensions of comparison in succeeding chapters.

## Institutional Reasoning and Discourse

Institutions have traditionally played a starring role in comparative analysis. Governmental bodies, in particular, as the organs through which policy is formally articulated and implemented, are the places one turns to first for evidence of political mobilization and policy change. Usually, the focus is on what institutions do rather than how they do it, or how they think,[62] but with growing interest in the political role of ideas—their origins, power, and dissemination—institutions have attracted renewed attention as sites for interpretive analysis. This is reflected in the work of the new institutionalists, who recognize the capacity of institutions to embody meaning, create social relationships and symbolic orders, and "set the limits on the very nature of rationality."[63] A concrete result of this shift is the need to look at the political and policy discourses used by institutions as objects of analysis. In a book about biotechnology, the discourses of risk assessment, bioethics, and intellectual property law have special relevance as instruments for framing issues, ordering new knowledge, and (re)allocating power. How these discourses vary across national lines and actor groups becomes part of the agenda of comparison.

Because genetic engineering transgresses some of the most deeply entrenched categories of Western thought, the institutions that promote and regulate biotechnology are particularly likely to be involved in the production of novel ideas, norms, and meanings. Religious, moral, practical, and aesthetic ideas, according to the anthropologist Clifford Geertz, must be "carried by powerful social groups to have powerful social effects; someone must revere them, celebrate them, defend them, impose them."[64] Biotechnology touches on all these kinds of ideas, as well as on scientific ones, which Geertz (perhaps because he was writing about Indonesia, but perhaps also because he was unconsciously reproducing modernity's foundational nature/culture divide) did not explicitly include in his list. Nor did he expressly mention the counterforces of reaction that are likely to be set loose by powerful new ideas when actors seek to defend or impose them.

Clustered around genetic engineering in each country, we find not only celebratory institutions (mostly those of science, industry, the mainstream media, and the state), but also institutions (mostly *not* those of the state) that fear, doubt, question, or decidedly oppose the workings of the new technologies. Between these poles stand an array of public and private institutions whose task is neither to praise nor to blame biotechnology but more cautiously to deliberate on its management. The claims, beliefs, discourses, and actions of all these institutions, and the strategies by which they acquire and maintain legitimacy, will constitute another important strand in our comparison.

### Actors' Identities

In a book dedicated as this one is to understanding how natural and social orders are coproduced, and how cultural commitments are rewritten on changing terrain, the emergence of new social identities and the actors who embody them attracts particular interest. We expect to gain insight into how societies cope with the novelty of science and technology by observing changes in existing actor categories, such as "expert" or "ethicist," that expand social roles or alter their meanings. In the context of biotechnology, relevant new actors include not only institutional presences such as IVF clinics and expert advisory bodies, but also professional groups such as bioethicists, and social actors such as surrogate mothers and patients' organizations. An unexpected actor category consists of what we may term "liminal agents"—like the pre-embryos and supernumerary embryos discussed in chapters 6 and 7 that channel the flow of politics no less effectively than the humans who speak for them. The visibility and influence of each actor type varies from country to country, as do their institutional resources and opportunities to participate in political debate. Identifying relevant actors and reflecting on the basis for their political standing is another dimension of comparison, indeed an essential one if we are to produce a meaningful ethnography of political culture.

The point here is not to duplicate the analytic style of social movement theorists, who also study the identity-shaping roles of political actors, but to supplement that work with a more fine-grained reading of cultural responses to developments in the life sciences and technologies. Not all societal adjustments rise to the level of protest movements nor arise from below, as in the formation of group resistance; to the contrary, many salient adjustments in actors' identities, with profound consequences for the day-to-day conduct of society, occur *within* elites, in the courts, the expert bodies that advise parliaments and presidents, and the professional classes that control much of the meaning making in advanced industrial societies. These are the groups, then, that can be observed enacting and performing some of the continuities of culture, with significant implications for convergence and divergence across national polities.

## The Field of Comparison: A Topography

Two axes of comparison structure the remainder of this book. It is, to begin with, a comparison of three advanced industrial states: Britain, Germany, and the United States. The European Union also requires attention, both as a source of autonomous biotechnology policy initiatives and because two of the selected countries are EU member states. In important ways, however, much of the politics of biotechnology has been driven even at the European level by national rather than supranational concerns. The book's design reflects this recognition. The second comparative axis cuts across several sites of debate within the field of biotechnology politics and policy, encompassing developments in agriculture as well as biomedicine. A brief word on the selection of the three countries and a slightly longer reflection on the use of biotechnology as a lens for comparison are now in order.

### Nations of Choice

The comparative method works best when the entities to be compared are different enough to present interesting contrasts, yet similar enough for the variations to be disciplined. The similarities among the science policy cultures of the three countries chosen for this study—Britain, Germany, the United States—are considerable and easily stated. All are economically and technologically advanced democratic nations, with long records of public sponsorship of research in biomedicine and the life sciences. All subscribe to neoliberal policies of privatization in the delivery of goods and services, although in each country support for market values is counterbalanced by other collective goals and interests. More specifically, commitments to free scientific inquiry and progress through technological innovation are offset in each national context by basic concerns for human dignity and autonomy, nondiscrimination and equality of opportunity, and the preservation of nature and the natural. In each country, too, well-organized representatives of civil society, serving both corporate and public interests, are in constant dialogue with governmental bodies on the appropriate directions and means for steering biotechnology.

Statistical information on public support for biotechnology in the three countries is more difficult to provide. The absence of a precise definition for "biotechnology" and the lack of international standards in data collection stand in the way of exact comparisons. Nonetheless, the Organization for Economic Cooperation and Development (OECD), a body dedicated to harmonizing trade and regulatory policies among the world's wealthiest nations, has compiled data that point to some interesting cross-national convergences and divergences. Table 1.1 compares selected indicators for Britain, Germany, and the United States. These show consistently high public expenditures in all three countries, but with the United States leading in the race to commercialization, as indicated by the number of field trials of genetically engineered traits in crop plants and in patent applications for biotechnology.[65]

The differences among the three countries may be less obvious, but they hold considerable interest for a study aimed at illuminating divergent models of democracy at play in the early twenty-first century. There is, to start with, the empirical observation that Britain, Germany, and the United States have responded to biotechnology with different assessments of what is acceptable or unacceptable, and different legislative, judicial, and policy instruments for correcting the perceived threats. That these reactions to some extent contradict earlier national responses to environmental risks only adds piquancy to the comparative project.

Despite their structural similarities, the three countries differ importantly in their legal and political traditions, with substantial implications for the conduct of politics. Particularly noteworthy are differences in the formality and pervasiveness of legal processes, the approaches to administrative decision making, the methods of engaging with expert advice, and the means of incorporating public perspectives into policy. Table 1.2 summarizes these basic contrasts. There are, as well, reasons to believe that the three countries vary interestingly along the dimensions of comparison outlined above: framing,

TABLE 1.1
National Profiles of Biotechnology

| Indicators | Country | | |
|---|---|---|---|
| | Britain (population: 59.5 million) | Germany (population: 82.1 million) | United States (population: 272.9 million) |
| Total public funding for biotechnology, 1997 (million PPP$) | 705.1 | 1048.2 | |
| R&D biotech/R&D overall, 1997 | 7.8% | 6.7% | |
| Percentage of total venture capital investments, 1999 | ~6 | ~18 | ~5 |
| Percentage of total scientific publications (1998) | 9.3 | 6.0 | 23.9 |
| Number of U.S.-issued biotechnology patents (2000) | 299 | 373 | 5,233 |
| GMO field trial-traits, 1995–2000 | 162 | 123 | 5136 |

Sources: Brigitte von Beuzekom, *Biotechnology Statistics in OECD Member Countries: Compendium of Existing National Statistics*, STI Working Papers 2001/6 (OECD, 2001); "Origin of US Biotechnology Patents," *Chemical and Engineering News* 79, 44 (October 29, 2001): 56.

TABLE 1.2
Comparative Political Systems

| | Britain | Germany | United States |
|---|---|---|---|
| Legal tradition | Common law; unwritten constitution | Civil law; written Basic Law | Common law; written constitution |
| Administrative style | Informal | Formal | Formal |
| Expert engagement | Informal; consultative | Formal; negotiating | Formal; technical |
| Public participation | By invitation; limited to recognized social interests | By appointment; party and institution-centered | Self-generated; open to interested and affected groups |

boundary work, institutional reasoning, and actor identities. A comparison along these lines is therefore likely to be specially productive from the standpoint of increasing our theoretical understanding of science, democracy, and political culture. The empirical studies in later chapters will put these intuitions to the test.

## The Lens of Biotechnology

Modern biotechnology did not spring full-blown from an instant of brilliant scientific inspiration. Nor did it instantly reveal its political potential. The collection of techniques dubbed "biotechnology" gradually took shape over many decades of research on the processes by which genetic information is stored and transmitted in living organisms.[66] The political and moral tensions surrounding biotechnology cannot be separated from the fact that there are two distinctive registers in which they can be discussed: that of pure science, harking back to the discovery of the structure of DNA, and that of industrial production, with its associations of efficiency, commodification, and control. In this section, we look first at the major steps in the evolution of biotechnology from university to industry, and then at the salient political themes that have grown up around the industrial and commercial uses of the life sciences.

## The Birth of an Industry

Biotechnology is founded on molecular biology and the possibilities it offers for regulating organisms through planned genetic manipulation.[67] The word "gene," so central to biotechnology, was coined in 1909 by the Danish botanist Wilhelm Johanssen. He wanted a "little word" for the hereditary units that determine the characteristics of organisms in accordance with the principles of inheritance outlined by Gregor Mendel in the 1860s.[68] The word needed a physical structure, and eventually a mechanism, to achieve its operational power. The most important step in this evolution was the 1953 discovery by James Watson and Francis Crick of the structure of deoxyribonucleic acid (DNA),[69] the basic genetic material of almost all living organisms.[70] So simple and elegant was their model that it captured the scientific and popular imagination unlike any discovery since the breakthroughs in atomic physics more than a generation before. It was greeted as a revelation, uncontaminated by any of the messiness, false turns, or dead ends of routine scientific practice. Crick's own reflections testify to the role reversal by which DNA gained greater power to shape the human imagination than the men who elucidated its structure: "Rather than believe that Watson and Crick made the DNA structure, I would rather stress that the structure made Watson and Crick. After all, I was almost totally unknown at the time, and Watson was regarded in most circles, as too bright to be really sound. But what I think is overlooked in such arguments is the intrinsic beauty of the DNA double helix. It is the molecule that has style, quite as much as the scientists."[71]

A DNA molecule, "as every schoolboy knows,"[72] looks like a double helix—two long, intertwined strands composed of four nucleotides. These are the bases adenine, guanine, thymine, and cytosine, usually abbreviated as A, G, T, and C, which are attached to a uniform "backbone" of sugar and phosphate molecules (see fig. 1.1). The key to the "stylishness" of DNA is the

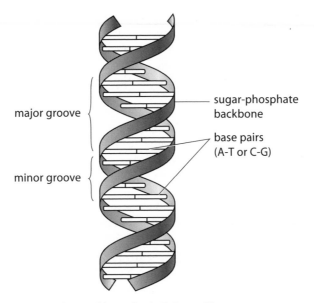

major groove

minor groove

sugar-phosphate backbone

base pairs
(A-T or C-G)

Figure 1.1. Structure of DNA (design by A. P. Jasanoff)

paired relationship of the four bases. For chemical reasons, adenine bonds only with thymine (A = T) and guanine with cytosine (G = C). Accordingly, the sequence of bases on either strand of a DNA chain automatically provides the corresponding information about the other strand; an A on one strand is always paired with a T on the other, and similarly a G with a C.

The brief paper in which Watson and Crick first described these findings contained a sentence whose verbal restraint was inversely proportional to its future economic and ethical significance: "It has not escaped our notice that the specific pairing we have postulated immediately suggests a possible copying mechanism for the genetic material."[73] This may have been studied nonchalance, although Crick attributed their initial refusal to say more to Watson's fear of being wrong about the proposed structure. Whatever the motivation, the remark proved prescient. DNA replication was soon shown to work by the unwinding of the two chains and the use of each half to guide the formation of a parallel chain. Because the position of each base on the new chain is fixed in relation to its opposite number on the old one, the process results in the creation of two identical copies of the original, with the base pairs lining up in exactly the same order on each.

The double helix was the product of pure university science, the brainchild of two gifted young men tinkering with a model in Cambridge University's famed Cavendish Laboratory and building on work done by other researchers using different tools to study DNA's structure. In this case, though, consequences for medicine and industry followed very quickly. First came the discovery that DNA chains could be cut at specific sites with the use of compounds called restriction enzymes, and that they could also be joined together by using enzymes called DNA ligases. Cut fragments consist at each end of unpaired bases forming "sticky ends," so called because they easily pair with their complementary bases. Twenty years after the decoding of the structure of DNA, two Stanford University scientists, Stanley Cohen and Herbert Boyer, perfected and patented a technique for systematically cloning, or reproducing, specific lengths of DNA.[74] Their method involved taking DNA fragments obtained through the use of restriction enzymes and inserting them into circular molecules of nonchromosomal bacterial DNA known as plasmids. These hybrid (or recombinant) plasmids could then be reinserted into *Escherichia coli* (*E. coli*), a bacterium widely used in scientific research, where they would replicate stably (see fig. 1.2). Subsequent developments made it possible to transfer biologically active foreign DNA into many different host environments, including plants and higher animals.

States, industry, and scientists themselves very soon recognized the economic potential of a new biotechnology founded, at the subcellular level, on the techniques of genetic engineering and, at higher levels of organization, on cell and tissue culture techniques. Biotechnology was targeted by the mid-1970s for a wide range of commercial applications, although progress was slower than expected. Nearly thirty years later, products still had not reached the market in very large numbers, but hopes for economic regeneration through biotechnology remained undimmed in states seeking to maintain positions of global dominance in a second, science-driven industrial revolution.[75]

In industry and university laboratories, genetic research divided into two streams, labeled "red" for biomedicine and "green" for agriculture and environment. Under the green heading, biotechnology was put to use in producing new viral pesticides and plant varieties resistant to pests, herbicides, and other environmental stresses. Biotechnological methods were used to engineer commercially desirable traits into farm animals, such as higher milk yield in cows, leaner meat in pigs, and a higher volume of edible flesh in fish. These manipulations often involved cross-species transfers of genetic material, resulting in "transgenic" organisms that could not have been produced through traditional breeding. Genetically engineered bacteria were experimentally created to serve a variety of environmentally beneficial functions, especially in the field of bioremediation. Overshadowing all these in media attention was the announcement in February 1997 that a sheep, facetiously named Dolly (for the American country and western singer Dolly Parton), had been cloned months earlier from an adult cell taken from another ewe.[76] From one perspective, this event was just another step in a long history of animal breeding; from another, it suddenly opened a window onto the disturbing possibility of intentionally producing replicas of human beings.[77]

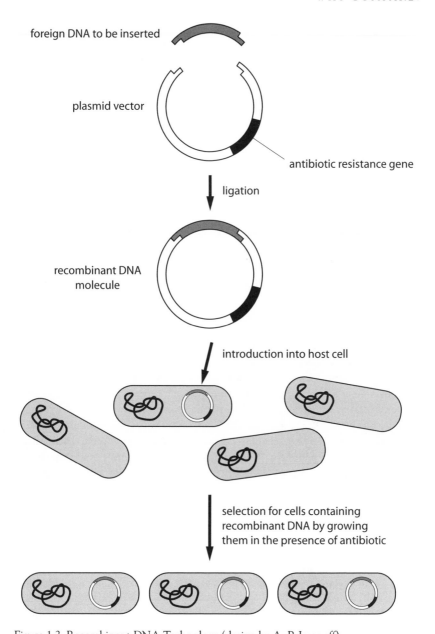

foreign DNA to be inserted

plasmid vector

antibiotic resistance gene

ligation

recombinant DNA
molecule

introduction into host cell

selection for cells containing
recombinant DNA by growing
them in the presence of antibiotic

Figure 1.2. Recombinant DNA Technology (design by A. P. Jasanoff)

Working on red biotechnology in the meantime, pharmaceutical companies quickly seized on the possibility of creating genetically engineered substitutes for scarce therapeutic agents, such as insulin for the treatment of diabetes and human growth hormone for congenital dwarfism. Techniques for mapping and sequencing genes, and eventually the whole human genome, laid the foundation for a promising market in diagnostic tests; work also began on individually targeted "designer molecules" for therapeutic purposes, rekindling the perennial hope of cures for cancer, autoimmune diseases, and various genetic disorders. DNA typing, a method of identifying people through characteristic "fingerprints" obtained from samples of bodily fluids, emerged as an invaluable tool for law enforcement and paternity testing. In related moves, techniques from molecular biology were applied to create human embryos outside the womb, opening the way to in vitro fertilization, embryonic stem cell research, and (especially after the birth of Dolly) possible reproductive human cloning. Let us turn briefly now to the political themes that have been articulated around these developments.

### Scientific Advances, Social Anxieties

If chemistry and physics underwrote state power through the twentieth century's two great conflagrations, it now seems biology's turn to define new roles for government. Life itself, as Michel Foucault compellingly argued, has become the new preoccupation of states, and the resulting biopolitics gives citizens a new arena on which to demand and contest the exercise of state power.[78] Certainly, there is no lack of public interest in the breakthroughs that biology promises. As the human conquest of nature chalks up new victories, those natural forces that remain outside human control seem increasingly more arbitrary and pointless. Who would not prevent if they could the devastating epidemics, the crops that fail, the pain of infertility, the unfairness of hereditary disease, the assaults of cancer, the decay of memory and reason, and the sadness of untimely death? These persistent troubles threaten the well-being of even the most prosperous modern societies. If collective defense and welfare goals remain intransigent problems, as the "war on terror" clearly demonstrates, then the mood of the moment seems all the more hospitable to state-supported advances in the life sciences, which promise citizens fulfillment on an intimate, personal scale, through longer, healthier, more liberated lives for themselves and, in time, their genetically tailored children.[79]

The life sciences for their part stand poised to serve both state ambition and private desire. Biology enjoyed its share of patronage even before the middle of the past century,[80] but the discovery of the structure of DNA and the astonishing cascade of developments in genetics and molecular biology in succeeding decades provided the basis for a much closer alliance between science and the state. It seemed that a whole range of social problems could be rewritten so as to take advantage of scientists' growing capacity to manipulate the molecular foundations of life. Genetic engineering promised a new efficiency in meeting global targets for medicine, agriculture, environmental protection, and public health that had frustrated generations of idealistic social engineers and their disappointed clients. In wealthy nations, improved genetic understanding was also seen as holding the keys to threatening behavioral problems, such as violence, addiction, and madness.[81] Advances in biological knowledge seemed to add point and meaning to the modernist project of rational, science-based problem solving at a moment when doubts about the goals and instruments of modernity were increasingly in evidence.[82] Biological science and technology projected a confident ability to take much that is mysterious, elusive, particular, and problematic in the human condition and bring it within the realms of order, prediction, uniformity, and control. The life sciences in short presented themselves as ideal instruments to states in late-modern crises of legitimation.

On a personal level, biological intervention holds out hopes for individualized medicine, conferring cures for currently incurable conditions. More enticingly, the notion of design has begun to permeate talk and imagination through such terms as designer genes, designer molecules, and even designer babies. Designs on nature—once thought to be the prerogative only of a divine creator—seem now well within the reach of human capability.

Yet, there are cross-currents that make this moment seem less propitious for a scientific and technological revolution of wide-ranging proportions. The prospect of tinkering with the human genome disturbs deep-seated notions of the integrity and inviolability of human nature. Questions of responsibility are yet more troubling. Pledges of progress sit uneasily with publics who have grown weary of unfulfilled promises, cynical about the good intentions of states, angry with the continuing gap between rich and poor, and newly conscious of the constructed and value-laden character of scientific knowledge.[83] While biologists celebrate their achievements in the laboratory and the field, and governments burnish their credibility in science's reflected glory, intellectuals, artists, and many ordinary citizens view the genetic revolution with much more contained enthusiasm. Biological hazards have captured the public imagination and fostered a kind of "genetic anxiety" comparable to the "nuclear fear" of an earlier time.[84] Some cite archetypal myths and contemporary disasters—Prometheus, Frankenstein, Nazi eugenics,[85] ozone depletion, climate change, "mad cow disease"—to argue that human foresight and institutions are doomed to lag behind human ingenuity in tampering with nature's secrets.[86] Others deplore the reductionism of the genetic vision of life and the associated dangers of heightened state control, diminished human dignity, and inevitable inroads upon the mystery and diversity of life.[87] Still others see the alliance among Western science, capital, and the state as the recipe for a new colonialism that will again appropriate indigenous resources and threaten the ecological and economic survival of the developing world.[88]

These tensions between the perils and promises of genetic engineering have created the testing ground on which we can observe the politics and policy of mature liberal democracies differentially at work. New theaters of action have opened up in which governments have had to put new administrative routines in place: not only to secure public funding for biotechnology, but also to assess its hazards, monitor its development, create markets for its products, and encourage ethical and responsible use of biological research. The threat of biological warfare acquired fresh reality after the post–September 11, 2001, anthrax attacks and the still unproven charges against Saddam Hussein's Iraq. In sum, life after the double helix has disclosed many missing elements in the governance of biotechnology and wide gaps among state, citizen, expert, and corporate perceptions of the risks, benefits, and moral ambiguities surrounding the life sciences. Normative problems inevitably follow. Whose views should control the governance of biotechnology? Are governmental attempts to steer biotechnology stifling or setting free new forms of democratic self-expression? How should we evaluate the performance of different nations with respect to governing biotechnology? Will the experiences of wealthy nations serve as model or warning with respect to science and social order in a globalizing world?

In subsequent chapters, I trace the trajectories of these concerns within the legal, political, and policy systems of Britain, Germany, the United States, and, at a supranational level, the European Union. Comparisons will cover the formal principles and processes of law and regulation, their application to specific cases, the practices of debate and dissent, and the formation of institutions and expert discourses by which states and publics assess the risks and benefits of biotechnology. Cases selected for these purposes are drawn from green as well as red biotechnology, even though the two areas have come to be associated with somewhat different scientific debates, ethical concerns, and political questions. The nature of biotechnology and the politics surrounding it justify this inclusive approach. Fundamental to *all* recent advances in biotechnology is a set of ideas and techniques that have reshaped people's understanding of what life is and helped in important ways to frame political thinking about the life sciences. Civil society responses to biotechnology accordingly cut across sector-specific policy concerns. Likewise, the institutional and political changes that beg for cross-national analysis encompass both major domains of industrial application. From a theoretical standpoint, then, more is to be gained from a cross-sectoral comparative strategy than by allowing the scope of analysis to be governed by the contours of specific policy domains.

## Sites of Reflection: A Schematic Roadmap

My objective in the remainder of this book is to explain as fully as possible why new developments in the life sciences were differently received into three national political systems, and what the implications of these stories are for the future democratic control of biotechnology. Prospects for supranational harmonization, both at the EU level and globally, will be deduced to some extent from the comparison of national cases. In keeping with constructivist and coproductionist understandings of the relationship between science and politics, or natural and social order, I will be interested throughout in comparing the framing and bounding of issues for decisionmaking, their uptake into governing institutions and their discursive regimes, and their impact on actors and social identities.

In this spirit, chapter 2 describes three "controlling narratives" that framed the course of policy development on genetic engineering in the three countries. These narratives characterize biotechnology as (1) a novel *process* for intervening in nature, (2) a source of new *products* for the benefit of humans and the environment, and (3) a state-sponsored *program* of standardization and control carrying profound implications for human dignity and freedom, and raising questions of constitutional significance. This chapter reviews the historical and social origins of these three competing narratives and seeks to understand why they were differently received and institutionalized within each national political system.

Chapter 3 takes up a central methodological problem confronting a comparative analysis of the United States with individual European nations: what to do about the role of the European Union. The problem looms large partly because the EU established itself during the period of this study as an independent presence in policymaking for science and technology, including biotechnology. Equally important, though, the EU's own political identity was concurrently under negotiation, so that policymaking for biotechnology became one of the channels through which the EU sought to constitute itself. This "problem" actually offers a theoretical entry point for looking at European biotechnology policy through a coproductionist lens. Periods of emergence during which new social and scientific orders are put in place offer a particularly rich site for examining the interplay and mutual reinforcement of scientific and social possibilities.[89] In this respect, key episodes in the formation of European biotechnology policy shed light on the construction of Europe itself.

Chapters 4 to 8 carry out something akin to a "multisited ethnography"[90] of three national political cultures as they each adjust to the ethical, legal, and social challenges posed by biotechnology. I use the term "ethnography" analogically here, mainly to suggest that political culture, like culture in general, has to be sought at varied sites and in successive episodes of enactment and performance, rather than as codified once and for all within static institutions or regulatory traditions. My method, however, is geared less toward conveying the descriptive thickness of individual events in the manner of Clifford Geertz, or to "following" objects, persons, and discursive trajectories in the manner of Bruno Latour or George Marcus, than to capturing important regularities in episodes of political challenge and democratic legitimation. Whereas current anthropology is often preoccupied with movement and flux, I am interested in the self-perpetuating normative commitments that give societies a claim to coherence and solidarity even in the face of shocks and change. The developments compared in these chapters accordingly meet most or all of the following criteria: they provoked substantial public debate and usually drew forth a discernible policy response at the national level; they gave rise to new institutional arrangements; they altered the existing terms of normative discourse or introduced new elements into it; they were associated with the formation of new or altered social identities.

Chapter 4 looks at the release of genetically modified organisms (GMOs) into the environment and the associated development of new approaches to regulation and risk assessment. This chapter shows that early attempts to settle the debate on environmental safety reopened under the pressures of commercial use, but controversies about genetically modified crops followed different national trajectories. This history helps to illuminate the transatlantic divergence between the U.S. insistence on "risk assessment based on sound science" and the EU nations' concern for "precautionary" policy. Chapter 5 considers the legal and political wrangling over the safety of genetically modified foods. Discursive differences between the U.K. and U.S. cases form the chapter's primary focus, with particular attention to the nature and organization of expertise developed to deal with food risks. Chapter 6 looks at the advent of new reproductive technologies based on in vitro fertilization and the initial steps taken by each nation to ensure that assisted reproduction is kept within the bounds of "the natural." In chapter 7 we revisit the consequences of these regimes in response to techniques such as cloning and research with embryonic stem cells; questions considered more or less settled for a decade were suddenly reopened as citizens and governments discovered unsuspected ethical quandaries lurking behind the naturalized façade of assisted reproduction.

Chapter 7 also builds a bridge to chapter 8 in that these two chapters deal, respectively, with the role of bioethics and patent law in regulating advances in biotechnology. In each domain, new institutions and formal concepts emerged side-by-side with new technological capabilities; changes in ethics and law not only facilitated public deliberation but also served governmental attempts to support genetic research and development. The instrumental role of these professional discourses in framing biotechnology policy (for example, by foregrounding some issues for concern and backgrounding others) is of particular interest. More specifically, chapter 7 examines how the notion of "bioethics" itself was interrogated and recast in each country within new institutions dedicated to promoting biotechnology. Chapter 8, by contrast, explores how existing legal institutions addressed the novel property claims constituted by and for the life sciences and considers how these developments helped to reinforce cultural conceptions of human agency, dignity, and personhood.

Chapter 9 looks at universities as crucially important sites in leading the revolution in the life sciences, but also—increasingly—as experiment stations for a new social contract between science and society. Reviewing national policies for technology transfer from universities to industry, the chapter asks how the entente between the university's historical commitment to interest-free inquiry and its new enrollment into commercial and industrial projects affects the relationship of knowledge to democratic politics.

Chapter 10 reflects on the implications of this cluster of developments for the widely perceived crisis of modernity in advanced industrial states.[91] The chapter's primary contribution is to theorize the role of science and technology in the formation of democratic political culture. The chapter argues that democratic theory in the era of the knowledge society must actively take on board the involvement of citizens in the production, use, and interpretation of knowledge for public purposes. To advance this aim, the chapter develops the concept of *civic epistemology* and compares how it has been actualized in the three countries.

Chapter 11 returns to the book's central normative questions. It asks how democratic institutions on the two sides of the Atlantic have responded to the problems of risk, ethics, and human agency raised by developments in the life sciences and biotechnology. A critical issue is whether the politics of biotechnology has reinforced the familiar modernist paradigm of scientific rationalization and control or whether one can discern here signs of a novel postmodern accommodation, founded on more fluid and less hierarchical (in short, more overtly experimental) relations among science, society, and the state.

# 2

---

## Controlling Narratives

### Framing Biotechnology

Biotechnology made its political debut in the 1970s with its copybook still unsmudged. Founded on the sciences of life, biotechnology promised release from the ambiguities and calamities that had marked the twentieth-century histories of physics and chemistry. By the end of World War II, physics was indelibly identified with the atomic weapons used on Hiroshima and Nagasaki; to this the cold war's unbridled arms race added the globally perceived threat of nuclear annihilation. Hopes that physics would solve the world's energy problems were repeatedly dashed. The 1957 fire at Britain's Windscale nuclear reactor, hastily built at the war's end to maintain U.K. competitiveness as a nuclear power, offered early evidence of risks from the unchecked spread of nuclear technology. Later attempts by governments to develop "atoms for peace" sparked citizen opposition in several countries. Bloody antinuclear protests erupted in Germany in the 1970s, while in the United States a lively antinuclear movement drew force from the near-disaster at the Three Mile Island power plant in 1979 and brought investment in nuclear power to a virtual standstill.[1] Less than a decade later, radioactive and political fallout from the 1986 Chernobyl accident ended for a time any expansionist hopes for nuclear energy in Europe.[2]

Chemistry, too, had to contend with its narratives of abuse and destruction. A century of warfare implicated chemicals in many barbaric forms, from the terrors of mustard gas in the trenches of World War I to the obscenity of Zyklon B in the gas chambers of Nazi Germany, and the devastation of napalm and Agent Orange in Vietnam. Wartime research also produced life-saving and life-enhancing innovations in pharmaceuticals, pesticides, and synthetic materials, but the indiscriminate use of chemicals in agriculture and manufacturing caused rising levels of pollution and grievous accidents throughout the world. In 1962 the biologist Rachel Carson wrote about the effects of persistent pesticides on bird life in *Silent Spring*, the evocative best-seller that helped to launch the modern American environmental movement.[3] In 1984 what some called the world's worst industrial accident occurred at a Union Carbide plant in the central Indian city of Bhopal, snuffing out thousands of lives and injuring tens of thousands others.[4] In succeeding years, the international community confronted yet more chemical crises, from the global effects of stratospheric ozone-depleting chlorofluorocarbons, persistent organic pollutants, and renewed threats of chemical warfare.

Biology, by contrast, remained mostly untouched by negative publicity in the 1970s, when the industrial application of genetic discoveries first began to look feasible. The world had never experienced biological horrors comparable to those of Auschwitz, Hiroshima, Bhopal, or Chernobyl. Indeed, if chemistry and physics represented the forcible conquest of nature, the life sciences seemed to offer more peaceful and cooperative scenarios. Unlocking the secrets of life promised the ability to perfect life itself.[5] Then, too, biotechnology offered an opportunity to close the door *before*, not after, the horse of technological innovation had bolted from the barn of social control. This time, human wisdom would finally learn how to keep ahead of human invention. There was hope that the risks of genetic engineering would be identified in advance, and protective measures would be adopted well before the harmful products of biological knowledge, like those of physics and chemistry, were dispersed into the distant corners of the world.

To be sure, elements of less benign narratives, and the ingredients from which controls on biotechnology would eventually be fashioned, were already circulating in science and popular culture by the late twentieth century. These included the themes of transgressing nature, courting the unknown, and inviting totalitarian control through biopower. Just as images of radiation and radioactivity structured public thought and expectation before the atomic bomb shattered the myth of the "pure" physics,[6] so fears of "playing god" long predated molecular biology and the discovery of DNA. Mary Shelley's *Frankenstein*, the literary locus classicus of modern fantasies of scientific overreaching, became an almost instant success after its publication in 1818.[7] Only five years later, a stage adaptation attended by Mary Shelley herself thrilled London audiences. Repeated cinematic versions, including one produced by Thomas Edison in 1910, gave the Frankenstein myth renewed artistic life in the twentieth century, while also testifying to its remarkable staying power.[8]

In the environmental arena, both fact and fiction lent credence to worries about human error and lack of foresight.[9] Conservation biologists and ecosystem ecologists, in particular, had many tales to tell of disaster caused by unthinking human interference.[10] A number of these featured the transplanting of species from one ecological niche to another in which the absence of controls—predators or other environmental factors—produced explosive growth. Thus the kudzu vine, imported from Japan into the American South in the 1920s to control erosion and produce forage, flourished like a weed in its all-too-hospitable new environment, earning the nickname "mile-a-minute vine." Wild rabbits introduced into Australia in 1859 to provide sport for gentlemen hunters multiplied uncontrollably into "a more nightmarish plague than ever the Egyptians invited," destroying crops, endangering native fauna and flora, and degrading the land.[11] Other threats were less publicly acknowledged. Only the scientists working with rDNA molecules in the early 1970s actually knew what kinds of hybrids they were building and could imagine how these might affect human health or the environment, but their concerns gradually seeped into wider consciousness. These experiences of uncertainty, past and future, lent weight to the discourse of risk that built up around public efforts to regulate new technologies in the latter half of the twentieth century.

If Frankenstein played on primal fears of supplanting divine order with man's imperfect understanding, Aldous Huxley's *Brave New World* broached themes more suited to the midcentury's secular, totalitarian experiments. This was a world of graded, standardized, and denatured human beings, manufactured like the orcs in J.R.R. Tolkien's epic fantasies to meet the needs of an all-powerful state. Turned into instruments of others' interest, people were deprived of uniqueness, autonomy, and free will. Yet the eugenic ideas that found nightmarish expression in Huxley's novel were enthusiastically embraced by Western progressives and intellectuals in the early twentieth century. Just five years before the publication of *Brave New World*, Supreme Court Justice Oliver Wendell Holmes, a giant of American jurisprudence, wrote in *Buck v. Bell*: "It is better for all the world if, instead of waiting to execute degenerate offspring for crime or to let them starve for their imbecility, society can prevent those who are manifestly unfit from continuing their kind. The principle that sustains compulsory vaccination is broad enough to cover cutting the Fallopian tubes. Three generations of imbeciles are enough."[12] Eugenic theories motivated the U.S. Immigration Restriction Act of 1924, which discriminated against Jews and people of southern Mediterranean origin. Not until the excesses of Nazi eugenics, culminating in the Holocaust, were these ideas substantially discredited as a basis for policy.[13] Indeed, in socialist Sweden, a forty-year program of forced sterilization based on eugenic principles ended only in 1976. Meanwhile, seemingly untouched by the Nazi experience, U.S. biomedical researchers' desire for knowledge ran ahead of ethical concerns for the protection of human subjects right into the 1960s.[14]

Politically, the field of potential concern and action remained inchoate well into the 1980s. Researchers saw, of course, that useful medical and agricultural products might one day flow in profusion from their discoveries, but neither they nor industry could guess as yet which research directions and marketing strategies would prove most productive. Beyond the organized interests of science, state, and industry, it was unclear which voices from the general polity might legitimately question the trajectory of the life sciences. The issues swirling around the new biotechnology still needed to be named and characterized, related to realms of prior experience, and be given meaning—in short, they had to be framed as targets of collective action.

In this chapter we turn to the national framings that would have important consequences for the political future of biotechnology. Controlling narratives constructed during the first decade or two of policy development provided resources for, and shaped the contours of, political participation and debate for a long time to come. Biotechnology, as already indicated, came to be conceptualized in three quite different ways in these early years: as a technoscientific *process*, as a stream of *products*, and as a *program* of governance and control. All three frames were seen as significant wherever biotechnology rose to political salience, but they were differently articulated from one country to another in response to contingent historical and political circumstances; in turn, the dominant framing helped support substantially different legislative and administrative arrangements in each national context, with downstream repercussions for science, industry, and democracy.

## A Not-So-Innocent Process

In the United States, questions about how to regulate genetic manipulation unfolded within a well-established culture of technological optimism, but also against a robust tradition of scientific activism and open debate about the social impacts of science and technology. In 1957, for example, seven eminent U.S. scientists were among the twenty-two participants (including two U.K. physicists, but none from Germany) at the first Pugwash Conference in Nova Scotia, Canada, to address the threat of thermonuclear war.[15] Concerned nuclear experts later provided key arguments around which the broader U.S. antinuclear movement coalesced.[16] In 1969, at the height of the Vietnam War, faculty and students at the Massachusetts Institute of Technology called for a reduction in the military uses of science and greater attention by researchers to the social and environmental applications of their work; their efforts led to the formation of the Union of Concerned Scientists, whose activities later encompassed the critical oversight of U.S. biotechnology policy.

Molecular biologists had these precedents to work with when questions about the safety of rDNA research first arose at the 1973 Gordon Research

Conference on Nucleic Acids. Conference cochairs Maxine Singer of the National Institutes of Health and Dieter Söll of Yale University wrote a letter to the National Academy of Sciences (NAS), America's scientific "brain bank," urging it to study the issue and recommend appropriate protective measures. Recognizing the significance of such concerns at a rapidly moving research frontier, the NAS put together a distinguished eleven-member study committee under the chairmanship of Paul Berg of Stanford University. Berg had already hosted a conference on rDNA and possible biohazards at the Asilomar conference center in California in 1973 and was therefore a logical choice to head the NAS effort. Nonetheless, thirty years and several social upheavals later, the Berg committee's composition looks astonishingly narrow: eleven male scientists of stellar credentials, all already active in rDNA experimentation.[17] Yet this elite group's framing ideas were to reverberate through the policy discourse on biotechnology for years, and not only in the United States.

Published in *Science* on July 26, 1974 (and almost simultaneously republished in the leading British journal *Nature*), the Berg committee's report is often cited as the starting point for serious public engagement with the control of genetic engineering.[18] The so-called Berg letter called for a voluntary moratorium on certain types of research until their risks were better understood and appropriate precautionary measures were designed. The signatories looked to NIH, their principal sponsor, to take the lead in creating the necessary safeguards:

> [T]he director of the National Institutes of Health is requested to give immediate consideration to establishing an advisory committee charged with (i) overseeing an experimental program to evaluate the potential biological and ecological hazards of the above types of recombinant DNA molecules; (ii) developing procedures which will minimize the spread of such molecules within human and other populations; and (iii) devising guidelines to be followed by investigators working with potentially hazardous recombinant DNA molecules.[19]

The Berg committee also proposed an international scientific meeting "to review scientific progress in this area and to further discuss appropriate ways to deal with the potential biohazards of recombinant DNA molecules." The follow-up conference, held at Asilomar, California in February 1975, was attended mainly by scientists but also by a small group of lawyers and journalists. It set out the principles that would guide the first round of regulation: rDNA research would be classified according to degrees of hazard (four were ultimately specified, designated P1–P4), and these would be kept within bounds through a cautiously tailored program of physical and biological containment.

The Berg study and the 1975 Asilomar conference won many accolades as exemplary displays of scientific responsibility and self-restraint. Years later, however, it is the restricted representation at those events, together with their emphasis on the regulation of scientists by and for scientists, that strikes us as most remarkable.[20] Those early stirrings of doubt and alarm among biologists supplied crucial conceptual and procedural raw materials for structures that continued to operate when the world of biotechnology had grown a great deal more complex. Among the features that stand out in hindsight, perhaps the most significant is the Asilomar scientists' preoccupation with "recombinant DNA molecules." Understandably, given the disciplinary identities of the leading participants, it was the process of altering molecules that garnered the most attention. Biologists had embarked on creating new living things, such as viruses and bacteria not previously encountered in nature. They saw that this was not a project to be lightly undertaken, but they also presumed that the route to greater understanding lay at the level at which they were conducting their ingenious experiments, that is, at the level of molecular manipulation and control. Molecules were small and relatively easy to understand, as well as inanimate, and thus safely removed from questions of politics or values. That biotechnology might one day destabilize basic elements of social order—kinship, for example, or farmers' rights to own and sow seeds—was very far from the thoughts of the field's founding fathers.

The risks that preoccupied scientists at Asilomar were *biological*, conceived in terms of possible harm to human health and the environment through the unchecked spread of undesired genes. Participating scientists worried about the introduction of dangerous traits, for antibiotic resistance or toxin formation for instance, into molecules that might prove unexpectedly hard to contain within the lab or within the altered organism. The new organisms they were fashioning joined together biologically active genetic material from different species. Who knew what could happen to such constructs if they migrated to the wrong environments? A viral cancer gene transferred to a bacterium normally found in the human gut, for example, might successfully replicate and end up increasing the incidence of human cancer.

Already at Asilomar scientists recognized that they could not go it alone in controlling these risks. Governmental participation would be essential in deciding how to regulate rDNA research, and a formula for the appropriate manner of supervision also began to emerge. Since risks were thought to originate at the molecular level, it made eminently good sense in the private world of rDNA researchers for NIH, the biomedical research community's leading governmental sponsor, to assume responsibility for regulatory oversight. As a grant-giving agency, NIH was most familiar with one instrument of control: peer review of scientists by other scientists to ensure the quality and integrity of their research. Donald Frederickson, then NIH director, simply adapted this tried and true device to the new issues of lab and environmental safety presented by rDNA research. An interdisciplinary Recombinant DNA Advisory

47

Committee (RAC), at first consisting only of scientists but soon modified to include a few lay members, was established to review all proposals for conducting rDNA research and to make sure they complied with applicable guidelines.

Bathed in the long afterglow of Asilomar, where scientists had visibly led government, the media, and the public in reflecting on the risks of their research, the NIH-RAC structure dominated the policy scene in the United States until well into the 1980s. A few scientists and public interest activists were dissatisfied with the Asilomar consensus as being insufficiently attentive to social and moral concerns.[21] Their views, however, remained distinctly in the minority. But as biotechnology inexorably moved from the cloistered scientific laboratory to the competitive marketplace, it became clear that the post-Asilomar settlement could not continue unamended.

## From Process to Products

The second influential narrative concerning the risks of biotechnology also emerged from the U.S. political context as researchers began experimenting with commercial uses of rDNA techniques. In this phase, the focus of regulatory debate shifted from the process of genetic engineering to the safety of its products. The U.S. Congress, the nation's most representative political institution, was once again notably absent from the list of actors who brought about this shift in policy framing, although congressional silence speaks perhaps more to the scientific community's skill in evading legislation than to the legislature's institutional inability to take action.[22] Instead, the Supreme Court, the White House, the executive agencies, and of course the research community provided key inputs. In the background, shadowy as yet but not mute, stood the private interests that hoped to gain from the commercialization of biotechnology.

### A Reward for Invention

Litigation, not surprisingly, played a formative role in shaping the politics of biotechnology in the United States. The first and most influential test of judicial attitudes toward the new technology occurred in a patent controversy that led all the way to the Supreme Court.[23]

Ananda Chakrabarty, a scientist working for General Electric, set the ball rolling by requesting a patent for a bacterium of the genus *pseudomonas*, which he had modified using molecular (but not rDNA) techniques. A patent examiner denied Chakrabarty's claim because the thing he sought to patent was a living organism and hence not within the legal definition of patentatble "subject matter." The Court of Customs and Patent Appeals reversed the examiner's decision, holding that "the fact that microorganisms . . . are alive . . . [is] without legal significance."[24] The Patent Office appealed, but a 5–4 majority of the Supreme Court sided with the appeals court, asserting that the important question was not whether the thing to be patented was living, but whether it was "a 'manufacture' or 'composition of matter' within the meaning of the statute."[25] On this narrow question, the majority saw no room for doubt. Chakrabarty's organism fell within the universe of man-made objects that did not previously exist in nature. Thomas Jefferson, the author of the patent law, had intended the act to be expansively construed, the Court observed, and Congress had confirmed that intention as recently as 1952 by decreeing that patents could be granted for "anything under the sun that is made by man."[26] Clearly, the Court concluded, Chakrabarty's bacterium met the test of being man-made.

The legal argument in *Chakrabarty* masked the politically more interesting question of where technology assessment should take place in a democratic society—in the legislature or in the courts. But this was a question the Supreme Court resolutely refused to take up as relevant to the disposition of the case. Uncontroversially, the majority stated that evaluating the desirability of genetic engineering was not a task for the judiciary, which possesses neither the competence to make policy nor the power to enforce its conclusions: "judicial fiat as to patentability will not deter the scientific mind from probing into the unknown any more than Canute could command the tides."[27] The meta-issue that the Court side-stepped was whether the character of invention itself had changed so radically in the intervening years as to render Jefferson's vision of intellectual property out of date. By treating the patentability of microorganisms as a purely technical issue governed by existing law, the Court effectively denied that the advent of molecular biology had caused any such fundamental shift to take place. *Chakrabarty* gave biotechnology a green light and thereby removed from Congress any incentive to assess the adequacy of the patent law in relation to new developments in the life sciences.

If the Court chose a narrow path, it was not for lack of alternatives. One of the many *amici curiae* (friends of the court) who filed briefs in *Chakrabarty* was the People's Business Commission (PBC), a watchdog organization founded by the activist author and technology critic Jeremy Rifkin. PBC attempted to situate the immediate patent dispute within a broader trajectory of technological change with far-reaching moral and political implications. Through techniques of genetic modification, the PBC argued, humans had for the first time gained the ability to "manufactur[e] life itself" and even to "influenc[e] human evolution."[28] This was a technology born of the long Western tradition of scientific reductionism, and it would enable scientists to manipulate and objectify nature still more completely. By eroding the distinction between living and inanimate matter, patent law would lend its considerable weight to consummating an ongoing process of denaturing nature. With this reasoning, the PBC hoped to recast the legal patent dispute as a conflict over the larger politics of technological governance—hence as an issue belonging properly to Congress rather than the courts. Reframing the debate in this way would have rebuilt the actor-network supporting commercial biotechnology so as to include the legislature and its work; Congress would have had to reenter the picture as an active node in the network of governance for the new technology.

Perhaps predictably, the judicial majority in *Chakrabarty* took no notice of the PBC's arguments. As far as the Court was concerned, the technology-promoting thrust of American patent law was thoroughly unambiguous and plainly applicable to the circumstances of this case; there was nothing here that necessitated any legislative second thoughts.[29] Interestingly, as we will see, positions similar to those advanced by the PBC fared differently when promoted by the Greens in Germany. In the United States, however, the shift of policy attention from the *process* of genetic engineering, which Rifkin and his associates deplored, to its *products*, which Ananda Chakrabarty and his employer sought to patent, was now definitively underway. It would gather momentum in the next few years.

### From Unplanned Release to Policy Coordination

Several events in the early 1980s highlighted the insufficiency of the NIH-RAC review process for controlling commercial biotechnology. The first challenge came when two University of California scientists, Steven Lindow and Nikolaos Panopoulos, sought permission to carry out a field test using the "Ice-Minus" bacterium, a member of the *pseudomonas* family that had been genetically modified to increase the frost resistance of plants. In its widely dispersed natural form, this bacterium contains an ice-nucleating protein that makes the plants on which it is found susceptible to frost damage at relatively higher temperatures. By deleting the gene that produces this protein, the researchers hoped to create a new "ice-minus" strain that would not encourage ice formation and, by displacing the more common ice-nucleating strain, would increase the sturdiness of many plants. California's lucrative strawberry production would be a particular beneficiary. The RAC members who reviewed the application initially asked for some modifications but decided unanimously after a second round of review that the experiment was safe.

The scientific community had traveled far indeed from its precautionary posture at the Asilomar conference. Experiments of the very kind now proposed by Lindow and Panopoulos, and matter-of-factly endorsed by RAC, had been banned by the 1976 NIH guidelines, barely a half-dozen years earlier, as too risky or uncertain to be undertaken. Not everyone found this scientific about-face satisfactory, and the Foundation on Economic Trends

(FET), another of Rifkin's organizations, sued to overturn the decision. In *FET v. Heckler*,[30] a federal court of appeals blocked the proposed field trial on the ground that NIH had not carried out the environmental impact assessment required by the U.S. National Environmental Policy Act (NEPA). The court was troubled by NIH's failure to explain why a type of experiment that had been explicitly ruled out under earlier guidelines could now be permitted to go forward with no more than a sentence of justification. The decision challenged the RAC's implicit assumption that, in assessing the safety of rDNA research, the accountability of scientists to other scientists was all that mattered. The law, according to *Heckler*, required a more open and deliberative process that would enable the public to judge for itself the quality of the experts' reasoning. NEPA, the court held, demanded a different, less internalist mode of justification in order to speak to public concerns about the environmental risks of commercial biotechnology.

Other problems soon followed. In one case, the molecular focus of the NIH guidelines created a technical barrier to regulation. In mid-1987 Gary Strobel, a plant pathologist at Montana State University, injected a genetically altered strain of the *Pseudomonas syringae* bacterium into fourteen elm trees without proper authorization. Eager to find a cure for Dutch elm disease, Strobel was openly impatient with what he saw as unnecessary bureaucratic restrictions on his research. An NIH special committee convened to review Strobel's behavior disqualified itself as lacking jurisdiction since his particular study did not involve "recombinant DNA molecules." As defined by the rDNA guidelines, recombinant DNA molecules are "either (i) molecules which are constructed outside living cells by joining natural or synthetic DNA segments to DNA molecules that can replicate in a living cell, or (ii) DNA molecules that result from the replication of those described in (i) above."[31] Strobel's study used a recombinant plasmid, but this construct did not replicate fully in the host organism that Strobel "deliberately released" into the environment; hence, the released organism did not contain rDNA molecules within the technical meaning of the guidelines and thus fell outside NIH's regulatory purview.

Private companies, too, faced difficulties in readying their products for testing in the field. Since their research had not necessarily been supported by government funding, NIH could not officially certify their activities as safe. Some companies turned for help to the Environmental Protection Agency (EPA), which had begun drafting rDNA regulations under existing legislation on toxic substances, but unresolved jurisdictional issues and conceptual ambiguities led to more controversy and growing doubt whether anybody at all was minding the store with respect to America's emerging market in agricultural biotechnology.[32] In 1986 President Reagan's Office of Science and Technology Policy (OSTP) stepped into the breach with a report entitled "Coordinated Framework for the Regulation of Biotechnology." OSTP identified three agencies that already had extensive legal jurisdiction over the new technology—the EPA for environmental applications and impacts, including pesticide regulation; the Food and Drug Administration (FDA) for new foods, animal drugs, and pharmaceuticals; and the U.S. Department of Agriculture (USDA) for new crops and animals. Under OSTP's leadership, a Biotechnology Science Coordinating Committee (BSCC) was established to develop a common interagency approach on regulatory matters. In addition, each of the lead regulatory agencies developed new institutional means for dealing with biotechnology. For example, EPA established a Biotechnology Science Advisory Committee (BSAC) to give advice on the scientific and technical aspects of regulation.

These institutional arrangements cemented a consensus across the U.S. government that biotechnological products—pesticides, drugs, foods, crops, and food additives—should not be treated any differently for regulatory purposes from similar products manufactured by biological or chemical processes that did not involve gene manipulation. Since biotechnology was framed as just another industrial process, lacking any special attributes or consequences, existing regulatory laws aimed at controlling specific classes of biotechnological products were deemed sufficient to control any new risks. OSTP and the three agencies cooperating in the Coordinated Framework persuaded Congress that regulations issued under prior laws would adequately clarify the necessary concepts and eliminate possible jurisdictional conflicts.

Under the Coordinated Framework, biotechnology ceased to be a matter for broad participatory politics and became instead an object of bureaucratic decision making under the guidance of technical experts. The forces of mainstream science and industry—not social activists, technology critics, or the environmental movement—were henceforth to be in the driver's seat in managing the emergent technology. Consistent with this framing of the issues, the Coordinated Framework cheerfully admitted the need for new scientific bodies while rejecting the need for new legal authority. The creation of an expert advisory committee (BSAC) for EPA and a scientific coordinating committee (BSCC) at the interagency level showed that federal regulators viewed the task ahead as primarily one of technical assessment, divorced from any unprecedented normative concerns, and were prepared to strengthen their institutional capacity accordingly.

### A Science of Precise Interventions

Scientists on the whole applauded these moves, keeping intact the spirit of the post-Asilomar accommodation between the life sciences and the state, but not its precautionary orientation. The *Heckler* opinion incensed many in the scientific community for whom RAC's vigorous approval of the experiment had been quite enough to ensure its reliability; a "technically illiterate" court, they believed, had no business second-guessing the expert committee's informed judgment.[33] Besides, with increasing familiarity, molecular biologists were losing their own earlier nervousness about genetic engineering. James Watson, an Asilomar conference participant, was a vocal, visible, and influential convert to the new position. In a 1977 *New Republic* article, he labeled the Asilomar conference "an exercise in the theater of the absurd" and asserted that the effort to assess and control genetic engineering was "a massive miscalculation in which we cried wolf without having seen or even heard one."[34] In place of "crying wolf," the very precision of the techniques by which genes could be manipulated began to cast a seductive spell on the imagination of molecular biologists. With such seemingly perfect control at their command, why should they allow their techniques to be coupled with unnamed fears and horrors put forward by ignorant or eccentric social activists? It was time, the scientists felt, to dissociate the process of genetic engineering from any unjustified imputation of risk.

The growing scientific consensus on the safety of genetic engineering was soon transmitted to the policy arena through the medium of expert advice. A high-level report produced by the National Research Council in 1989 shored up the U.S. government's position that commercial biotechnology posed no special risks to human health or the environment.[35] On each major issue where splits had developed among the leading federal regulatory agencies, the NRC report sided with the FDA and USDA, the agencies that took the more relaxed view of biotechnology's hazards.[36] Specifically, the NRC report concluded that

> (i) the *product* of genetic modification and selection constitutes the primary
> basis for decisions . . . , and not the *process* by which the product was obtained;
> (ii) although knowledge about the process used to produce a genetically modified
> organism is important . . . , the nature of the process is not useful for determining
> the amount of oversight; and (iii) organisms modified by modern molecular and
> cellular methods are governed by the same physical and biological laws as are
> organisms produced by classical methods.[37]

The message was unambiguous: mere use of rDNA techniques did not make a harmless product dangerous; nor, conversely, were organisms produced by "classical methods" safe simply because they were not genetically engineered. For policy purposes, biotechnology was henceforth to be regarded as a supplier of familiar classes of products requiring familiar types of review—not as a unique technological process threatening society with uncertain or incalculable harm.

In contrast to the Berg committee's anxious, cautionary tone, the NRC experts just fifteen years later sounded a fanfare of optimism about genetic engineering, stressing how much was already known about its possible impacts. The molecular theme still dominated, but it was tuned now to the register of reassurance. The report took pains to point out that molecular methods, whether used on plants or microorganisms, are highly precise and lead to modifications that can be fully characterized and understood.[38] This precision, the committee felt, provided adequate safeguards against unpredictable behavior by genetically modified organisms. Regulatory attention was thus diverted away from the process of genetic engineering by stabilizing—or black-boxing—a particular interpretation of that very process, one stressing its precision, predictability, and near-perfect amenability to control.

Opposed to molecular biology's powerful rhetorics of precision and knowledgeability were the alleged unreason, ignorance, and sheer bureaucratic incompetence of the policymakers who wished to place hurdles on the road to technological innovation. The confrontation between science's precision claims and regulators' alleged resistance became embedded within U.S. political discourse. Referring to the ice-minus case as recently as 2002, Henry I. Miller, a former FDA regulator and noted biotechnology enthusiast, invoked both themes to explain how a promising invention had been stopped in its tracks: "Using very precise biotechnology techniques called 'gene splicing,' the researchers removed the gene for the ice nucleation protein and planned field tests with ice-minus bacteria. . . . Nonetheless, the field trial was subjected to an interminable and burdensome review just because the organism was gene-spliced (and even though it contained no new genetic material, but had merely had part of one gene deleted)."[39] Based on Miller's analysis, the EPA had clearly erred in failing to follow the scientists' conclusion that it was the product, not the process, that mattered for regulatory review.

## European Articulations: Process or Program?

In Europe the product narrative met a less friendly reception, although most European molecular biologists shared their American colleagues' confidence that genetic engineering per se presented no substantial threats to health, safety, or the environment. The persistence of a process-based regulatory approach in Europe had to do in part with the institutionalization of policy authority in the European Commission (as discussed in chapter 3) and in technologically advanced EU member states. In part, it had to do with the rise of a third, and equally powerful, characterization of the risks of biotechnology—one that focused not only on the physical but also the social and political dimensions of the changes that the new technology might set in motion. Biotechnology, according to this third view, was neither simply a novel scientific process nor a cornucopia of new products, but a *program* of state-sponsored activity that threatened important individual and collective values with its emphasis on reduction and manipulation. A comparison of British and German policymaking in the decade of the 1980s clarifies and sharpens the contrasts between the European and American approaches.

### Britain: The Persistence of Process

Britain, where Francis Crick together with James Watson cracked the genetic code and shared the Nobel Prize for their discovery, provided from the start a seemingly idyllic home for genetic research and development. With a head start in the basic sciences, a powerful economic base in the chemical industry, and a still embryonic environmental and consumer movement, British politics might have been expected to downplay the risks of biotechnology more than any other state's. But Britain's responses to Asilomar and to subsequent European legislation in fact produced a more cautious initial framing than that adopted by U.S. science advisers and policymakers. The divergence began almost immediately after Asilomar. Laboratory work involving what was termed "genetic manipulation" in Britain was controlled through regulations under the Health and Safety at Work Act of 1974. Implementing this law was a relatively unusual "tripartite" structure of advisory committees that included representation from all three of Britain's most powerful social partners: labor, industry, and local as well as national government.[40] The Genetic Manipulation Advisory Group (GMAG), the committee set up to review applications for rDNA research, was therefore also constituted as a tripartite body. Labor's presence on GMAG implicitly acknowledged that rDNA research could carry consequences—to the workforce, for example—that were not reducible to the scientifically determinable risks of harm to human beings and the environment.

In 1984, as research moved nearer to commercialization, GMAG was reconstituted as the Advisory Committee on Genetic Manipulation, later renamed the Advisory Committee on Genetic Modification (ACGM),[41] to the Health and Safety Commission and its operational arm, the Health and Safety Executive (HSE). Biotechnological work with environmental implications was further reviewed by the Department of the Environment (DOE), which obtained expert advice from its own interim Advisory Committee on Introductions. By the late 1980s, however, it became clear in Britain as in the United States that many biotechnological activities, including large-scale industrial production and deliberate releases into the environment, could not easily be controlled through existing regulatory structures.[42]

Developments within the European Community provided additional impetus for new legislative controls. In April 1990 the Community adopted two directives relating to biotechnology: one on contained experiments, which was largely consistent with existing British policy, and one on deliberate release of GMOs, which required active implementation. The U.K. government introduced into the Environmental Protection Act of 1990 (the so-called

Green Bill) a new Part VI specifically dealing with GMOs. Meanwhile, environmental and health and safety authorities decided to consolidate their separate expert committees into a single new committee to review applications for releasing GMOs into the environment. The resulting interdepartmental Advisory Committee on Releases to the Environment (ACRE) held its first meetings in July 1990.

Debate on the Green Bill provided a focal point for environmentalists to demand formal participation in decisions about GMOs. Somewhat to the activists' surprise, the government responded favorably and agreed to include an environmental representative on ACRE. The person selected for this position was Julie Hill, a member of the Green Alliance, an environmental lobbying group spun off from the Liberal Party that had been active in commenting on the Green Bill. Supported by funds from government and industry, as well as individual memberships, the Green Alliance was known for its middle-of-the-road views on biotechnology, favoring tighter regulation but no outright bans.[43] Hill herself brought no particular scientific or technical expertise to her new role, but she did bring the British liberal left's concern for greater transparency and openness in official decisionmaking.

Within Britain's consensual and incrementalist policy culture, Hill's appointment marked at once a departure from tradition and a concession to longstanding political practice. Asking a Green Alliance representative to sit on ACRE affirmed that the hitherto unorganized environmental interest in biotechnology was significant enough to be represented in future negotiations over safety. It was a relatively radical move in the British context, where segments of society that have neither strong party affiliations nor demonstrated technical expertise are not normally granted representation on key decisionmaking bodies. But ACRE, like ACGM before it, was formed under the aegis of the Health and Safety Commission, at the time the most participatory of Britain's regulatory agencies. This institutional context perhaps made it easier for ACRE to accommodate a new interest (environmentalism) than it would have been for more traditional expert committees, such as those attached to the notoriously pro-industry Ministry of Agriculture, Fisheries and Food (MAFF).[44]

Britain's traumatic encounter with "mad cow disease" or BSE (bovine spongiform encephalopathy) was already brewing out of public view, and a contrast of MAFF's handling of expertise in that case with ACRE's greater inclusiveness is instructive. To head its first Working Party on BSE, MAFF appointed Sir Richard Southwood, former chairman of the Royal Commission on Environmental Pollution (RCEP) and vice-chancellor designate of Oxford University, and very much a "safe pair of hands" in British parlance. Southwood had to overcome some resistance on MAFF's part even to appoint a human health expert to his committee, since both MAFF and the Health Ministry agreed that "the public should not be given a false impression that a health risk in man is likely."[45] Needless to say, there was no question of including consumer or even grower and producer interests on a committee that was supposed to conduct its work, even in the face of an unprecedented crisis, without arousing public interest or concern.

Environmentalists gained a seat on ACRE at a time when the Conservative government, undergoing a crisis of confidence toward the end of Margaret Thatcher's leadership, was seeking to expand its hold among moderate environmentalists. For British government and industry, the Green Alliance may well have represented environmentalism with a human face—a voice of civilized dissent that could be internalized without seriously jeopardizing the development of biotechnology. In constructing an appropriate advisory committee on the risks of commercial biotechnology, the government simultaneously constructed—or coproduced—rules of the game for the kind of "green participation" that it was prepared to live with.

In Britain as in the United States, a well-timed report by a prestigious expert body helped reinforce the government's legal and institutional arrangements. The Royal Commission on Environmental Pollution, a standing body that advises the government on environmental matters, decided that the time was ripe for a thorough evaluation of deliberate release, looking at both the possible consequences of releasing GMOs and the procedures for identifying, assessing, and mitigating their risks.[46] RCEP's assessment, however, differed markedly from that of the molecular biologists on the NRC committee in the United States.

Issued in 1989, contemporaneously with the NRC report, the royal commission's review was both more inclusive in its treatment of impacts and more open in admitting uncertainty. Thus, instead of dwelling on benign past experiences and the precision of molecular techniques, as did the NRC, the British experts emphasized how much was still unknown—and hence how little could be predicted with assurance—about the likely behavior of GMOs in the environment. With respect to genetically modified plants, for example, the RCEP report considered a broader range of possible risks than the NRC and seemed unwilling to dismiss any of the scenarios as wholly improbable. Similarly, the historical experience with the introduction of exotics was deemed highly relevant to the release of GMOs into environments where they were not native.[47] With respect to herbicide resistance, the commission considered not only the biological likelihood that resistant genes might spread to weedy species, but also the secondary social consequence that the genetic engineering of plants resistant to herbicides might lead to intensified use of environmentally damaging herbicides.[48]

As we will see in chapter 4, the theme of nature's unpredictability echoed through official British policy. Was this a legacy of Britain's colonial past, in which the global movement of plants and animals had played such a constitutive, though largely unpremeditated and unpredictable, role?[49] We will return to Britain's institutionalized experiences of public knowledge-making and knowledge use in our discussion of civic epistemology in chapter 10. For now, the important point is that British authorities accepted seemingly without hesitation the Royal Commission's recommendation that, to start with, all GMO releases should be subject to regulatory scrutiny. Put differently, this amounted to accepting the principle, denied in America, that the *process* of genetic modification should define the scope of policy action. Officials at both DOE and HSE acknowledged that it might eventually be possible to establish risk categories that would either exempt some products from evaluation or subject them to reduced oversight.[50] But any such relaxation, they indicated, would have to rest on actual experience, that is, on empirically observed data from prior releases rather than on an in-principle determination that the GM process poses no risks.

### Germany: Programmatic Fears

Public debate about biotechnology was slower to gain momentum in Germany than in the United States (and slower still in Britain), but when it did, it drew on a particular reservoir of national memories, experiences, and political activism with no close analogues in any other Western country. Only ten years before the discovery of DNA's structure, German geneticists along with other biomedical scientists were watching, and sometimes actively abetting, a state consciously dedicated to the proposition that all men are *not* created equal.[51] Influential European thinkers saw science's complicity in the Nazi Holocaust as a logical extension of the fatal modern alliance between the ordering and controlling state and its obedient instruments for subjugating nature.[52] Auschwitz, the noted East German playwright Heiner Müller declared, was "the last stage of the Enlightenment," the altar of a rationality which "as the only binding criterion reduces man to his material worth."[53] Reestablished on democratic principles after the war, and officially cleansed of its most gruesome doctrines, the German federal state nevertheless represented for many of its citizens an entity whose good faith in the sponsorship of technoscience was still unproven and not at all to be taken on trust.

None of these worries were openly on display in the early 1960s, when the federal government first targeted the biological sciences for state support, thus setting in motion the official policy history of genetic engineering. Biotechnology received a particular boost with the creation in 1972 of the Federal Ministry for Research and Technology (BMFT),[54] whose main mission was to channel funding toward designated "key technologies." When German scientists, like their counterparts elsewhere in the world, responded to Asilomar's call for controlling the risks of genetic manipulation, BMFT took on the role played by NIH in the United States. German policy ideas were largely derivative at this stage, and political unrest about the life sciences was yet to surface. Under BMFT's supervision, a restricted, ad hoc committee of experts drafted guidelines closely modeled on NIH's; neither labor nor industry interests were included at first, although both were later represented in a twelve-member commission formed to implement the guidelines.[55] Through the early 1980s, regulatory debate was confined within this carefully constructed expert committee, thus ensuring a relatively narrow focus on the physical risks of rDNA research and correspondingly muted attention to its wider social and political ramifications.

Mere absence of debate, however, as we have learned from steadily accumulating studies of technological controversies, is not a reliable indicator of public acceptance.[56] In the Germany of the mid-1980s, the rise of new social movements and the waning of earlier controversies over nuclear power set the stage for a more participatory politics of biotechnology.[57] The Green Party, first elected to the Bundestag in March 1983 and searching to refine its political agenda, soon created a working group on genetic technology. In the same year, an alliance between the Greens and the Social Democrats led to the formation of a parliamentary Commission of Enquiry (Enquete-Kommission) to examine, in the words of its official title, the "prospects and risks of genetic engineering" (*Chancen und Risiken der Gentechnologie*).[58] The seventeen-member commission was chaired by the Social Democrat Wolf-Michael Catenhusen and consisted of nine members of the Bundestag (four Christian Democrats, three Social Democrats, one Free Democrat, one Green) and eight nonpolitical experts.

The Greens, though massively outnumbered, compensated for their political weakness with the purity of conviction. At their February 1986 national meeting, the party issued a starkly uncompromising declaration on gene technology and its use in human reproduction.[59] The action was intended to challenge impending moves in state policy, which the Greens hoped to derail with their own ringing opposition. Arguing that biotechnology would increase inequality, harm the Third World, ignore the basic causes of ill health, undermine social solidarity, and promote biological weapons—all contrary to promises of increased health and safety—the Greens categorically rejected biotechnology in any of its possible guises. It was a very public shot across the bows, foreshadowing the party's oppositional stance within the Enquiry Commission.

The four-hundred-page report that the commission issued in 1987 was much more moderate, concluding chiefly that new legislation was needed to control biotechnology properly. But as the inquiry process for the first time subjected state policy on biotechnology to systematic, institutionalized criticism, two radically divergent views emerged concerning the nature of the problem confronting policymakers. The Greens and the Social Democrats argued that the risks of biotechnology were sufficiently unsettling—uncertain, reductionist, manipulative, potentially catastrophic, perhaps irreversible—to require a new political dispensation for their management and control.[60] Key to this new order should be a more pronounced voice for the public, institutionalized through new forms of participation. The Christian Democrats insisted, in tones reminiscent of official U.S. policy, that biotechnology was fully amenable to control through normal channels of assessment by technical experts.

As a prominent advisory body to the Bundestag and as a microcosm of German party politics, the Enquiry Commission provided a high-profile forum for airing these differences. Through its representative Heidemarie Dann, the Green constituency in the Bundestag presented a procedurally daring "special vote" dissenting from the commission majority's conclusions. It was highly irregular for a parliamentary minority to try to introduce such a measure into an official report to the Bundestag (and yet more to disclose it to the press); strikingly, too, at three times the authorized number of pages, the Green dissent violated all of the commission's own guidelines on the length, timing, and disclosure of submissions. The very first sentences flung down a strident challenge to the commission's modus operandi and conclusions: "When dealing with a technology, pregnant with such far-reaching consequences and affecting the very basis of life, *it is essential to hold a wide-ranging, public and essentially open debate on its development, before the development itself is allowed to proceed. The majority of the Commission endeavored to prevent rather than promote such a debate* (emphasis in original)."[61]

The approach the Greens espoused, knowing full well that there was no hope of its adoption by the legislative majority, was aimed at reversing the burden of proof on commercial biotechnology. Safety and feasibility were not alone to determine the direction of R&D; the Greens asked instead that alternative methods be explored before any practical application of genetic engineering was attempted in any field, including medicine, agriculture, food production, and environmental protection. Only in this way, they argued, would the new technology be made truly responsive to human needs. We observe here the beginnings of a demand for social accountability that would gain strength in the later European politics of biotechnology—and the politics of science and technology more broadly—but at the time the Greens' case for a showing of need made only modest headway.

Despite evident annoyance with the Greens' tactics and their conclusions, the commission majority decided to incorporate the full text of the "special vote" into the final report so as not to grant the authors "any pretext for representing themselves as a persecuted minority."[62] The decision spoke volumes about the Greens' perceived political and moral strength, even when their wholly idealistic policy position stood no chance of gaining majority approval. The publication of the special vote put into circulation at the highest levels of political discourse an exceptionally wide-ranging thematization of biotechnology's social and political, as well as physical, risks. Arguments that had sunk with hardly a trace in an obscure *amicus* brief to the U.S. Supreme Court in *Diamond v. Chakrabarty* here achieved political standing, if not anything like political acceptance.

Green opposition to biotechnology played itself out not only within the Enquiry Commission but also in the parallel and in some ways less predictable venue of litigation. In an unusual lawsuit against Hoechst chemical company, German environmentalists in Hessen challenged a planned facility for the production of genetically engineered insulin on the ground that the state—conceived under German law as having an affirmative duty to protect its citizens—had not as yet adequately guaranteed the safety of biotechnology. Existing laws, they argued, did not provide a sufficient basis for controlling risks whose characteristics required explicit legislative authorization, just as nuclear power had done a decade earlier. The administrative court of Hessen accepted this representation of biotechnology's uniqueness and ordered a halt to industrial biotechnology until a suitable legal framework was in place. It was a significant decision in German constitutional law because it reinforced the proposition that the state could be held accountable not only for the overextension of its delegated power (*Übermass*) but also for its underuse (*Untermass*). This conception of an underconscientious state has no exact parallel in Anglo-American law.

Within a year, the German parliament set aside the inconvenient roadblock created by the Hessen decision by enacting the 1990 Genetic Engineering Law (Gentechnikgesetz), which incorporated many but not all of the Enquiry Commission's recommendations. In particular, no moratorium was declared on field releases of GMOs. Overall, however, the German law imposed significantly more stringent reporting and administrative requirements on the biotechnology industry, even for the lowest risk (P-1) experiments, than were in effect at the time in the United States. Evidently, the Greens' aggressive political dissent had made a difference.

Critics immediately denounced the law for repudiating the demands for greater participation in the government's insulated, bureaucratic-technocratic structures of control. By combining the functions of protection (*Schutz*) and promotion (*Förderung*) within a single law, the state had asserted its capacity to undertake these potentially conflicting tasks without compromising the values or rights of its citizens, but dissenters questioned whether this optimism was justified. As a partial concession to public concerns, the German law opened up participation on the government's key advisory committee (as Britain had also done) and created a new public hearing process for reviewing deliberate release applications. These innovations addressed the themes of social and political risk articulated through the parliamentary inquiry before the law's enactment. In practice, however, the procedural reforms proved short-lived, as will be seen in chapter 4.

## Repertoires of Political Action

Three very different framings of biotechnology as a regulatory problem emerged in the United States, Britain, and Germany in the 1980s. How can we explain these divergences? There are, to begin with, no simple connections between the intensity of public risk perceptions and the breadth or stringency of the regulatory response. The United States, despite its history of environmental activism and scientific controversy, emerged as the place where the risks of genetic engineering were defined in the narrowest terms, associated only with classes of products, and where new legislative and institutional proposals were deflected with little dissent. In Germany, an exceptionally heightened perception of risk did indeed coincide with tough policy measures, but German activists perceived what was at stake in the development of genetic engineering in vastly different terms from U.S. environmentalists. For the Greens, the matter at issue was the relationship of science and technology to society, not simply the risks of biotechnology as contemplated by experts in the life sciences. In Britain, where public debate was most muted at the start, the official regulatory posture remained more cautious throughout than in the United States. Here, we begin to address some of the reasons for these discrepancies, whose implications and ramifications will occupy us in later chapters.

The most striking feature of the early national debates on commercial biotechnology is that they were carried out in such different institutional settings. Three aspects of this institutionalization merit particular attention: the role of scientists in framing the issues for public policy; the discourses of persuasion employed by the key participants; and the constitutional position of the state in relation to science and technology.

### Scientists and Experts

In her analysis of possible structural relations between scientists and the state, the political scientist Etel Solingen offered a prediction that perfectly accounts for the success of U.S. rDNA researchers in setting the policy agenda at Asilomar and beyond. There will be "happy convergence" between the goals of the state and its scientific communities, Solingen proposed, when there is "a high degree of consensus between state structures and scientists, who enjoy internal freedom of inquiry and relatively comfortable material rewards."[63] Such a consensus marked the early development of American biotechnology. Molecular biologists, acting in their own interest but not unmindful of the public good, supplied some of the key formulas that legitimated government policies favoring relatively unconstrained research and development. In particular, the National Research Council's 1989 report added considerable authority to the adoption of a product-based (hence more pro-industry) rather than process-based regulatory framing. In return, molecular biology enjoyed continued high support coupled with relatively low state supervision, until conflicts over stem cell research erupted during the Bush administration.

But we need a subtler touch than Solingen's to explain why scientists—despite their similar structural positions—played such different roles in framing British and German biotechnology policy. One notes to start with that the "independent," university-based research community whose ideas triumphed at Asilomar made less of a difference as an interest group in European than in U.S. policy. Britain's Sidney Brenner, for instance, was a lead figure in early regulatory developments,[64] but his significance diminished as the debate moved to the commercial applications of biotechnology. Scientists continued to make key inputs into policy in both Britain and Germany, but only in the guise of "experts"—that is, not only, or even primarily, as spokespersons for the nation's knowledge elite ("scientists"), but as men and women deemed specifically competent to advise government on the design of policy. The socially constructed role of "expert," whether in Britain's RCEP or Germany's parliamentary Enquiry Commission, was thus quite different from the role of scientists on the Berg study committee, at the Asilomar meeting, or on RAC in the United States. The former were consulted on the appropriate way to regulate science; the latter spoke out on how to tailor regulation so as to accommodate science's best interests. In Europe the state set the parameters of the debate and then drew on science; in the United States, scientists actively set the agenda and the state endorsed it. In other words, scientists intervened or were enrolled into the political process according to different national conceptions of their relationship with the state and society. The work that scientists did to safeguard particular regulatory framings and methods of oversight differed correspondingly. We will return to these points repeatedly in subsequent chapters.

## Discourses of Persuasion

Actors in the three countries differed strikingly in the rhetorical and persuasive strategies they employed to frame biotechnology. The importance of litigation in defining the U.S. policy agenda is not surprising. More notable is the fact that *Chakrabarty* retreated from the strains of democratic populism embraced by the courts in the previous decade of environmental litigation. In the patenting case, legal authority was invoked to shut off, not open up, new grounds for concern about technological risk. The Supreme Court limited its analysis, both rhetorically and conceptually, to the narrow basis of statutory construction. In so doing, the Court implicitly ruled out the possibility that the very concept of "invention" had changed so radically in the life sciences that it no longer made sense to apply historical precedents. Instead, likening their own role to that of the legendary King Canute who vainly commanded the tides to hold still, the

Justices in the majority naturalized the process of invention, declaring it unstoppable like the tides of the ocean. By these moves, *Chakrabarty* reinforced traditional American narratives of emancipation and progress through science and technology. Simultaneously, the Court helped constitute the emerging political consensus that the market, rather than the law, was the right instrument for controlling the inventiveness of biotechnology.

In Britain, political mobilization on biotechnology was virtually nonexistent in the early 1980s, but the legislative amendments proposed to implement the EC directives (the "Green Bill") briefly provided an entry point for environmentalists. Having established their credentials in the political arena, however, British "greens" were incorporated into a key decisionmaking forum, ACRE, which was governed by altogether different deliberative rules. Here, seated among accredited technical experts, but not fully on a par with them, the environmentalists' capacity to influence policy became congruent with the ability of their designated representative (the Green Alliance's Julie Hill in the first instance) to critique the conduct of risk assessment. Not surprisingly, Hill's contributions leaned less toward the intricacies of technical analysis and more toward associated questions of governance, such as the transparency of decision making.

The electoral success of the German Greens gave them a voice in the legislative process that was not available to biotechnology's critics in either Britain or the United States. Through the vehicle of the Enquiry Commission, this minority party parlayed a procedural advantage into an amplified political presence. Unconstrained by the technical discourses of law or risk assessment, the Greens undertook a fundamental critique of modern science's alliance with the state that went well beyond the PBC's unsuccessful brief in *Chakrabarty*. At the heart of the Green protest was a rejection of a particular view of science and technology, one they saw as increasingly divorced from any pure commitments to observation and theory, but driven by commercial needs toward laboratory experimentation and intervention. In a revealing if naïve aside, the Greens applauded the famed American geneticist Barbara McClintock for her holistic approach to studying genes:

> Using a mainly observational technique she discovered the "jumping genes" as early as the forties and interpreted them as a phenomenon which at the very least would modify the central core of inheritance dogma. Over twenty years later molecular biologists confirmed this work by other methods. Also, McClintock did not need to perform any genetic engineering manipulations for her work and no new and potentially dangerous organisms were produced by it.[65]

Science, according to the Greens, admits of many possible forms of inquiry into the same set of problems. If modern molecular biology is not the only way to understand heredity, as they claimed, then the state's decision to foster that particular disciplinary approach was not simply a way of furthering constitutionally protected inquiry. The choice could only be seen as political and economic, hence open to political critique. This argument also surfaced in the Greens' discussion of the state's constitutional role in relation to scientific inquiry.

### Science, State, and the Constitutional Order

A third theme that figured in the framing of biotechnology in all three countries was the constitutional relationship of science and the state. Whether or not this issue was openly broached in political debate, different conceptions about it underlay the developments reviewed in this chapter. These contrasts, briefly outlined here, add specificity to the theoretical argument about coproduction sketched out in chapter 1 and further developed in the chapters that follow.

The only mention of science in the U.S. Constitution is in Article I, which grants Congress the power to "promote the Progress of Science and Useful Arts by Securing for limited Times the exclusive Right to their respective Writings and Discoveries."[66] This provision authorized the original Patent Act, whose meaning the Supreme Court construed so generously in *Chakrabarty*. The decision, moreover, was perfectly consistent with the practice of American courts in upholding the freedom of scientific inquiry whenever it does not threaten the institutional power of the courts to declare what counts as right knowledge.[67] More generally, the Court embraced a strictly instrumental conception of the state's role in promoting science and technology. It showed no inclination to explore the deeper threats to human self-conception (such as the shift from an autonomous to a manipulated view of human nature) that could be posed by biotechnology; nor were the Justices concerned with the possible implications of such threats for a constitutional order based on notions of individual autonomy and liberty.[68]

Britain, with its unwritten constitution, is the nation where one would least have expected an explicit, formal discussion of the science-state relationship. Yet features of the British constitutional order invisibly worked their way into the early accommodation with commercial biotechnology. The scientist-philosopher Michael Polanyi had celebrated the "republic of science" in the 1970s, calling attention to its absence of internal hierarchy and its freedom from all restraints but that of internal peer criticism.[69] But Polanyi ignored the political economy of British science, conducted in state-sponsored institutions and oriented toward achieving national goals in defense and medicine. He also seemed oblivious to the historical traditions that conditioned British scientists to accept a conceptual separation between "pure" science, however intertwined with national interests, and the taint of industrial exploitation. The "republic" of British science maintained its independence, in fact, by remaining more detached from industrial and commercial ties than much American science, but also by never questioning its deep financial, political, and cultural ties to the state. Significantly, there was no talk in Britain of "rights" to free inquiry or to the fruits of one's inventions in connection with the early regulatory initiatives on biotechnology. The sequestering of advice, and of possible dissent, within the state's expert advisory structures at once ratified the principle of ministerial discretion in matters of governance, including the governance of science, and reaffirmed the presumption that scientists, like all responsible members of the British elite, are servants of the state.

It was Germany, with its still-young Basic Law and sharp distrust of power, that most explicitly addressed the state's constitutional responsibility in relation to new and potentially hazardous technologies. The parliamentary Enquiry Commission asked whether the Basic Law either compelled the legislature to prohibit genetic engineering or prevented it from restricting the technique's use. Answering both questions in the negative, the commission noted that freedom of research is guaranteed by the Basic Law and is underwritten by a conception of "science as a process of discovery and a search for truth."[70] Basic and applied research were equally protected; the commission saw no significant differences from the standpoint of constitutional rights.

Again, the Greens protested. In their view, genetic technologies were converting the biological sciences into instruments of control and manipulation. This slippage from "basic" to "applied" science demanded, they argued, a thorough reconsideration of whether genetic engineering as actually practiced remained worthy of the constitutional protection given to free inquiry. Logically, the reasons for protecting inquiry, as integral to the unconstrained development of human personality and autonomy, could not apply if inquiry was directed only toward applied and instrumental ends. Indeed, in this latter context, technoscience itself might function as a regulatory device shaping human possibilities; its limits would then have to be clarified and democratically deliberated just like those of any other exercise of state power. This direct confrontation with constitutional issues remained a hallmark of Germany's public engagement with biotechnology, despite the apparent ineffectuality of the Greens in translating these ideas into legislative policy.

## Conclusion

Modern biotechnology's second wave, the phase that led from the early breakthroughs in molecular biology to marketing the first products of that enterprise, encountered broadly similar constellations of interests in Britain, Germany, and the United States. In each country, the political dynamics of biotechnology were shaped by firm national commitments to basic science, economically powerful industries, and state authorities eager to demonstrate their support for a winning technology. These forces, however, were institutionalized into the political process in critically different ways, offering the actors who sought to shape policy access to substantially different strategic and discursive resources. The consequence, as we have seen, was that different initial problem framings emerged as salient across the three countries; in each, the dominant frame was consistent with national traditions of scientific advice giving, citizen participation, and historical—we may even say constitutional—accommodations between science and the state.

This is not to say that there was anything inevitable or fixed for all time about the ways in which genetic engineering occupied each country's political imagination in the 1980s. Cross-national commonalities ran deep, in commitments to economic growth, technological progress, social equality, and human autonomy. Key actors, notably scientists and private companies, but in time also consumers, environmentalists, and health activists, pursued their projects and strategies across national borders and hence either rejected or opposed locally specific framings of policy issues. In a period of increasing global communication, scientific discoveries were widely reported and the debates around them flowed easily across territorial boundaries. The forces of bureaucratic harmonization and political integration, especially within Europe, added more layers of complexity to the country-specific initiatives described in this chapter. Most of these factors would have predicted a gradual weakening of national particularities and the emergence of an international political consensus around biotechnology. Yet, as we will see, the normative differences captured in the divergent framings remained alive and repeatedly rose to prominence as new issues came into focus on national political agendas. National differences did not simply fail to die away. They sometimes intensified. To deepen our understanding of these dynamics, we turn next to one of the principal sites of transnational policy formulation, the European Union.

# 3

## A Question of Europe

While nation-states were busy incorporating the life sciences into their policy agendas, a political world beyond nations was also in the making. Globalization was the watchword of the late twentieth century, subsuming a host of flows and movements that weakened the nation-state and gave rise to new supranational organs of power, protest, and politics.[1] Of these, by far the most important for our purposes is the European Union.[2] Beginning with the six nations of the European Coal and Steel Community (ECSC) in 1951, the EU grew to fifteen member states by 1995, and as of 1998 it began contemplating a further enlargement to add at least thirteen more. In January 2002 twelve of the fifteen EU members adopted a new common currency, the euro, brought into circulation in a remarkably swift, though not uniformly smooth, transition orchestrated by the European Central Bank.[3] On February 27, 2002, delegates from all of the member states and candidate countries gathered in Brussels for a first meeting of the "Convention for the Future of Europe." Inconclusively finishing its work in July 2003, the convention and its immediate aftermath led some cynics to write off the occasion as just another European talk show or, worse, an exercise in futility.[4] More sanguine observers saw the events as the opening gambit in what EU Commission President Romano Prodi dramatically designated in 2002 as the "the birth of Europe as a political entity."[5] In the end, the 325-page document agreed to by the twenty-five member states in June 2004 was less a work of inspiration than of complex integration—and it still had to be ratified by every state.

What was the nature of this emerging entity, and how was that nature reflected in EU politics and policy with regard to the life sciences? The question needs an answer because it goes to the heart of cross-national comparison. We have to assess the EU's role in relation to the two member states—Britain and Germany—that are at the center of this study. Did the rising importance of European institutions render national governments irrelevant or at best subservient to European policy, or did the EU fail in its efforts to coordinate policy across divergent national cultures? Or, resisting such binary formulations, was the nature of European influence altogether more subtle, and did it relate in unexpected ways to the changing complexion of European politics during this period of ambiguous, contested *super*state formation?[6]

Although it is easy to pose the question about the EU's influence on biotechnology policy, efforts to answer it lead to analytic difficulties. As we will see in this chapter, the character of the Union and the shape of its life science policies were both in flux during the period of this study, and attempts to define the latter were in some ways reflective of the former. Seemingly technical questions about how the EU should promote biotechnology in the member states turned out, in other words, to be unanswerable without also taking on board the deeper question of what kind of European Union there should be. Complicating the analyst's task is the ambiguity of Europe itself as a geographical, political, and cultural construct. In the period covered by this book, none of these dimensions of Europeanness was by any means settled. That instability poses an unresolvable linguistic as well as conceptual challenge, inasmuch as Europe has to be invoked as an agent with fixed attributes and capacity to act, even though its actions are the very means by which new European identities are being constituted.

This chapter looks, then, at some twenty-five years of European biotechnology policy as both shaping and shaped by European politics. Each word in that phrase—European biotechnology policy—is contested and fluid; each moves through history attached to changing discursive and material practices. And yet each is also fixed enough to be easily understood by a heterogeneous collection of actors spread across many countries and interest groups. To grasp both its continuity and its contingency, we look at several key moments in the formation of this construct, from roughly 1980 to just after the turn of the twenty-first century, assessing the ways in which each term (European, biotechnology, policy) has evolved through that period, but also asking how each has shored up the meaning of the other two—and of Europeanness more broadly. In this sense, the formation of European biotechnology policy is a story of coproduction.[7] It is at once about the evolutionary transformation of Europe as an economic, political, and cultural union and about the consolidation, within Europe, of the technoscientific sector known as biotechnology.

The periodization of European integration and of European biotechnology policy follows, needless to say, no perfect pattern of cause and effect nor any neat temporal sequence. The artificiality of fitting these two complex and interlinked developments into a linear historical narrative can only be overcome through iteration and a more detailed accounting of parts of the story (such as Europe's role in food safety or patent policy) in other chapters under other topical headings. Here, I begin with a brief overview of the chief institutional and political transformations that affected the discourse on biotechnology within Europe. I then focus on several controversial moments in the articulation of biotechnology policy, linking them as far as possible to Europe's evolution as a supranational political entity. In each of these episodes, as we will see, a primary challenge has been to find appropriate discursive and institutional means with which to address the problems that Europe confronts. European policy has struggled to find a balancing point among several overlapping and at times contradictory pressures: the ambiguous relationship of the EU to its member states; differing attitudes toward the EU among those states; the ever-active push of transatlantic competition with the United States; and the legitimacy problems confronted by Brussels in relation to European citizens.

The issue of legitimacy relates back in turn to the kind of union Europe hopes to be, and indeed is permitted to be. For it is a curious political formation, whose constitutive laws contain no provisions for a direct response by the center to the demands of an organic European public. The principle of subsidiarity, moreover, locates many of the details of politics back in the member states, by demanding that implementation be delegated to the smallest and lowest appropriate level of governance. Adding to the mix, some of the nations pushing hardest for European integration, including France and Germany, were in 2004 among the worst laggards in implementing European legislation. Areas of noncompliance included some of direct relevance to this study, such as intellectual property rights and protection for biological inventions.[8] The framing discourses of European biotechnology policy reflect all these tensions about the meaning and force of a united Europe.

## An Unprecedented Union

In the spring of 2002, the walls of many European Commission offices in Brussels sported a jaunty poster in the primary colors of a child's paint box or a late Matisse collage. Under the heading "9 May 2002 Europe Day" were displayed, in spare graphic design, a blue euro symbol, a pointy yellow star, and a red hand; the message below read "The Euro: The European Union in your hand." A fitting slogan indeed for a political union rising on the foundation stones of a single market! Personal prosperity was here equated with European unity, while also implying that the single currency would place the future of the EU in the hands of its economically empowered citizens. But the annual Europe Day posters also served an inward-looking function. Asked about it, a commission colleague wryly observed, "Well, you know every institution needs its own day to celebrate. This is ours."[9]

Her comment, made only partly in jest, points to a basic dilemma confronting the EU's administrative institutions. Their power and influence have grown beyond anything that founding figures such as Robert Schuman or Jean Monnet could have imagined a half-century ago. But is the EU supposed to function as a *technocratic* intergovernmental organization comparable to the World Bank or the International Monetary Fund, which are (allegedly) answerable only for the goodness of their policy outcomes, or is it a *political* supranational union, accountable under democratic principles to an increasingly Europeanized public and society? If the latter, then Europe has a problem, for—despite repeated invocation of fundamental values such as peace, precaution, and solidarity—the EU, unlike its several member states, has no constitutional status and no reserves of shared linguistic, cultural, or religious loyalty to legitimate its exercises of authority. Born out of agonizing conflict, the EU is inscribed upon a palimpsest of histories that confound easy claims of European unity.

How can the nascent Europe achieve something of the ideological coherence of a nation-state without an "imagined community"[10] of European citizens to give meaning to its efforts? Formal structures, after all, are not enough to build a polity, especially when the *demos* is irreducibly multiethnic and heterogeneous, any more than imagination alone is enough to hold nations together without the solider appurtenances of power and order—a law code, an army, an educational system, a tax base.[11] There is, too, a tension between two quite different ways of imagining a more political superstate. Should the EU adopt an active-integrationist posture, seeking to eliminate cross-state divergences in policy framings, or is it better off playing a passive-preservationist game, aimed at protecting deep-seated national values that give rise to such cross-cultural differences?[12] Some aspects of European policy for the life sciences reflect insufficiently worked out answers to these questions.

Commemorative symbols offer one way to fill the vacuum of identity, but symbols can go only so far in the absence of other devices for building solidarity. Even something so minor on its face as designating an official day for Europe reflects the commission's ambivalence on this point, or perhaps simply its lack of self-reflection.[13] The web page for Europe Day states in its opening statement, "Today, the 9th of May has become a European symbol (Europe Day) which along with the single currency (the euro), the flag and the anthem, identifies the political entity of the European Union." Even the designers of the page concede that this is a day without meaning for most of their intended audience. A link to the question "What is Europe Day?" leads to the disarmingly honest answer, "Probably very few people in Europe know that on 9 May 1950 the first move was made towards the creation of what is now known as the European Union."[14] Even the posters reflect the artificiality of the construct they represent, for they are united mainly in their brash coloration; the euro symbol, connoting a primarily economic consolidation, features in three of the posters since 1996.

Europeanization, the process leading to this awkward way-station between common market and uncommon political union, can be outlined in three broad historical strokes, each of which is intimately tied up with the formulation of European biotechnology policy. The first centers on the formal legal and institutional development of Europe, from the ECSC to the EU and beyond, entailing changes in Europe's self-definition as an actor on the international stage and, more particularly, in its economic and political relations with that other great democratic confederation, the United States of America. The second concerns the internal bureaucratic politics of the European Commission, as reflected in the changing rules and dynamics of its policy process. The third and most political strand has to do with the gradual emergence of a series of agenda items that are properly defined as European and cannot be attributed exclusively to the concerns of individual member states. Each provides some of the context for the policy debates on biotechnology that will occupy us later in this chapter.

### European Transformations: Progress at a Price

The history of the EU began in the shattered aftermath of World War II, when Winston Churchill and other European leaders broached the idea of a "United States of Europe" to end once and for all the hostilities among their countries. Their aspirations were partly realized in 1951 through the signing of the Treaty of Paris establishing the coal and steel union, but it was not until the Treaties of Rome in 1957, forming the European Economic Community (EEC) and the European Atomic Energy Community (Euratom), that the structure of a new supranational government began to solidify. This was the year in which a Parliamentary Assembly, predecessor to the European Parliament (EP), was established. Even then, the union comprised only six members;[15] the United Kingdom, despite Churchill's early enthusiasm for unity, remained outside the EEC for another fifteen years.

Significant institutional developments in Europe can be listed almost month by month after the formation of the EEC, but the most important further consolidations came in the mid-1980s and later: the 1985 Single European Act creating a single market for movements of goods, services, capital, and persons; the 1992 Maastricht Treaty establishing the European Union; the 1997 Amsterdam Treaty strengthening cooperation on employment, security, the environment, and foreign affairs, and increasing the power of the European Parliament; and the 2001 Nice Treaty laying out procedures for the admission of countries from central and eastern Europe. There were shocks and opt-outs along the way. Danish voters famously—and astonishingly—rejected the Maastricht treaty in 1992, coming around only after special provisions were made for Denmark on issues like the single currency and EU citizenship. Britain did not adopt the Social Charter on worker rights and, along with Sweden and Denmark, initially decided to stay out of the European Monetary Union. And in June 2001 Ireland, the sole country requiring a national referendum to ratify the Nice Treaty, dealt the EU an unexpected blow when no more than a third of its voters turned out, only to reject the accord by about 54 to 41 percent.

These dramatic gestures of repudiation are the most obvious signs of continuing disquiet among European politicians, citizens, and academics about the nature of the EU and the basis for its legitimacy. Internationalization is generally seen as a threat to democracy, because it distances the exercise of power from traditional forms of popular control,[16] and the EU's indeterminate status exacerbates this concern. On the one hand, all recognize that the EU is not a simple intergovernmental organization, deriving its powers under principles of international law from the states that authorized its creation. Membership in the EU requires a well-understood ceding of sovereignty on selected matters by member states to European institutions.[17] On the other hand, the EU is also not a federated superstate legitimated through direct representation of citizens or sustained by thick ties of language, ethnicity, traditions, or a common civil society and public sphere. Only the European Parliament, with an unwieldy 736 members after the 2009 election, is elected by direct universal suffrage, and it alone among Community institutions meets and deliberates in public. The Council of the European Union, the EU's lead decision-making body, is answerable to the member states and not directly to citizens, although, since Maastricht and Amsterdam, it shares responsibility with the EP through an expanded and simplified codecision procedure. How then do EU decisions aspire to legitimacy, let alone achieve it?

On a formal level, the changing role and significance of the European Parliament has been an indicator of the EU's evolving relation to its possible publics.[18] Until the Maastricht Treaty, the EP was a purely consultative body, exercising no real legislative power. After 1992 the codecision procedure spelled out in Article 251 of the EC Treaty permitted the EP to participate with the Council on a more equal footing in approving legislation; preparing legislative proposals, however, remains the Commission's exclusive responsibility. Among the areas where the parliament's powers have increased are environment and research, with consequences we will return to below. The EP also approves the president of the European Commission and the remaining commissioners for five-year terms and has the power to dissolve the Commission through a censure motion. Although this power was not formally exercised in the 1990s, the threat of censure caused the Commission headed by President Jacques Santer to resign collectively in March 1999, following widely publicized allegations of fraud, mismanagement, and nepotism.

TABLE 3.1
European Parliament Elections: Voter Turnout, 1979–2004(in percent)

|         | 1979 | 1984 | 1989 | 1994 | 1999 | 2004 |
|---------|------|------|------|------|------|------|
| EU      | 63   | 61   | 58.5 | 56.8 | 49.4 | 44.2 |
| Germany | 65.7 | 56.8 | 62.4 | 60   | 45.2 | 43   |
| UK      | 31.6 | 32.6 | 36.2 | 36.4 | 24.0 | 38.9 |

*Source:* UK Office of the European Parliament, http://www.europarl.org.uk/guide/textonly/ Gelecttx.htm (visited July 2004).

The EP's large gains within the Community power structure, however, were not sufficient to secure it an enthusiastic public following. Voter apathy grew steadily from 1979, when the first European parliamentary elections were held, and turnout dropped to its lowest level of 44.2 percent in the sixth election of June 2004. While the decline in voting was Europe-wide, absolute percentages varied enormously country by country, from a 2004 high of 90.8 percent in Belgium (where voting is compulsory) to a record low of 24 percent in Britain in 1999; in general, Germany hovered roughly around the European average while Britain consistently registered among the lowest levels of voter interest. Divergent framings of the meaning of European integration partly explain the cross-national variation.[19] Across the board, these dropping figures fueled perceptions of a "democratic deficit" (or, in German terms, *Öffentlichkeitsdefizit*, public deficit) at the European level. But the European Parliament was not the syndrome's only victim; the European Commission suffered as well.

### Performative Politics: Improving Governance

When former Italian Prime Minister Romano Prodi took office as European Commission president in September 1999, he knew he had a massive job of confidence rebuilding on his hands. The general problem of disenchantment faced by many democratically elected Western governments, particularly after the end of the cold war, was magnified in the case of EU institutions by a complex and cumbersome institutional structure, an opaque administrative process, and the cynicism provoked by the abrupt downfall of the preceding Santer Commission. To these were added the complications of the EU's most ambitious planned enlargement in 2004, which threatened to ratchet up existing problems of legitimacy beyond the threshold of institutional resilience. The very practical challenges of this historical moment took shape, moreover, against a backdrop of growing theoretical confusion about the EU's political identity. Unique in its history, structure, reach, and ambitions, the EU seemed to correspond to no well-defined niche in democratic theory; instead, its acts of self-realization proved to be theory-forcing for political analysts, much in the way that complex technological systems sometimes preexist and presage the theories that account for their success.[20]

One of the Prodi Commission's actions is especially relevant because it prompted debate on the nature of European democracy and carried longer-term implications for European policies affecting biotechnology. This was the White Paper on Governance,[21] issued in July 2001, which aimed to reform the process of decisionmaking in the Commission. The document, answering one of Prodi's personal initiatives, laid out two principal goals: to improve the openness, transparency, and accountability of Commission decisions, and to improve their coherence and efficiency. Some observers, however, discerned a third, unstated motive: to redress the balance of power within the EU in favor of the Commission, wounded by attacks from the press and the public as the EU's least democratic branch.[22]

Both the White Paper's content and the comments it generated helped crystallize troubling questions about EU legitimacy that were percolating beneath the surface even before *governance* was tagged as an explicit item for discussion. The political scientist Philippe Schmitter, a state theorist by specialty, put his finger on the problem, asking, "What is there to legitimize in the European Union . . . and how might this be accomplished?"[23] Schmitter noted that, although charges of a "democratic deficit" were leveled against the EU, it was inappropriate to hold this *objet politique non-identifié* (unidentified political object[24]) to the standards and measures of democratic performance normally applied to nation-states. Neil Walker, a constitutional lawyer, refined the problem by disentangling three strands of legitimacy analysis—regime, polity, and performance legitimacy—their centers of definition located, respectively, in a political entity's formal organization and structure, its relations with citizens, and its delivery of good policies. Walker suggested that the Commission's White Paper was pursuing all three goals at once, although in a discourse that seemed intent more on veiling than making transparent its quasi-constitutional ambitions.[25]

Of particular interest to us is a part of the White Paper that dealt specifically with confidence in the expert advice offered to the Commission. Biotechnology and food crises were singled out for explicit mention as the Commission acknowledged that "a better informed public increasingly questions the content and independence of the expert advice that is given." Among the problems the Commission identified were the need for a wider range of disciplined inputs "beyond the purely scientific," and the need to counter the opacity of the existing system of expert committees and their modus operandi.[26]

### Politics in Search of a Polity?

These initiatives and the reactions to them raise another fundamental question about the EU at the millennial turning point: who are the Commission's audiences, real or imagined, and who are the consumers of the improved governance measures that the 2001 White Paper outlined in schematic yet controversial form? Measured in numbers, the White Paper caused hardly a tremor in the daily conduct of European politics. The comment period ending in March 2002 produced only 250 submissions, although some were substantial; scholars also entered the debate, as in an essay collection produced by lawyers and political scientists at the European University Institute in Florence.[27] The White Paper, besides, was not an end in itself but the start of an effort to follow through on some thirty action points over the course of 2002. The concept of governance had clearly arrived to stay.

Critics observed, however, that the Commission's view of its intended audience remained extremely narrow. To be sure, one of the five principles of good governance that the Commission endorsed was participation: "The quality, relevance and effectiveness of EU policies depends on ensuring wide participation throughout the policy chain—from conception to implementation. Improved participation is likely to create more confidence in the end result and in the institutions which deliver policies."[28] By including a fuller range of participants, the Commission implied that its practices would become more closely aligned with those of national governments, showing similar concern for democracy and the rule of law. But as Paul Magnette noted, the text ran counter to the Commission's apparent intent, in that it limited participation largely to organized sectoral actors, and even that chiefly during the consultative or predecision process.[29] One could legitimately ask whether such elite stakeholder-oriented participatory strategies were in any sense an effective antidote to European civil society's apathy toward the EU and its works. Whatever else, this approach did not cohere with an active-integrationist approach toward building Europe on harmonized lines featuring convergent public values.

While top-down initiatives such as the White Paper cannot generate a polity out of whole cloth, the impact of such a document, and the action points derived from it, could be more significant if there were in some sense a polity already in the making prepared to respond to Commission initiatives. European social scientists have debated whether an identifiable European public and public sphere do exist, and, if so, how they can be characterized or even measured. Qualitatively speaking, there is reason to consider the "European public" a reality, even if it lacks two of the most basic instruments of the public sphere as traditionally conceived: a common language and common media of communication.[30]

Jürgen Habermas, the foremost theorist of the public sphere, never posited the need for a common language as a prerequisite for democratic deliberation. The German sociologist Klaus Eder similarly rejects the language argument. Eder has suggested three reasons for regarding united Europe at its fiftieth anniversary as possessing its own emergent society, occupied with public debate on genuinely European questions. Phenomenologically, he notes, several problems and scandals of the late twentieth century heightened European citizens' concern about matters originating outside their own national borders and yet affecting their security and well-being. These included the "mad cow" crisis precipitated by the discovery of tainted British beef, which came to light in 1996;[31] the debate over EU border security and the Schengen process for cross-border travel; and the corruption charges against the Santer Commission. In these and similar cases, Eder argues, publics began to use EU forums to critique both transnational (EU) institutions and institutions of other member states. This process in turn gave rise to actor-networks that criss-cross national boundaries, even though not all social strata (blue-collar and agricultural workers, for instance) are as yet equally enrolled into such interactions.[32]

These events and debates point to the open-ended character of European integration following the cold war and extending into the twenty-first century. Issues, institutions, discourses, actors, and political communications are all identifiable at the transnational European level, as are some incipient markers of a European cultural identity, but little is set in stone, and some of Europe's founding premises seem more contested fifty years on than at the creation of the first economic union of the ECSC. How have Europe's initiatives in the field of biotechnology affected, or been affected by, the unsteady processes of political integration? That is the topic of the remainder of this chapter.

## Biotechnology Policy: Fields of Discourse and Practice

The European Commission identified biotechnology as a key area for policy intervention as far back as the mid-1970s, but its actions picked up substantially in speed and diversity in the 1980s. Herbert Gottweis has identified several "discursive codes" used to justify a European presence in this field in the early years: among them, a technological race with the United States, a solution to structural problems in health care and agriculture, and consolidation of "a strong and unified Europe."[33] These were not merely arbitrary linguistic choices but strategic uses of language to *produce*, in effect, problems that European institutions could then position themselves to address. The focus on biotechnology as a field of risks as well as opportunities is consistent with the political theorist Yaron Ezrahi's conception of the liberal democratic state, which constantly seeks to legitimate itself through instrumental uses of science and technology.[34] States put on their technological performances in the court of public opinion, and citizens, acting as "attestive witnesses" in Ezrahi's terms, grant or withhold their assent in accordance with their assessment of the state's competence.

Unlike the atomic bomb, polio vaccine, or Apollo missions, however, the new genetic sciences and their attendant technologies resisted producing spectacular successes (or, as in the case of the moon landing, hugely success-ful spectacles[35]) through the latter quarter of the twentieth century. Accord-ingly, the would-be European superstate had to invent not only the discourse of problem-solving but also the ways in which it could make the urgency of its task, as well as possible markers of its success, visible to its intended audi-ences. Biotechnology, in other words, posed much the same kind of challenge that modernizing states had contended with in the late nineteenth century in defining and coping with a whole range of new social problems. At that time, the social and human sciences evolved rapidly to underwrite state efforts to counter such evils as poverty, unemployment, and occupational injuries. Numbers and statistics, in particular, proved to be inordinately important tools for making society visible to itself—and visible, moreover, as a collec-tion of at risk or suffering groups whom states could protect or make whole through appropriately targeted policies.[36]

The European Commission's efforts to represent biotechnology policy as rational and beneficial called for similar demonstrations, but the process turned out to be anything but linear and tidy. Obstacles to rationalization emerged both internally, within the European institutional framework, and externally, as a result of exogenous events and the shifting character of Euro-pean political sensibilities. Conflicts arose around three sites of contestation—the initial framing of regulatory policy, the definition of the "public" and its interests, and the role of ethics in science and technology policy. As a result of these divisions, European influence mainly served to keep certain issues alive on the policy agenda, and to tag these items as suitable for "European" resolu-tion, presumably for the benefit of a correspondingly "European" public. The EU was less successful in bringing about convergence on a variety of key ques-tions and hence was relegated to a largely passive preservationist role.

### Early Steps: Horizontal and Vertical Frames

European institutions can influence policy in the member states by formal as well as informal means. Formally, the EU commands several instruments through which it intervenes in national science and technology policy. Most important is the directive, which requires member states to accommodate their domestic legislation to a common European regulatory norm. Other points of intervention include European research funding, under the so-called Framework Programmes, and various technical harmonizing activities of European expert committees. The informal choices are by definition less clear on the surface, but potentially no less important in the long run. These include internal struggles over where to lodge lead responsibility for policy development on specific issues, as well as attendant choices among alternative ways of framing the debate and implementing policies. Since the early 1980s, the EU has employed all these means to shape the European policy discourse, and policy, for biotechnology.

The first stage of European activity covered more or less the decade of the 1980s, ending in 1990 with the adoption of three major directives pertaining to biotechnology: Council Directive 90/220/EEC on the deliberate release of genetically modified organisms; Council Directive 90/219/EEC on the contained use of genetically modified microorganisms; and Council Directive 90/679/EEC on the protection of workers from the risks related to exposure to biological agents at work. Of these, the first, on deliberate release, most centrally addressed the framing questions described in chapter 2. Should biotechnology be represented for European regulatory purposes as a technological *process*, demanding special concern because of its intrinsic properties (the U.K. and German position), or as an in-itself harmless tool for the manufacture of *products* that could be assessed according to already extant regulatory principles (the U.S. position)? The directive adopted the process frame. Some saw this as a victory for prudence and precaution; others—let us call them the antiprocess interests—saw it as a signal defeat for policy coordination inside the Commission.

In Brussels, the task of coordinating biotechnology policy was distributed by the mid-1980s among a number of bodies of varying significance and leadership ability: the Biotechnology Steering Committee (BSC), formed in 1984, its chairman the director-general of DG XII, the Directorate-General for Science, Research and Development;[37] BSC's small but highly proactive secretariat, the Concertation Unit for Biotechnology in Europe (CUBE), created in 1984 and dissolved in 1992; and the Biotechnology Regulatory Interservice Committee, spun off from the BSC and in operation from 1985 to 1989. Superficially boring and bureaucratic, these arrangements were the administrative façade for what insiders saw as a power struggle among the directorates, in which DG XII (Research) ultimately lost to DG XI, the Directorate-General for Environment and Nuclear Safety.[38]

According to Mark Cantley, the maverick and visionary former head of CUBE, DG XII's failure consisted in the Research Directorate's inability to sell to its sister directorates the image of biotechnology as a cross-sectoral, knowledge-based engine of innovation, revolutionary in its promise for human betterment, and unstoppable in its inevitable advance. Cantley's own perspectives were significantly shaped by several years of surveying biotechnology from the vantage point of the Forecasting and Assessment in Science and Technology (FAST)[39] Program. This modest program, with a budget of 4.4 million ECUs from 1978 to 1982, enjoyed the advantages of what Cantley, in retrospect, termed "holy poverty":

> The FAST mandate was beautiful. It actually told us to highlight prospects, products and potential conflicts affecting the long-term development of the Community. So we had an explicit mandate to look at the long-term and not to hide the policy conflicts. So it's totally counter-cultural from most bureaucracies, where

your obligation is to conceal the conflicts. We were actually armed with a formal mandate from Parliament and Council to highlight the prospects, problems and potential conflicts likely to affect the long-term development of the Community.[40]

But freedom came at a price. The holistic framing of biotechnology that FAST proposed from its privileged outsider position could not survive the interservice competition of the later 1980s.

In 1982 FAST issued a report on Community strategies for scientific research and development in several key areas, including biotechnology, that were identified with the future of Europe.[41] The chapter on "bio-society," Cantley recalled, "had quite an effect." Gottweis represents this chapter as an ambitious attempt to construct an "out of control" European society in need of ordering through biotechnological means.[42] Doubtless these resonances were not wholly absent, as connotations of rationalization and rectifying never can be from projects using the life sciences to improve on nature.[43] In advocating the concept of bio-society, however, the authors of the FAST report were concerned more with the need for European coordination and synthesis than with pathologizing European society and correcting its deficits and errors. Three levels of integration were explicitly foreseen in order to fulfill Europe's grand designs on nature: political, in that this was to be a community-wide initiative; scientific, in that biotechnology was seen as implicating the "whole range of the life sciences, their foundations and their applications";[44] and societal, in that the program aimed at "the conscious management of self-organizing systems for the sustenance and enrichment of human life and purposes."[45] But the institutional commitment needed to translate FAST's highly integrative vision of a European bio-society into concrete policies never materialized. What were the reasons?

Looking back, we can observe two basic contradictions in the positions that FAST and the opponents of the "process" framework sought to advance within the Commission. The first was simply the effort to integrate policies for promoting biotechnology horizontally across an institution that was organized vertically among directorates, each of which had its own network of clients and supporters fanning out across the member states. In Cantley's frustrated assessment:

> Time and again, one comes back to the fact that the Commission is built in seventeen vertical slices—seventeen Commissioners,[46] who are nominated for a fixed period, who are unsackable, who have no electoral constraints inviting them to "hang together, or, hang, separately." So they are effectively uncontrollable and uninterested in horizontal cooperation except in some very short-term and direct bilateral bargains. . . . And that made it exceedingly difficult to conduct a coherent strategy for biotech, so we didn't realize in FAST, how potentially impossible the concept of a Community strategy for biotechnology actually was. We learned gradually.

To be sure, interagency rivalries are scarcely unique to the European Commission. In the 1980s, however, the task of coordination was substantially more daunting in the European Commission than in the United States, the birthplace of the "product" framework for regulating biotechnology. U.S. regulatory agencies, unlike the European DGs, did not have to account to a multiplicity of national ministries driven by disparate political imperatives. White House leadership, exercised through the Office of Science and Technology Policy, helped pull together the agencies most concerned with biotechnology, and the Food and Drug Administration in particular was prepared to lead the effort to build an interagency consensus. Further, coordination was helped along in the United States at this juncture by a critical regulatory vacuum around commercialization, prior interagency cooperation, support from a prestigious scientific body (the National Academy of Sciences), and organized scientific and industrial lobbies. The deregulatory philosophy of the Reagan years also did its part by badly demoralizing the Environmental Protection Agency, the most proactive and precautionary of the agencies responsible for controlling biotechnology, and potentially the most sympathetic to the "process" framework. In Europe, by contrast, the other DGs did not wish to accept DG XII's leadership, and BSC was weakened by a rotating chairmanship, insufficiently frequent meetings, low and overly technical levels of civil servant staffing, and continuing tension among the DGs.

A more subtle contradiction can also be discerned in the framing of biotechnology by the antiprocess interests in the Commission. As far as benefits were concerned, CUBE and others in DG XII put forward an ambitiously holistic view of biotechnology. Proponents noted that dramatic and fundamental shifts were taking place in the life sciences as a whole. Sooner or later, these discoveries and their applications would revolutionize all the fields in which biology played an important part, including medicine, agriculture, environment, and Third World development. Biotechnology, these advocates argued, should therefore be fostered through a single, coordinated program of research and development. This after all was the basis for demanding a unified "Community strategy" for the field. When it came to regulation, however, the self-same defenders of policy integration insisted on a more decentralized, disaggregated approach based on the primacy of sector-specific regulation. The argument was that risks—unlike benefits—lay not in the techniques of genetic modification, but in its specific applications. These more or less familiar risks could be targeted and adequately managed through sector-specific controls. Not all European actors, it later proved, were willing to settle for this asymmetrical representation of integrated benefits and fragmented risks.

As policy formulation moved forward in the late 1980s, splits developed between DG XI (Environment), which took charge of drafting the two most important directives on biotechnology, and DG XII (Research), as represented by CUBE. In preparing Directive 90/220/EEC on deliberate release, DG XI adopted the process framework, extending legislative control to all genetically modified organisms rather than to specific products within the remit of specific directorates. In effect, the directive overcame the strictures of vertical organization by bringing GMOs under a single, horizontally applicable regulatory mandate addressing a new technological means of production. Critics saw this as a capitulation to German Green interests, which had successfully stigmatized genetic modification as unnecessarily hazardous and a possible driver of nonsustainable agricultural practices.[47] The Environment Directorate understandably rejected the charge of Green capture. Officials instead saw the directive as a necessary first step toward building a common framework for environmental protection, given vastly uneven statutory controls across the Community. Such a baseline, they argued, would provide a platform for harmonizing national, and in some cases European, regulations already in place for product sectors such as drugs, pesticides, and plants—thereby eventually bringing European and U.S. frameworks into closer alignment.[48]

At DG XII, Cantley and his colleague Ken Sargeant reacted with dismay to what they saw as a growing discrepancy between the European and U.S. regulatory frameworks. They believed that European policy had taken a wrong turn in targeting the *technology* of genetic manipulation rather than the *actual risks* they believed should be the focus of regulation. An internal memorandum of the period noted, in language reminiscent of contemporaneous U.S. policy, that risks "derive from dangerous organisms and their products, rather than from the technologies through which these are manipulated." It identified the drawbacks of using a process-based approach in the following terms:

(i) increasing the regulatory burden on our industry, thus weakening its competitiveness;
(ii) inducing industry to displace to more favourable regulatory environments (e.g., carrying out their research or making their investments outside the Community);
(iii) shifting the debate to abstract issues such as attempts to define the boundaries of what is called "natural," and disputes over whether genetically manipulated organisms are intrinsically more dangerous; such as have hampered developments in the US in recent years.[49]

But attempts to translate these views into Commission policy foundered for the time in the dynamics of interservice politics. With the departure of Commissioner (Viscount) Etienne Davignon in 1985, direction of the Commission's R&D policy shifted to less experienced and less universally respected hands. By the late 1980s CUBE and Cantley himself came under suspicion of being too closely affiliated with industrial interests, a view that gained perhaps unfair credibility when the Senior Advisory Group on Biotechnology (SAGB), a newly constituted trade association, recruited Brian Ager, Cantley's deputy, as its director.[50] By late 1992, under fire from environmental interests and lacking firm support from above, CUBE was dissolved as part of an administrative reorganization.

Changes in EU law and politics over the next decade arguably weakened the stark distinction between the product-based U.S. and the process-based European regulatory approaches. In particular, EU Directive 2001/18/EC repealed the 1990 directive on deliberate release and introduced a regime in which products derived from GMOs, such as paste or ketchup from genetically modified tomatoes, were covered by relevant "vertical" or sectoral provisions addressing product categories, such as the 1997 Novel Foods regulation.[51] At the same time, the new legislation reaffirmed some important transatlantic differences and left room for state-by-state variation within the EU. The new directive's preamble and text explicitly recognized the controlling role of the precautionary principle, a source of growing contention with the United States, in implementing European rules on GMOs. The attempt to harmonize GMO approvals across member states also ran into complications as countries continued to assert their right to ban genetically modified crops approved at the European level.[52] By the end of 2003 the United States had initiated a case before the World Trade Organization based partly on the European Commission's failure to get its members to accept the safety assessments produced by the EC's own scientific advisory panels.

In evaluating these developments from the standpoint of comparative analysis, one cannot overemphasize the contingency of the EU itself during the 1990s. In particular, this decade saw the political emergence of the European Parliament as a more powerful force pursuant to the Maastricht treaty, and with that, at least temporarily, a consolidation and elevation of the Green presence in European legislative politics. Institutionally, the EP, partly in consequence of its limited in-house technical capacity, proved receptive to an informal network of European and American ecologists and activists who emphasized uncertainty and urged a precautionary stance on GM technology.[53] The EP, too, was the vehicle through which Ken Collins, chairman of the Environment Committee, initiated a request to the Commission to consider building socioeconomic impact assessment into the regulation of some agricultural products, including some produced through biotechnology. Officially termed the "fourth criterion" or "fourth hurdle,"[54] the Commission's draft guideline was voted down by the Parliament in 1990, but, although unsuccessful, the episode provided early warning that European consumers and politicians cared about more than biological risk in the introduction of products seen as socially destabilizing, like genetically engineered bovine growth hormone. What seemed only a minor stirring of the air at the time would rise to gale force later on with the transatlantic fight over GM foods.

Subsequently, the formation of the European Environment Agency helped disseminate precautionary perspectives across the spectrum of EU environmental legislation.[55] Even "normal politics" at the Commission continued to produce its own contingent outcomes. Thus, Chancellor Schröder's red-green coalition underpinned the German presidency of the Commission during a key period in 1999 and influenced the drafting of the 2001 deliberate release directive. The new directive's continued strong emphasis on stepwise evaluation of GMOs, its explicit incorporation of the precautionary principle, and the room it left for independent ethical judgments in EU member states can be imputed in part to this fortuitous conjunction.[56] These observations underscore this book's overall claim that biotechnology helped make European politics, particularly by keeping alive an interest in social and economic impacts and a commitment to precaution, just as much as Europe controlled the making of biotechnology policy during this formative period in the EU's political history.

### Constructing the European Public

At quiet times in mature democracies, the work of management proceeds with mechanical impersonality, as if people and politics scarcely mattered. It is far different in times of crisis, when a political apparatus feels the need to justify its utility and efficiency to the citizens it seeks to serve. The European Commission has scarcely known a quiet moment in this sense. Its task of self-projection is continuous, as we have already seen, and it is all the more delicate because there is no preformed European polity whose allegiance European institutions can automatically count on. Policymaking in Brussels therefore remains a deeply (though not transparently) political undertaking, not only because its legitimacy is frequently questioned, but because it helps to constitute the very politics to which it then responds. The FAST report's reference to bio-society can be seen in retrospect as one dimension of this political dynamic: the construction of specifically European societal expectations of the life sciences, or designs on nature, which the Commission's biotechnology program could then plausibly seek to address.

But how could Europe's needs, wants, and deficiencies with respect to biotechnology be compellingly represented? One strategy was to stress competition with the United States, in which Europe could be shown to be lagging. Commission officials, as we have seen, privately deplored Europe's failure to adopt the U.S. framing of biotechnology and develop coherent policy around it. Many steps were taken to remedy this perceived gap, from commissioning ad hoc research reports on Europe's competitive position[57] to more formal transatlantic bridge-building. In 1990 an EC-US Task Force on Biotechnology Research was formed to encourage communication among scientists in the two regions; joint workshops were held virtually each year on issues such as environmental biotechnology, database interoperability, and bioinformatics.[58] In 2000 an EU-US Biotechnology Consultative Forum was established to reflect on the benefits and risks of modern biotechnology.

Defining policy in opposition to competition from abroad is not a sure-fire recipe for placating constituents at home, as politicians in the globalizing world have discovered to their sorrow. In policy as in politics, there is no substitute for a committed domestic constituency satisfied with the handling of immediately recognizable local problems. Municipal elections from Boston to Baghdad are still won and lost on mundane records of policing, road maintenance, garbage removal, or the delivery of essential utilities. This, too, the European bureaucracy grasped intuitively in seeking justification for the EU's biotechnology policy. As early as 1987 Cantley, in a published article, named four reasons for a coordinated European policy for biotechnology. One was "unremitting global economic competition"; the others were biotechnology's vast potential for further scientific breakthroughs, its promise of reducing hunger and disease, and, most interestingly, its capacity to provoke "popular apprehension about non-understood science and socially or culturally unacceptable innovation."[59] Implicit in the last reason was a characterization of the European public as deficient in scientific understanding and ethically confused or uncertain. These were local problems that EU authorities could reasonably seek to address in the public interest, but more than speculation was needed to cement the need for remedial action at the European level.

Like modernizing states of the past, the Commission had already discovered the need to sample and monitor its public. Since 1973 the Commission has conducted regular sampling of public opinion in the member states to support its policies as well as evaluate their success. The instrument used for this purpose is the *Eurobarometer*, a twice yearly survey that collects, classifies, and most significantly compares "what Europeans think" about a variety of issues central to the European policy agenda.[60] In early 2002, for example, the standard survey measured citizen reactions to the euro, the new currency launched in January of that year, as well as the public's knowledge of and trust in European institutions. Added in the late 1980s, the *Flash Eurobarometer* conducts targeted, smaller-scale surveys designed to produce early warning of changes in European public opinion. These instruments are not merely objective tools of policy and politics. They are ontological ordering devices: in sampling European opinion they help to constitute the very thing that they seek to represent[61]—a united European polity, even though its component parts can be accessed, sampled, and surveyed only through the constituent nation-states.

It is of no small interest, then, to find biotechnology surfacing as a survey topic in the *Eurobarometer*. Questions regarding European attitudes toward, expectations of, and, after 1993, knowledge of biotechnology were asked four times in the 1990s (in 1991, 1993, 1996, and 1999) and again in 2002.[62]

TABLE 3.2
British, German, and European Attitudes toward Biotechnology, 2002

| Country | Genetic Tests | Clone human cells | Enzymes | Crops | Food |
|---|---|---|---|---|---|
| UK | + + | + | + | + | − |
| Germany (national) | + | + | + | + | − |
| Europe | + + | + + | + | + | − |

*Source:* George Gaskell, Nick Allum, and Sally Stares, "Europeans and Biotechnology in 2002," Report to the EC Directorate General for Research, 2nd Edition, March 21, 2003.

*Note:* + + = strong support; + = weak support; − = weak opposition.

The assumptions embedded in these surveys and the debates that they, in turn, helped generate are critically important in evaluating the European Commission's presence in the politics of biotechnology.

Although contained within an opinion poll, the biotechnology questions not only sampled current and future public attitudes but also assessed what people know about genetics and whom they would turn to for information. Under the heading of attitudes, people were asked how they perceive the potential of biotechnology in relation to other new technologies, what they expect from biotechnology, and how acceptable they consider its products and applications. The results indicated striking attitudinal discrepancies toward different applications of biotechnology among socio-economically similar member states. For example, respondents were asked to say what they think about when modern biotechnology is mentioned. They were then asked to state, on a three-point scale, whether their responses to these aspects of biotechnology were positive (3), neutral (2), or negative (1). Differences among Britain, Germany, and the European national averages on five major dimensions of biotechnology that the survey identified are displayed in table 3.2. They show that reactions to "red," or pharmaceutical, biotechnology (but not cloning) were more positive than toward GM foods; they also indicate less positive responses to environmental genetic testing in Germany than in the United Kingdom.

To test citizens' knowledge, the survey asked respondents to answer a "biotechnology quiz," and also to state which sources of information they trusted. In 2002, the survey included ten propositions calling for true-false answers (see table 3.3); of these, nine had also appeared in the 1996 survey, and all ten in the 1999 survey. One major conclusion was that citizen's level of knowledge about genetics was low, as evidenced by 36 percent of Europeans (41 percent of Germans) agreeing that "ordinary tomatoes do not contain genes, while genetically modified ones do." The Commission also drew other

TABLE 3.3

Eurobarometer Biotechnology Quiz—1996–2002 (Showing European average correct answers in percentages)

| | % Correct | | |
|---|---|---|---|
| Proposition | 1996 | 1999 | 2002 |
| There are bacteria that live from waste water. | 83 | 83 | 84 |
| Ordinary tomatoes do not contain genes, while genetically modified tomatoes do. | 35 | 35 | 36 |
| The cloning of living things produces genetically identical offspring. | 46 | 64 | 66 |
| By eating a genetically modified fruit, a person's genes could also become modified. | 48 | 42 | 49 |
| It is the father's [1999]/mother's [2002] genes that determine whether a child is a girl. | N/A | 44 | 53 |
| Yeast for brewing beer consists of living organisms. | 68 | 66 | 63 |
| It is possible to find out in the first few months of pregnancy whether a child will have Down's syndrome. | 81 | 79 | 79 |
| Genetically modified animals are always bigger than ordinary ones. | 36 | 34 | 38 |
| More than half of human genes are identical to those of chimpanzees. | 51 | 48 | 52 |
| It is impossible to transfer animal genes into plants. | 27 | 26 | 26 |

Source: George Gaskell, Nick Allum, and Sally Stares, "Europeans and Biotechnology in 2002," Report to the EC Directorate General for Research, 2nd Edition, March 21, 2003.

conclusions: knowledge had not substantially changed, let alone increased. In 1999, Europeans trusted consumer (55 percent), medical (53 percent), and environmental (45 percent) organizations more than universities (26 percent) and national public authorities (15 percent); Europeans do not fear technology in general but are not highly enthusiastic about biotechnology—and these attitudes vary significantly across Europe.[63]

The portrait of the European citizen that emerged from these surveys— as an ignorant, distrustful, risk-averse biotechnology skeptic—was immediately challenged by scholars who took issue with the *Eurobarometer*'s epistemological presumptions. Critics questioned the survey's implicit acceptance of a fact-value distinction that other research on risk has rendered problematic or untenable.[64] "Risk" and "trust" were treated in the *Eurobarometer* as if they could be assessed independently of one another: risk, in the survery designers' view, was a "natural," mathematically calculable balance of probabilities, whereas trust was a matter of opinion, falling in the domain of values. Against this implicit partitioning, critics noted that qualitative research has consistently shown close ties between perceptions of risk and historically and culturally conditioned expectations about the trustworthiness of governing institutions.[65] Risk and trust accordingly are interdependent, not independent, variables. A five-country study of the public perception of GMOs, also funded by the European Commission, found that responses in all five were ambivalent rather than starkly negative or positive, and that in this respect public reactions were consistent, not divergent, across member states. Focus group participants asked why technological choices were not discussed in advance, why risks appeared to be incompletely assessed, and who would be in charge if failures occurred. In contrast to the ignorant and doubting citizen "found" by the *Eurobarometer*, this comparative study "found" a reflective, actively questioning, and politically engaged European citizen.[66]

Such variation suggests that study respondents to some extent show investigators the face that the study is designed to illuminate. Publics and their understanding, in short, are only imperfectly captured in studies designed to characterize them. These observations in turn point to the need for a more ambitious and inclusive theorization of the knowledgeable citizen in a democratic polity, a figure I have elsewhere referred to as *homo sciens*, as distinct from the *homo economicus* who underpins economic theory.[67] We will return to that theoretical project in the chapter on civic epistemology.

A second issue concerned the definition and meaning of "ignorance." The *Eurobarometer* analysts equated lack of knowledge with factual errors made by survey respondents: a key example, widely reported, was that one-third of the surveyed population did not know that ordinary tomatoes have genes but assumed that genes were found only in GM tomatoes. Work in the public understanding of science has questioned both the content and the relevance of such observations. On the substantive side, answers may reflect a more complex understanding of science and technology than such reductionist questions can hope to capture; for example, the public may correctly believe that genes in tomatoes do not matter for either consumption or policy (hence, for pragmatic purposes, do not "exist") unless they have been artificially imported into the tomatoes for specific purposes through technological means.[68] Aggregate numbers, too, tell only a part of the story. For example, more women than men correctly answered that the father's genes determine the child's gender; it is interesting to speculate why, when women on the whole did worse than men on the quiz. With respect to the survey's relevance, many question whether lack of factual knowledge has any bearing on public attitudes toward technology or the public's willingness to support scientific research in areas seen as beneficial.[69] In the United States, for example, health research has enjoyed consistently strong public and political support in spite of high levels of ignorance regarding scientific facts.[70]

Irrespective of one's take on these issues, one can see that the *Eurobarometer* surveys and other sociological instruments gave rise to an active cross-national debate about the characteristics of the European citizen in relation to biotechnology. Is that citizen ignorant or informed, fearful or confident, subservient to or emancipated from national cultural and political framings? While that debate remains largely sequestered among elites—university researchers, Commission staff and analysts, and national policymakers—it is nonetheless a visible locus of European political discourses and identities in the making.

### European Ethics for European Science?

Debates about biotechnology in the 1990s were not isolated political events but were woven into a field of discourse that included intra-European disputes over the "mad cow" crisis, transatlantic trade wars over products such as hormone-treated beef, and ongoing state-society conflicts over environmental issues such as the disposal of the Brent Spar oil platform in the North Sea. Policymakers were tempted to characterize all these as conflicts over values: over the meaning of nature and the natural, the protection of local customs in food and agriculture, and the preference for precaution over risk-taking. If policy conflicts could be attributed to divergent values, then a logical response for nations as well as the European superstate was to ask for better analysis and management of such value differences—in short, for more expertise in ethics. The rise of ethics on the European agenda is a response to these concerns and is closely tied to EU policymaking for the life sciences.

More than any other European politician, François Mitterand led the way in raising the political profile of bioethics. His 1983 presidential decree established a standing National Consultative Committee on ethics and the life sciences.[71] Mitterand acted for France, but the term "ethics" was also circulating in wider European policy discourse by the mid-1980s. The 1984 FAST report listed "the examination of societal dimensions"—including ethical implications, risk, and acceptability—as one of four focal problems under the heading "bio-society."[72] That report also identified questions that did not become politically significant until fifteen years later, when circumstances rendered them more visible (see chapter 7): "Is a chromosomal inheritance of minimum standard to become a right? Do spare, deep-frozen blastocysts have rights? Should human embryo tissue be cultivated for transplant purposes, such as skin or brain tissue repair?"[73]

By November 1991 the European Commission was sufficiently concerned about its expertise in ethics to establish an independent Group of Advisers on the Ethical Implications of Biotechnology (GAEIB). It was not so much a bold new initiative as a self-conscious response to the times. A senior Commission official observed, "Ethics is a fashion now."[74] The concept of "ethics," as the initial membership list made clear, was loosely defined and fitted to circumstances within the individual member states: for example, Britain was represented by Baroness Mary Warnock, a distinguished moral philosopher and former chair of the U.K. Commission of Inquiry on Human Fertilisation and Embryology,[75] whereas Germany was represented by Hans F. Zacher, then president of the Max Planck Institute for Patent Law in Munich.[76] Ideas of personhood and property from philosophy and law were indiscriminately thrown together in constructing an authoritative European presence in bioethics. The GAEIB was succeeded by a European Group on Ethics (EGE) on Science and New Technologies, established in December 1997 and reappointed in April 2001. Illustrating the rising prestige of ethics, the EGE's secretariat became a part of the Commission's Group of Policy Advisers, which works directly with the president and commissioners to identify cross-cutting issues with long time horizons.

The list of ethical opinions handed down during the 1990s suggests that EGE and its predecessor, the GAEIB, were conceived in instrumental terms, to neutralize the potentially divisive consequences of Commission policy regarding the life sciences. Between them, the two groups addressed virtually every controversial issue of the era—beginning with bovine somatotropin (bovine growth hormone) in March 1993, and continuing into the twenty-first century with such topics as gene therapy, patenting of biotechnological inventions, labeling GM foods, cloning, embryo research, and stem cells. The advisers' practices changed as well, toward more direct communication with their imagined or hoped for publics; beginning in 1998, all EGE opinions were accompanied by a press release summarizing the report's contents. But in focusing on the legitimating function of these bodies, one should not lose sight of another dimension: their quiet participation in the politics of European identity-building.

The EU's ethics advisers frequently observe that their work has to strike a delicate balance, between the universal aim of facilitating and promoting the sciences, on the one hand, and maintaining respect for the pluralism of European values on the other. The Commission's ultimate goal in science policy is to create a single European Research Area, permitting the free movement of researchers, projects, funds, materials, and topics across national boundaries. An important barrier, however, is the discrepant regulation on ethical grounds of certain types of research among the member states. GAEIB's 1997 opinion on the EU's Fifth Research Framework Programme exemplifies the resulting tension. Article 6 of the Programme required that "fundamental ethical principles" be adhered to in EU-funded research. The ethics group agreed on the need for an ethical assessment to comply with Article 6 while also stating, somewhat paradoxically, that "[a]lthough the concept of Europe is based on respect for national differences, it relies on strong common values."[77] The committee located these presumably shared values in a variety of international texts.

The problems noted by the GAEIB and EGE are mirrored within the Commission, where the in-house ethics office has the task of mapping and negotiating differences among member states on such thorny issues as the use of human embryos in research. These splits become visible at politically sensitive moments, as for instance during the debate on the Sixth Research Framework Programme (FP6) in May 2002. EU Framework Programmes are adopted through a codecision procedure that requires agreement between the Council and the European Parliament. Trouble on the FP6 began when the Parliament tabled a forbidding 170 amendments in preparation for the second reading, threatening to derail the negotiations. Contested issues included what some parliamentarians saw as excessive support for genomics compared with conventional biology, a tilt toward larger and more established facilities through "centers of excellence," and, most important, a controversial amendment that authorized the creation of less than fourteen-day-old embryos for research. Unless the conflicts on these points had been ironed out, FP6 could not have been approved during the then-current Spanish presidency; without an agreement, the conciliation process would have dragged on into the fall, compromising the funding of projects in January 2003.[78]

The EP research committee proposed to resolve the ethical dispute with a clause that would have excluded three research areas from funding: human reproductive cloning, germline gene modification, and the creation of embryos for research. While several member states supported this position, those with more permissive research environments, including the United Kingdom, saw the solution as too restrictive. In the ensuing Trilogue among Council, Commission, and Parliament, a compromise emerged that permitted the FP6 to be approved in timely fashion. The Commission declared that research funds would not be used for the three aforementioned purposes, although research could continue on spare embryos from in vitro fertilization.[79]

This ad hoc solution in effect created a common ethical standard for FP6, but in a way that limited its legal force and political visibility—and it also left unresolved the question broached at the 2001 bioethics meeting at Genshagen (see prologue): whether, in research ethics, there are indeed values that can reasonably be called "European." This pragmatic outcome was entirely consistent with Europe's status as a work in progress. Extreme positions were avoided, controversial science received an amber light, and EU funding was made available for scientists from both permissive and restrictive research cultures provided they complied with minimalist rules of the game. It was, in short, an affirmation of the passive-preservationist vision of Europe.

## Conclusion

Since the late 1970s, the European Community and its successor, the European Union, have grappled with questions of how to promote competitive research and development in the life sciences within the European region. Contrary to many conventional accounts of policymaking, this has not been a rational, linear process, in which decision makers first identify the goals they hope to attain and then find effective means of meeting them. Instead, as we have seen, biotechnology policy became a site of interpretive politics, in which important elements of European identity were debated along with the goals and strategies of European research. The project of unifying Europe as a political space advanced through efforts to define the objects and practices of European scientific inquiry. Similarly, to answer what Europe should strive to achieve in the field of biotechnology, it proved necessary to address what kind of union Europe was—or wanted to be—both in relation to its member states and as a player on the world stage.

This account makes it impossible to evaluate Europe's contributions to biotechnology policy through any objective assessment of successes and failures. Given the fluidity of its aims, and indeed its self-conception, asking how well the EU has performed in this arena makes little sense. It is more useful to specify in detail the various policy debates the EU has generated, and how it has pursued these institutionally in Brussels and elsewhere. Nor can we unambiguously assess how far the EU has influenced or been influenced by the policies of its member states. Again, the more interesting question is how the attempt to define a coherent approach to biotechnology has affected the dynamics of European integration, with repercussions for the nature of Europe's political union.

Under this heading, Europe's distinctive framing of the risks of biotechnology in the 1990s emerges as one step in a larger project of creating order in science and society, or coproduction. The absence of well-developed regulatory regimes, a relatively passive industry, Green influence in the Parliament and DG XI, and a holistic discourse about the life sciences all combined to produce a process-oriented policy framing that initially set European policy apart from directions taken in the United States. It was genetic manipulation that Europe targeted for concern in the 1990 directives for contained use and deliberate release, and even though these were revised ten years later, the groundwork had been laid for a more precautionary, socially oriented biotechnology policy than that obtaining on the other side of the Atlantic. Later changes in the EU's founding treaties only strengthened the commitment to precaution in regulatory matters;[80] by the turn of the century, this term stood for a distinctively European policy orientation, much as risk and sound science correspondingly did for U.S. policy.

To stop at framing, however, would be to miss another critically important dimension of European biotechnology policy, namely, its role in creating issues, publics, and values that can be designated as truly European. Like nineteenth-century nation-states, the EU has found it necessary to specify the problems it wants to solve in order to legitimate its political existence. It has mobilized for this purpose many of the instruments invented by modern governments to make their achievements visible to attestive publics: talk of crisis, expert committees, special reports, and new statistical measures, for example. Only the single, spectacular demonstration of success in curing a widely recognized problem, such as a common hereditary disease, has continued to elude Brussels. But the already complex task of defining a persuasive policy agenda was made more complicated in the European case by the absence of a polity to which the EU's ruling institutions are clearly accountable. The *Eurobarometer*'s biotechnology questions, as well as later efforts to map and characterize ethical issues in biological research, can be seen, in a sense, as a response to this problem of a missing or phantom polity. In however groping and awkward a way, the anonymous, impersonal, and still evolving European superstate was led by these means to constitute and to converse with its citizens. The flaws in these initiatives should not detract from their importance in launching a debate of European proportions. Limited to elites at the start, these conversations nonetheless helped to articulate democratic concerns about the social purposes of biotechnology, the representation of uncertainties in regulation, and the kinds of scientific knowledge and understanding that can appropriately be expected from citizens of technologically advanced societies. Europe, by the turn of the century, had become a space to watch for the deliberative politics of science and technology—and we can attribute this result, at least in part, to the complex developments surrounding European biotechnology policy.

# 4

## Unsettled Settlements

Between 1975 and 1995, biotechnology moved from a research enterprise that left even its most committed practitioners unsure of themselves to a global industry promising revolutionary benefits in return for allegedly well-understood and manageable risks. This shift occurred almost simultaneously and with remarkable speed throughout Europe and North America (see appendix). To facilitate commercialization, the United States, Britain, and Germany—and the European Community (later the EU)—all adapted their laws and regulations to control both laboratory research with genetically modified organisms and their planned, or in official language "deliberate," releases into the environment. Within barely a decade, environmental consequences that were once considered speculative and impossible to assess came to be regarded within policy circles as amenable to rational, scientific evaluation. By 1990 it appeared that, for genetically modified crops, apocalyptic visions and the rhetoric of science fiction could be set aside in favor of objective expert discourses and routine bureaucratic approvals.

These changes in the status of agricultural biotechnology were all the more unexpected because, at the time of commercialization, the risks of industrial-scale application remained largely hypothetical. Scientists and companies seemed confident that no serious harm would befall ecosystems or human health if corn and cotton crops were fitted out with herbicide- or insect-resistant genes, or if fruit farmers sprayed their orchards and berry plants with gene-deleted bacteria designed to prevent frost formation. Yet, unlike toxic chemicals, the products of the new biotechnology had not been in circulation long enough to manifest a wide range of beneficial or adverse effects. There was no storehouse of precedents that policymakers could reach into for historically documented evidence concerning the widespread use of laboratory-crafted organisms. As regulators in different policy systems approved the environmental release of GMOs, they were therefore obliged to find other credible ways of demonstrating the technology's safety.

Scientific and administrative hurdles stood in the way. On the scientific side, regulators and researchers had to agree on what needed to be known for policy purposes—in other words, they had to produce a robust and relevant body of regulatory science.[1] On the administrative side, systems of oversight and management had to be constructed to provide regulators with expert guidance and assure the public that commercialization would take place under adequate supervision. Cross-national divergences soon appeared. We saw earlier that attempts to conceptualize the "problem of biotechnology" for policy purposes initially led to three different interpretive frames in the United States, Europe, Britain, and Germany: as a collection of *products*, as a potentially hazardous technological *process*, and as a threatening *program* of state-sponsored control of society through technoscience. In this chapter we will see how these framing choices influenced the scientific practices, assessment principles, and management structures that each nation developed for releasing GMOs into the environment—and why, in each case, national efforts failed to silence further controversy.

The chapter, then, tells two stories. The first is that of normalization, a common theme in modernity. Vague, unnamed, and unbounded fears were specified and made tractable, or so it seemed, through evolving systems of framing, classification, calculation, and control. Producing a new state of technologically improved, or designed, nature demanded heroic efforts of legitimation on the part of scientists, producers, and regulatory institutions. To see how these ordering mechanisms were put together, we need to step back and examine the particular kinds of disorder that the proponents of biotechnology in each country were trying to discipline and control. We therefore begin with vignettes of three national controversies involving the deliberate release of GMOs. The manner in which they were framed reflected, and in a sense reaffirmed, each nation's particular style of controlling risk. In the United States, regulators claimed the authority of science to support their conclusions with regard to product safety; in Britain, by contrast, regulators relied on the more embodied concept of expert judgment to certify the safety of GM as a process; and in Germany, legitimacy was sought through targeted institutional and procedural reforms establishing new forms of dialogue between citizens and the programmatic state. But it was the fragility of each accommodation that proved in the end to be most unexpected. By the late-1990s debates reopened on issues that industry and government hoped had been definitively laid to rest.

The second part of the chapter, then, deals with the less common story of *de*normalization. It gradually became clear in each country that the political acceptability of agricultural biotechnology depends as much on the trust-worthiness of the supporting social and institutional arrangements as on the abstractions of scientific risk assessment. New debates and controversies revealed different fault lines in the consensus on framing achieved through the first round of normalization. The actors, the locus of controversy, and the terms in which disagreements were expressed all diverged cross-nationally, calling attention to the intensely culture-bound character of technology's public acceptance. These conflicts, in turn, elicited further social and scientific experiments, which round out the chapter. I conclude by relating these developments to underlying theoretical issues of coproduction and the relations of science, state, and citizens.

## The Greening of Biotechnology: Three Tales

Agricultural biotechnology has sought to establish its claim on public acceptance by explicitly distancing itself from the risky, dirty, polluting, and inefficient industries of the industrial era. It is a *green* technology, a point that industry ceaselessly documents in its web-based and televised promotional materials, as well as in glossy brochures and annual reports. These documents conjure up images of a fertile earth and its abundant fruits, often featuring sunny shots of families with young children. The theme of order is front and center: a favorite visual motif is unbroken rows of grain receding into the deep distance. The time lines frequently purveyed by biotech companies reconnect the agricultural enterprise to the remote past of human-nature interactions, untouched by the grime of the industrial revolution, and predating by millennia the controversial intrusion of modern reductionist science so deplored by the German Greens. For example, the Biotechnology Industry Organization (BIO)—a lobbying association of more than a thousand members formed in the late 1990s to press the cause of the agricultural and pharmaceutical sectors—lists the first three achievements of biotechnology as the brewing of beer by Sumerians in 1750 B.C., the use of moldy soybean curds as an antibiotic by the Chinese in 500 B.C., and the use of powdered chrysanthemum as an insecticide, again by the Chinese, in A.D. 100. Another BIO timeline starts off the march of biotechnology at 8000 B.C., with the first domestication of crops and livestock.[2]

This is wishful thinking carried to high art. It impresses pop culture into serving politics and merges advertising with history. To get people to believe in these representations, though, requires more than electronic hand-waving, and not only media consultants and public relations firms but, more importantly, science and government have to play their part in securing public acceptance of promises of safety. The first phase of orchestrating the preconditions for the environmental release of GMOs produced its share of discords as well as harmonies. These were articulated differently in the three national contexts. Their settlement drew on different traditions of enrolling science into decision making, as well as different procedural repertoires for building public trust. The resulting policies, too, were different, in ways that reflected local political circumstances. In turn, these early ordering moves laid the foundations for future expressions of uncertainty and discontent.

### United States: Science Speaks

The operative term driving U.S. politics on GMOs was risk, but that small word marks more the beginning than the endpoint of analysis. When the German sociologist Ulrich Beck wrote his highly influential monograph *Risk Society*,[3] he imagined risk as a transforming force reshaping social relations throughout the industrial world. People everywhere, Beck argued, were at risk from their own creative powers, materially transformed into hazardous technologies, and these risks could strike one down regardless of one's social or economic standing. Class in the traditional sense offered no defenses against, for example, the ozone hole, climate change, or nuclear catastrophe; the proliferation of risk created its own moral classifications of the potentially damned and the potentially saved. In U.S. social science and policy analysis, however, risk had in the 1980s as now a different flavor from risk as conceptualized in European social thought. If the world's most powerful nation saw itself as a "risk society" at all, it was only as a prelude to controlling yet better the threats resulting from novel methods of production. Not for U.S. decision makers was a sociological account that portrayed people as helpless victims of their own inventions.

Taming risks, however, requires work.[4] Who does this work and by what rules of the game? The answers, we have seen, were not fixed in U.S. law or policy when new GMOs were first readied for release into the environment. Indeed, GM techniques and the regulations authorizing their use evolved almost in unison during the late 1970s and early 1980s. In 1976, for example, the National Institutes of Health issued the first guidelines for recombinant DNA research. A year later, Steven Lindow, a graduate student at the University of Wisconsin, discovered that a mutant strain of the bacterium *Pseudomonas syringae* could inhibit frost formation on plants.[5] That year, 1977, some sixteen bills to regulate rDNA research were unsuccessfully introduced in the U.S. Congress. Vehemently opposed by both science and industry, federal legislation never materialized, and supervision of genetic research continued to rest in the hands of a grant-making agency—NIH—while the results of that research moved ever closer to the commercial market.

By the early 1980s Lindow, by then at the University of California, Berkeley, had refined his discoveries and was ready to test them outside the laboratory. His team had identified the gene responsible for producing an ice-nucleating protein in the "normal" strain of the *P. syringae* bacterium, called the "ice-plus" strain, and found a means of deleting this gene to create a frost-inhibiting "ice-minus" strain. The Berkeley researchers, including Lindow and his colleague Nikolaos Panopoulos, were now ready to shift their attention from engineering bacteria to engineering the conditions—both natural and social—for their deployment in the environment. Spraying the gene-deleted ice-minus bacterium on plants, they hypothesized, would displace the naturally occurring bacterial population and raise the frost resistance of the treated plants. It remained only to get official sanction to field-test their idea, and the institution from which these university researchers could most naturally seek approval was NIH.

From the researchers' standpoint, each step they took followed routinely from the one before, in a normal progression from basic research to product testing. From the standpoint of the social reception of these events, however, the Lindow-Panopoulos initiative, and equivalent moves by industry, breached several important institutional and conceptual boundaries and posed unprecedented problems. These boundary-crossings became progressively more unmanageable. The process began quietly enough with NIH's Recombinant DNA Advisory Committee, which was formally responsible for reviewing the application. NIH advisers had satisfied themselves by this time that the fears expressed at the 1975 Asilomar conference had been groundless, including even worries about GMO releases into the environment. Single gene deletion was too specific an intervention, its likely results too predictable, to occasion much eyebrow-raising among researchers using rDNA techniques. RAC approval, with scientists passing judgment on other scientists, was therefore unsurprising. But as we saw in chapter 2, review by disciplinary peers alone did not satisfy what opponents claimed was a legal mandate: the need to conduct a public environmental assessment under the National Environmental Policy Act. The ensuing litigation, which led to a decision in favor of the plaintiffs,[6] established that expert deliberations, however open and thorough, were no substitute for the public review of NIH's risk assessment principles contemplated by NEPA.

In taking the ice-minus bacteria from the lab to the field, researchers crossed more than the line between peer-reviewed basic science and its regulated applications. They also moved from a world of controlled experiment (science) to a world of messy experience (agriculture), from technical discourse to political debate, and from the relatively sheltered preserves of academic science to a space of higher economic stakes and public scrutiny.[7] Alliances, tactics, modes of expression and of action all changed as the context for assessing the scientists' work shifted. Disagreement, which thus far had been contained within professional circles, spilled into less rule-bound channels. The Berkeley team eventually gained the federal approval it had asked for, but the community of Tulelake, California, where the test was to be conducted, staged protests. An initial planting of three thousand treated potato plants was vandalized, although the trial was later repeated without opposition.[8] Controversy continued with the discovery that Advanced Genetic Sciences (AGS), an Oakland-based private company that had funded Lindow, had released the frost-preventing bacteria without proper authorization in a roof-top experiment on its own premises. Subsequently, AGS sought and received approval from the U.S. Environmental Protection Agency to conduct a study similar to Lindow's using strawberry plants. Legal maneuvers by Jeremy Rifkin and his associates to block the AGS test failed. Vandals destroyed much of the AGS site in April 1987, but in May company scientists dressed in eye-catching hazardous materials suits went forward with the intended release, earning predictable coverage from the national press.[9]

Some interpreted these events as the expected growing pains of a nascent technology. Popular anxiety, on this account, simply reflected the novel, unknown, and "dread" character of genetic engineering—attributes that social psychologists at the time commonly associated with elevated levels of public concern.[10] Things predictably grew calmer, on this same account, as communities became more familiar with genetic engineering, courts ceased encouraging irresponsible figures like Rifkin, and the media stopped retailing highly colored stories of improbable hazards. All these normalizing moves brought public perception back in line with the rational risk calculations made by experts; better information and more exposure acted as antidotes to the "sociology of error," that is, to collective responses based on a wrong assessment of the facts. The decrease in controversy "proved"—with only minor hiccups along the way—what scientists had claimed all along: that genetic engineering of crops and plants was safe, and would be seen to be so. Once conflict died down, the U.S. regulatory scheme for agricultural biotechnology came to be seen by many, especially within the United States, as a model of how scientific judgment could tame the uncertainties of technological innovation.

There are three problems with this happy reading of the ice-minus story, all of which loom as significant in the light of later events in the United States and Europe. First, the apparent closure of controversy was achieved in a period of American deregulation that reduced the type and intensity of scrutiny given to products of agricultural biotechnology. By early 1987, for example, RAC had decided to relax a number of restrictions to make studies like Lindow's significantly easier to conduct: RAC would not review tests already approved by other federal agencies, demand preauthorization of field tests for gene-deleted microorganisms, or require physical containment for organisms determined to pose low risk.[11] For the moment, regulators in and outside the United States interpreted this lowering of skeptical oversight as evidence that the research was acceptably safe, but the stability of this conclusion strongly depended on the credibility of the U.S. regulatory process as a whole. *That* would prove in time to be less robust than biotechnology advocates had hoped.

Second, the field tests did not so much resolve the scientific questions as displace them—from the capacity of the bacteria to reduce ice nucleation under field-test conditions to their possible longer-term effects on the environment, which the field test by definition could not assess. EPA imposed monitoring requirements to address the latter issue, but some denied the need for such studies, since the mutant ice-minus strain exists in nature and is therefore a "known" entity with respect to its biological properties. Henry Miller, a former official of the Food and Drug Administration (FDA) and research fellow at the conservative Hoover Institution, as well as an outspoken foe of regulation, was especially caustic: "Even after EPA finally granted its approval for testing the 'ice-minus' microorganisms in the field, the agency conducted elaborate, expensive, intrusive—and predictably worthless—monitoring of the field trials. (Monitoring for what, one wonders—the harmless bacteria mutating into pit bulls?)"[12]

Miller's polemic papers over an important point: environmental release entailed questions that could not be answered by molecular biologists alone. The precision of gene splicing had seduced these scientists into believing that their manipulations were highly specific, and therefore wholly manageable, but molecular methods could not by themselves predict how the altered organisms would behave in an uncontrolled environment, such as an open field. An intellectual line of contestation was drawn between those (mostly molecular biologists) who insisted on the precision of genetic engineering as sufficient evidence of safety, and those (mostly ecologists) who saw the technique's application as introducing uncertainties that could not be resolved in the current state of knowledge.[13] The former viewed field testing as unnecessary so long as the GM construct involved no hazardous manipulation; the latter considered the field tests as essential in the scale-up of GM crops from lab to commercial production. The dispute between these two views remained alive, in and outside the United States, despite the best efforts of the proponents of biotechnology to quell it. Arguments that "science" had shown biotechnology products to be "safe" downplayed the fact that science did not speak with one voice on this issue.

The third point relates to the demand side of biotechnology. The ice-minus experiment did not prove to be a commercial success. Though Miller cites EPA regulation as the primary culprit, there were other compelling reasons. By the mid-1980s work was already underway on engineering pest resistance into plants, a technology that was to have, under the primary regulatory jurisdiction of the U.S. Department of Agriculture, wide commercial success with little of the hullabaloo produced by the ice-minus episode. Questions arose about whether the frost-resistant properties conferred by the ice-minus strain were significant enough to merit substantial economic investment.[14]

Biotechnology's success in the marketplace ultimately depended on demand, and in the United States, it was not ice-minus but genetically engineered, pest-resistant corn, cotton, and soybeans that eventually met the test of marketability.[15] These products targeted the needs of large-scale growers and catered to the safety concerns of these well-satisfied clients. Ignored in industry's calculus of expansion were many other actors who, at other times and in other places, would exercise their voice in the biotechnology debates: small farmers, organic producers, supermarkets, the food industry, environmentalists, consumers, and of course concerned biologists from multiple disciplinary backgrounds. The molecular biologists' perceptions of risk and safety proved in the long run too restrictive to meet the concerns of this heterogeneous, but interested, population.

## Britain: Expertise Governs

Britain in the mid-1980s was a passive place for environmentalism. The British public displayed little of America's heightened concern for chemicals; even nuclear power ignited no protests comparable to those in Germany and the United States. Margaret Thatcher's Tory government scoffed at the perceived excesses of European green politics and remained unremittingly skeptical toward most claims of environmental degradation. Even the threat to the stratospheric ozone layer, discussed in U.S. scientific circles since the 1970s, was ignored at first by a prime minister who had been trained as a chemist. Her ministers adopted a no-nonsense, "show us the bodies" approach to scientific evidence, which was at odds with the more precautionary approach favored in other EC countries.

All this changed in the run-up to the 1988 British election, when a confident but also politically savvy Thatcher observed the rising green sentiment among the electorate and the aim of the Social and Liberal Democrats to turn the environment into a campaign issue. In a speech to the Royal Society on September 27, 1988, Thatcher surprised and pleased environmentalists by acknowledging that "we have unwittingly begun a massive experiment with the system of this planet itself."[16] Lecturing scientists on the need for better management and closer cooperation with industry, she also noted the need for more research on environmental issues. Follow-through was slow, but a year later the nomination of a new environment secretary, Christopher Patten (replacing the notably anti-environmental Nicholas Ridley), and the introduction of an omnibus "Green Bill" signaled some progress on the Conservatives' new environmental agenda.

Regulatory procedures for agricultural biotechnology were modified during this peaceful period, and they followed in the main British traditions of decisionmaking by experts. The first authorized releases of a genetically engineered microorganism in the United Kingdom illustrate the point. By the late 1980s research on biological pest control, using a baculovirus as the vector, was underway at the Institute of Virology and Environmental Microbiology (IVEM) in Oxford, a unit of the Natural Environment Research Council (NERC). David Bishop, IVEM director from 1985 to 1995, was determined not to repeat the mistakes of his counterparts in the United States. He hoped to avoid the traps they had fallen into by proceeding in small, incremental steps, each time collecting data to enable the next move. Others, he suggested, had been less cautious: "A lot of research is like trying to run before learning to walk, before learning to crawl, before learning to focus your eyes."[17] In IVEM's case, one solution was to use an enfeebled strain of the baculovirus by removing the gene that produced its protective coat protein. The modified organism would be less persistent in the environment and would thereby pose a smaller risk of escaping the researchers' control.

Bishop was extremely sensitive to the public relations side of IVEM's research. Prior consultation with environmental groups, notices in local papers, and a video explaining the nature of the research were among the means he used to reassure the public about the baculovirus release planned for the spring of 1989. The strategy apparently worked at the time of the first field test. IVEM received only two written requests for more information, with no follow-up from the concerned citizens. Newspapers and magazines did not report the event, their silence starkly contrasting with the media blitz around Lindow and AGS on the other side of the Atlantic. Bishop's scientific colleagues more or less reluctantly admired his handling of a potentially controversial situation, some praising it as "exemplary,"[18] but others expressing annoyance at his self-promotion in making "such a [public] meal of it."[19]

Behind the business-as-usual façade, questions continued to swirl, but, like the test virus itself, these did not spill into the open. Field test applications were approved by the Health and Safety Executive, acting on advice from the Advisory Committee on Releases to the Environment. The government's mandate was limited to reviewing the safety of the field test; in turn, the test itself was designed to illuminate only the questions that ecologists deemed important for evaluating safety—the modified organism's survival, persistence, and dispersal, and possible gene flow between it and other populations. But could contained releases such as IVEM's offer reliable insights into large-scale commercial use, industry's ultimate goal? The Oxford experiments, after all, were conducted under rigorous containment conditions, none of which could be maintained during full-scale commercial application: an enfeebled GMO strain, prior testing with limited numbers of target species, use of physical barriers, attentive monitoring, and eventual disinfection of the test site. Given these discrepancies between the real world and field tests, environmentalists wondered whether testing might not convey a misleading impression of the safety of GMOs.[20] The government's experts, however, were not asked to wrestle with deeper questions about the tests' correspondence to actual conditions of use—a problem that runs through all attempts to predict the efficacy of new technologies[21]—let alone to question the ultimate purposes of field testing particular organisms. Beneath the umbrella of expert reassurance, the seeds of doubt and uncertainty continued to germinate in secret.

### Germany: Procedure Rules

In Germany 1990 was a watershed year for biotechnology. In that year, the German parliament passed a new genetic engineering law (Gentechnikgesetz, GenTG) and the EC adopted its two major Europe-wide biotechnology directives. Up to this time, German geneticists had operated, much like their American counterparts in the mid-1980s, under the supervision of an expert committee, the Central Commission for Biological Safety (Zentrale Kommission für die Biologische Sicherheit, ZKBS). Constituted in 1981, the twelve-member body was originally composed of eight biologists and, reflecting Germany's corporatist traditions, one representative each of unions, industry, environmental groups, and research organizations.[22] Its activities were conducted largely out of the public eye. The first commission report—a small, stapled-together, mimeographed booklet, clearly not intended for public consumption—was issued in 1988 and covered the twenty-six meetings held during the previous seven-year period.[23]

The political circumstances leading to the adoption of the GenTG ensured that the new law would have to grapple more seriously with questions of federalism and participation. The allocation of regulatory authority was a perennially sore point in a governmental system founded on a careful division of power between the center and the states (Länder). Participation, especially in technical decision making, was an increasingly salient theme in German politics following the student uprisings of 1968, the anti nuclear protests of the 1970s and 1980s,[24] and the formation of the Green Party and its entry into the Bundestag in 1983.[25] Not surprisingly, much of the detailed negotiation on the GenTG reflected these driving concerns. On the side of federalism, the Länder joined with Klaus Töpfer's Environment Ministry in pressing for a more decentralized approach to licensing facilities and approving releases.[26] Both researchers and industry, however, favored the more centralized, one-stop approach through the ZKBS, which ultimately prevailed. On the side of participation, activists succeeded in altering the composition of the ZKBS: membership was raised from twelve to fifteen to provide a stronger voice for ecology and environmental protection, and over time the commission took more steps to make available the results of its deliberations. Paralleling Britain's ACRE, which advised the Health and Safety Executive, the ZKBS continued to operate under the jurisdiction of a health rather than an environment ministry. Chancellor Helmut Kohl was allegedly reluctant to transfer regulatory responsibility for an important industrial sector to his possibly too independent environment minister, Klaus Töpfer.[27] Keeping the ZKBS within the Health Ministry offered a practical solution.

Green activism was also responsible for the insertion of two public hearing requirements into the law, applicable to the construction and operation of genetic engineering facilities and the release of GMOs, respectively.[28] These provisions resembled the hearing requirement in earlier federal legislation on air pollution control, but in applying to institutions conducting basic research on GMOs, the deliberate release provision marked a departure from the prior focus on industrial hazards. It demanded that basic researchers account to the pubic for an aspect of their scientific aims and methods. In addition, paragraph 16 of the law required that releases should produce no unjustified harmful effects on humans, animals, plants, or the environment and property. Together, these changes soon proved consequential.

By the late 1980s scientists in Peter Meyer's research group at the Max-Planck-Institute for Plant Breeding Research (Züchtungsforschung) in Köln had planned and partly conducted a series of experiments designed to test the properties of genetically modified petunia plants. In the first phase, the plants were modified using a corn gene, which activated an enzyme that turned the normally white petunias a deep salmon red color; following the usual custom in such studies, the transgenic plants were also fitted out with a "marker" gene conferring resistance to the antibiotic drug kanamycin.[29] The results were published in the influential science journal *Nature* and attracted considerable media attention (although not, to begin with, in Germany) as a cute example of the funny things scientists do with genetic engineering.[30] In the second phase, researchers wanted to study the behavior of a class of "jumping genes" (transposons), which they expected would selectively turn off the color-producing gene, thereby creating variegated or pale pink flowers. To get meaningful results, Meyer's group planned a study requiring thirty thousand genetically modified plants to be grown in an open field. They were confident, on the basis of both the published literature and their own pilot studies, that no plants would survive from one growing season to another. Accordingly, they were not at all worried about safety.

This would be the second release of GMOs in Germany, but the first conducted in accordance with the procedures envisaged by the GenTG. The earlier release, also originating with Meyer's group, had proceeded without a hitch under the auspices of the ZKBS. The second study required a public hearing in addition to committee approval. That process proved more unruly and discursively undisciplined than the well-intentioned but politically inexperienced scientists had bargained for. Ten hours were consumed in discussing the design and worth of the experiment in order to satisfy the balancing of risks and benefits called for by paragraphs 1 and 16 of the GenTG. To Meyer's and others' deep dismay, environmental activists insisted on addressing procedural aspects of the hearing process instead of focusing on the study's scientific substance. The intervenors demanded, for instance, that many of the reference papers, which had been submitted in the original English, should be translated into German to facilitate access.[31] Critics, in short, used the occasion to pursue their broader agenda of throwing impediments in the way of what they saw, or at any rate publicly characterized, as unnatural experimentation with nature. All this was a shock to responsible scientists who had gone out of their way to assuage public concern, and they found the activists' behavior both objectionable and contrary to their understanding of the spirit of the law.

The open hearing not only breached scientists' understanding of the appropriate line between substantive and formal arguments, but it also contested their view of the kinds of uncertainties the public had a right to be concerned about. The initial petunia studies had produced some unexpected results. Transgenic plants that were stably colored in the greenhouse refused to stay stable in the outdoor environment, where they turned unexpectedly pale or variegated. Molecular analysis showed that this unpredictability was due not to transposons having excised the inserted genes (the mechanism the scientists wanted to study) but to environmental factors, such as above average heat and light that summer, as well as the age of the seeds from which the crosses had been made.[32] Such serendipitous observations are what science thrives on; they raise the curtain on new vistas of inquiry. In Meyer's view, while scientists owed the public a demonstration of safety, there was no reason, and indeed no basis, for his group to account for unexpected experimental findings. He sharply distinguished the issue of safety, where everything had to work (and apparently did work) according to plan, from the issue of experimental results, where surprise was a legitimate—indeed, for scientists, a most desirable—outcome:

> **PM:** . . . those results were very surprising, but of course, quite interesting for us. Now, all these results, of course, had nothing to do with the safety evaluation of the experiment and when we started the experiment, we said to the public that we would like to perform these experiments because we expect transposons to create variable phenotypes and we wanted to isolate variable phenotypes and we expect that they will be caused by transposons and we need a large number.[33]

The Max-Planck scientists saw no particular contradiction in a study whose results were at once "expected" and "surprising." These were, in Meyer's view, "two different things":

> **PM:** So the results were as expected. The petunias don't spread; they don't survive.
> **SJ:** In terms of safety, the results—or ethics?
> **PM:** That's what I mean. And these are two different things. We always say OK we would like to do an experiment and we can guarantee from our present knowledge,

as good as we can guarantee, that there will not be any danger, escape, or however you want to call this, [undecipherable] plants. That was part number one, which we thought was what we had to show to the public as best as we could. And the second part was, OK, we want to do an experiment and every experiment, of course, is open in its result, otherwise no funding agency would fund it. But of course we said that is what we expect from the experiment, and it didn't turn out as we expected, and maybe it was a mistake to tell the public, quite frankly, that we were quite happy that we found something which we didn't expect because those are very often the most interesting things. They lead you to new observations . . . [34]

Was Meyer's confidence in the safety studies warranted? In retrospect, we can only say that the premises on which those studies were designed were never carefully probed, although, as Brian Wynne and other science studies scholars have shown, such untested and unstudied assumptions about the social and natural worlds may be thoroughly unfounded.[35] Real-life examples from the U.S. experience with StarLink corn and Prodigene (see chapter 5) still lay in the future. Further, the Max-Planck researchers' focus at the time was on the escape and uncontrolled propagation of the GM plants, and not on questions that transcended the specific experiment and later became controversial, such as the appropriateness of using antibiotic resistance genes as markers. German and international biotechnology critics, at any rate, remained dissatisfied. Although the Max-Planck Institute experiments survived regulatory scrutiny and went forward over initial public objection, they left residual traces of illegitimacy. A decade later, the study was still being cited as an example of the "weird science" of genetic engineering by activist groups such as the Pesticide Action Network North America.[36] Numbers may also tell a part of the story. By 2000 the ZKBS had received a total of 118 applications for deliberate release. Of these, only 3 involved GM petunias.[37]

For the moment, however, German policy on genetic engineering appeared to have reached a workable, if tense, compromise. Public interest groups had gained a new procedural forum in which they could question the goals and premises of genetic engineering. Their questions upset researchers, challenging the scientists' preconceived notions about how far the public should be allowed to go in interrogating science—but in the end, after all the hassles, the research was permitted to go on.

## Unraveling: Normalization Breaks Down

Proponents of biotechnology, regulatory harmonization, and technological progress had reason for complacency in the early 1990s. In three leading industrial democracies—the United States, Britain, and Germany—public distrust and angst about a potentially disruptive new technology appeared to have been confronted and calmed, or in Britain's case avoided through careful management of science's relations with the public. Perceptive observers, however, might have detected clouds forming on the horizon. A tell-tale sign was that, although a consensus of sorts had been achieved, the basis varied from country to country. In the United States science was deemed to have answered, or at least to be capable of answering, all the relevant questions in a regulatory system firmly focused on the risks of particular biotechnological products. In Britain, by contrast, reassurance fell not to science in the abstract, but to a cadre of experienced, managerial scientists like David Bishop and the members of ACRE; it was these experts who diagnosed the public's needs and sought to satisfy them before any hint of trouble. Experts, in short, were entrusted with managing not merely risks but also the publics exposed to them. And in Germany novel procedures were concocted to deal with that country's particular insecurity about the abuse of science, opening a direct and unmediated dialogue between scientists and the public. But attempts to implement these procedures revealed deep conflicts about the very meaning of publicly evaluating the methods and goals of research. These cross-national differences and internal contradictions refused to stay buried and eventually led to renewed controversy.

### United States: Science Confounds Science

The institutional structure for evaluating GMO releases in the United States was designed to provide scientifically reliable answers to questions of risk. The lead regulatory agencies responsible for the oversight of biotechnology strengthened their advisory capacities to meet the challenges of this new industrial technology. At the same time, they moved to limit the range of concerns that could properly be voiced during regulatory assessment. When Monsanto began marketing genetically engineered bovine growth hormone, for instance, critics quickly discovered that there was no place in the federal government where they could raise their questions about risks to small farmers or damage to the welfare of treated cattle, let alone economic arguments against generating further surpluses in an already heavily subsidized industry, or ethical concerns about the instrumental use of dairy cattle as machines for high-intensity milk production. These views and values had to find expression in other than official channels—for example, in political cartoons or Internet postings. For purposes of governance, risks were narrowly defined as threats of harm to human health and the environment, and these in turn were felt to be the preserve of scientific analysis.

Almost imperceptibly, the U.S. discourse of regulating agricultural biotechnology began to equate *risk* assessment with *scientific* assessment. Public officials asserted that the only way to manage the threats of biotechnology was through risk assessment based on "sound science." Science, U.S.

107

administrators and politicians agreed, did not justify any serious worries about the release or consumption of GMOs. For the most part, the public seemed to go along with this assessment. No significant disputes arose as higher and higher percentages of key crops were replaced with transgenic variants. By 1998, 20.5 million hectares were sown with GM crops, up from 11 million in 1997 and 1.7 million in 1996. As a British expert body observed, "These are extremely high adoption rates for a new technology by agricultural standards."[38]

American agricultural biotechnology, then, came to depend on science in two respects: for inventiveness, leading to new products (Wynne calls this "innovation science"), and for regulatory purposes (in my terms, "regulatory science"). Though regulatory science derives constant legitimation from the label "science," sociologically it is a vastly different kind of activity from basic research, at least as that is ideally conceived.[39] An important difference, as Peter Meyer's experiences in Germany also showed, is that regulatory science needs to stay black-boxed, to deny its provisional or indeterminate status, if it is to be credible. Unusually prone to deconstruction in adversarial and political settings, such science depends on institutional closure mechanisms, such as authoritative expert advice, to keep challenges within bounds.[40] Ordinary science by contrast makes advances through uncertainty, provisionality, and surprise. One might expect the fluid and labile character of the latter to threaten the closure-seeking propensity of the former. And indeed, in the United states, the "science" relevant to the safety assessment of transgenic crops refused to stay black-boxed. Two episodes that occurred in 1999 and 2002 were important and illustrative.

In May 1999 John Losey and his colleagues, all entomologists at Cornell University, reported in *Nature* the results of studies they had done on the effects of a transgenic corn species on the monarch butterfly.[41] Known as Bt-corn because it contains genes from the bacterium *Bacillus thuringiensis*, this GM corn variety produces a toxin that is deadly to a common agricultural pest, the European corn borer. Losey's group dusted milkweed leaves with pollen from Bt-corn and fed these to monarch caterpillars, nearly half of which died within days. These results were not just surprising but potentially explosive. The monarch, with its distinctive orange and black coloring and its remarkable migratory habits, is one of America's most distinctive and beloved butterfly species. It is also a so-called nontarget species for Bt-corn, since monarch larvae do not feed on corn but on milkweed. Losey's experiments seemed to show that Bt-corn was dangerous not only to a designated pest, the corn borer, but also to a species that no one had any intention of harming.

The short *Nature* article had consequences beyond anything the authors had imagined. The biotechnology industry went into high gear in attempting to undermine the study's significance, commissioning counter-studies and aggressively marketing "information" to the public. The object was not to discredit Losey's competence or credibility so much as to make his study seem irrelevant to assessing the risks of Bt-corn. Among the most active players was Monsanto, the world's leading supplier of transgenic crops.[42] Monsanto posted on its "Biotech Knowledge Center" website the argument that "this experiment was conducted in a laboratory, not in the natural habitat of the Monarch butterfly."[43] With these words, the company in effect endorsed and strategically deployed an argument that academic and social critics had long leveled against industrial and governmental claims of safety: that the best test of a product's behavior in the real world is its actual behavior in that world.[44] Lab or field studies, however carefully designed, can do no more than approximate the complexity of actual use in real-world conditions. Indeed, studies done to verify Losey et al.'s conclusions suggested that, although some forms of Bt-corn were toxic to monarch butterflies, these varieties were not the ones in widespread use in U.S. agriculture.

Industry's attempts to refute any logical connection between the monarch study and commercial uses of Bt-corn may have succeeded at the level of professional scientific debate, but they worked substantially less well as a public relations strategy. Environmental and antiglobalization groups found in the potentially threatened monarch an irresistible symbol of the larger problems that they wished to bring to public attention: inadequate environmental testing and monitoring of GM crops, risks to nontarget species, damage to biodiversity, and corporate recklessness. Monarch images and costumes prominently figured in the 1999 riots against globalization in Seattle, Washington. Clearly, activists were unwilling to concede the basic point that "science" had adequately established the safety of products such as Bt-corn. Expert judgment had neither addressed nor answered the foundations of global public concern.

A rather uglier dispute erupted in the spring of 2002, with potentially longer-term consequences for the credibility of science. At issue again was a piece in *Nature*, this time written by Ignacio Chapela, a Mexican scientist, and his student David Quist, both biologists at the University of California, Berkeley.[45] Quist and Chapela reported that they had found evidence of cauliflower mosaic virus genes, commonly used as a promoter in industrially produced transgenic corn, in native ("criolla") strains grown in Mexico's Oaxaca region. Their experiments suggested that genes from bioengineered corn had migrated into native corn even in remote areas, with a likelihood of higher penetration in more accessible regions. The findings were particularly troubling in view of Mexico's having imposed a moratorium on planting transgenic corn in 1998. They were also politically sensitive, given the long history of corn cultivation in Mexico and that nation's strong commitment to protecting the genetic diversity of its native species.

Gene flow, or the transfer of genes from one population to another, has been one of the most hotly debated issues surrounding agricultural biotechnology. Opponents of large-scale commercial applications point to the possibility of gene flow as a threat to biodiversity, whereas proponents either minimize the probability of such transfers occurring at all or deny that it would be a threat even if it did occur.[46] Unexpected findings such as Chapela's were bound to fuel that ongoing controversy, but the vehemence of the reactions exceeded many expectations. Initial responses included a number of highly critical letters to *Nature* charging that the results were an artifact of poor experimental methods combined with an unfortunate rush to publish. Chapela and Quist had used the technique of inverse polymerase chain reaction (iPCR) to study their samples. This widely used method allows scientists to amplify and analyze small quantities of DNA, but it is also prone to contamination and can produce false positives. As the furor over the paper mounted, critics from industry joined the chorus, alleging that the authors had allowed their politics to override their science: they were behaving, in short, as "activists," not "scientists." Chapela, his detractors noted, had a history of political activism. He had opposed a five-year deal between the Swiss pharmaceutical and agrochemical company Novartis (later Syngenta) and Berkeley's Department of Plant and Microbial Biology, in which the company had agreed to pay the department twenty-five million dollars for research in return for benefits such as first rights to patents on potential discoveries. How could a person who had fought this deal conduct a neutral inquiry into the environmental consequences of agricultural biotechnology? In return, supporters of Chapela and Quist accused the attackers of illegitimate political motives and hidden connections to industry.[47]

Matters came to a head on April 4, 2002, when Philip Campbell, the respected editor of *Nature*, took the unprecedented step of withdrawing the journal's support for the contested article. It was not a retraction but something distinctly odder. Along with two letters critical of Chapela's results, Campbell published a note in the journal's online version stating that "the evidence available is not sufficient to justify the publication of the original paper." New data submitted by the authors had failed to establish "beyond reasonable doubt" that transgenes had been integrated into native corn genomes. Since the authors still stood by their original findings, however, Campbell felt it best "simply to make these circumstances clear, to publish the criticisms, the authors' response and new data, and to allow our readers to judge the science for themselves."[48]

In effect, Campbell's action opened the door to an unprecedented form of postpublication peer review. It was as if the first round of review, favorable to the authors, had only served as the "field test" of initial editorial scrutiny and approval. Now that the results were in full-blown public circulation, they had also, according to *Nature*'s editor, laid themselves open to a kind of scaling up, to extended peer review. But the consequence arguably was to subject an article on GM crops to a greater degree of scrutiny than the crops themselves had undergone in their passage from lab to field to commercial cultivation.

We will revisit the university-industry links that figured so prominently in the Mexican corn controversy in chapter 9. For present purposes, the more important point is the curious way in which this controversy at once undermined and sustained the status of *science* in the biotechnology debate. At one level, most observers agreed that science had been shown to be political, indeed that it was unavoidably so. At another level, appeals to science continued, especially in the comments of critics who portrayed Chapela and Quist as having fallen short of well-recognized canons of objectivity and good scientific practice. What lent irony to these charges was that the entire episode had disclosed just how fluid and unsystematic were the methods for investigating the environmental behavior of transgenic crops. Even *Nature's* peer review practices with regard to such studies were shown to be subject to contingent pressures and flexible interpretations.[49] Philip Campbell's admirably honest invitation to readers to make their own assessments of "the science" only heightened the irony by abdicating editorial omniscience while retaining power to exercise it, and by openly admitting the subjectivity of scientific judgment under conditions of uncertainty.

### Britain: Weakening Expertise

If the authority of science weakened in the United States under the stresses of supporting agricultural biotechnology, then it was the culturally sanctioned concept of expertise that came under comparable pressure in Britain. As we will see in the next chapter, it was not genetic modification but bovine spongiform encephalopathy (BSE), or "mad cow disease," that posed the most visible threat to expert authority in the 1990s. Yet well before the BSE crisis grabbed the center of political attention, British scientists, policymakers, and members of the concerned public had begun to question the meaning of expertise in relation to something so complex and hard to pin down as the risks of deliberate release. Some signs of the waning of expert authority were highly visible, while others remained more circumscribed, but it was clear by the end of that politically troubled decade that public policies for biotechnology would have to find new ways of engaging with, and reassuring, an increasingly skittish consuming public.

The fate of David Bishop, the cautious Oxford virologist who once made "such a meal" of his public relations, provides one instructive angle of vision. In the spring of 1994, field tests conducted by the Institute of Virology came under intense scrutiny because the terms of the experiments had changed dramatically. Interested now in concrete products, the institute had moved to test a viral pesticide with an inserted scorpion gene; this manipulation allowed the virus to produce a toxin that would quickly and efficiently kill plant pest caterpillars—more so, it was hoped, than conventional chemical pesticides. The GM virus, the *Autographa californica* NPV (AcNPV), was to

111

be tested on the cabbage looper, prompting joking references to cabbage patches, but also serious scientific and public concerns. In particular, Bishop's scientific colleagues in Oxford and elsewhere voiced a number of objections that were picked up and disseminated by the national media.

Openly on the table was whether Bishop, widely viewed as a strong and opinionated research leader, had adequately considered the risks of release, but additional questions revolved around who actually was responsible for making binding judgments on such issues. There were several technical questions that opponents of the release felt had been inadequately addressed. How could earlier trials conducted with enfeebled viral species be used to justify the release of an organism that was fully biologically active, and "unnatural" on top of it?[50] Scientists also questioned the likelihood of restricting the AcNPV to a particular target (the cabbage looper) when tests suggested that it could affect up to one hundred species of butterflies and moths. The fact that this was a nonnative virus, and that the release site was within close proximity of Wytham Woods, a "treasure trove" of lepidopteran biodiversity, only made matters worse in the critics' view. In an unusual move, some scientists even considered legal action to block the release, although it eventually went forward with approval from ACRE and the relevant government ministries.[51]

Other themes emerged as the trials were conducted, all consistent with Britain's initial framing of biotechnology as a process deserving special concern. The first was that of unintended consequences. Early reports from the pesticide trial suggested that some stocks of the engineered virus had become contaminated with the nonengineered wild type, making the experimental results uninterpretable. Although not specifically pertinent to the earlier debate on risk, the episode underscored the unpredictability of working with GMOs and contributed to doubts about the commercial viability of such products.[52] The second theme was accountability, as reports circulated that Bishop's team had failed to disclose potentially damaging data about the range of species affected by the engineered virus.[53] The third, and for us most interesting, theme was the dissatisfaction some scientists expressed with the culture of regulatory expertise that had permitted the AcNPV trials to proceed. Foreshadowing the criticism unleashed by the BSE crisis, Steve Jones, professor of genetics at University College London, took government scientists to task for their "'nanny knows best' attitude." He also voiced an empiricist's impatience with the sloppiness of the approval process: "They say the virus is not going to escape. If you look at the proposal that's clearly not true. They can by no means guarantee that this sodding virus is going to stay there."[54]

Less than a year later, in March 1995, David Bishop, the man at the center of the controversy, was abruptly dismissed from his post by the Natural Environment Research Council, under whose aegis he had functioned for eleven years as IVEM director. Publicly, NERC denied that the AcNPV trials had anything to do with the firing and even asserted that the trials would continue.[55] Bishop's dismissal was attributed to "structural redundancy" as a result of changes in NERC's research mission. Privately, people cited complaints from Bishop's coworkers about the nondisclosure and even distortion of data in the 1994 viral release application.[56] Perhaps fittingly, the man who decided Bishop's fate was John Krebs, NERC's chief executive, who would later become the first head Britain's Food Standards Agency, formed to restore public confidence in the wake of the BSE scandal.

The scorpion gene episode, capped by Bishop's sudden departure, was the most public controversy in the early development of British agricultural biotechnology. The anxieties it disclosed, however, were not limited to that single event, and geneticists were not the only ones worried about proceeding too far too fast with an untried technological process. Similar concerns surfaced in the work of the British Government Panel on Sustainable Development, headed by veteran diplomat and environmentalist Sir Crispin Tickell, a highly regarded member of Britain's "great and good." The panel selected biotechnology for its second report, issued in January 1996, months before the explosive turn taken by the BSE case.[57] Tickell himself wrote up the issue, nervous at first about going in over his depth, but also convinced (in part through representations from respected environmental groups such as the Green Alliance) that the government had paid insufficient attention to problems such as monitoring deliberate releases. The panel's objective, he said, was to set off a firecracker, and it succeeded in catching the government off guard: "It touched a raw nerve."[58] Within months the government responded, more extensively than the panel had expected, and, while continuing to assert its commitment to biotechnology, conceded that most of the problems identified in the panel report were well founded.[59] In a foretaste of things to come, the government agreed that its risk appraisal, emergency measures, and liability provisions might not adequately address all of the panel's concerns about agricultural biotechnology.

Political upheavals over the next few months and years gave new urgency to the precautionary trends illustrated by these events. The BSE crisis of early 1996 (of which more in chapter 5) helped open the door to the Labour Party's triumphant return to power under Tony Blair in 1997, with a mission to reform government and make it more transparent and accountable. Health and environmental regulation emerged as primary sites of institutional innovation, offering comparative analysts unparalleled opportunities to ask what changed and what remained the same in Britain's institutional ways of decision making. Following a review in 1999, the government decided it needed a broader range of inputs into its strategic framework for both green (agricultural) and red (pharmaceutical) biotechnology. Consequently, a new twenty-member Agriculture and Environment Biotechnology Commission (AEBC) was appointed in June 2000 to offer strategic advice and to work more closely with two other newly created committees, the Food Standards Agency and the Human Genetics Commission.

The new committee structures broadened earlier understandings of expertise by drawing a wider spectrum of opinion into the advisory process. AEBC members included academics and practitioners, scientists and ethicists, farmers and industrialists. This widening of the opinion network inevitably shifted the terms of debate on GM plants, particularly by placing the topic of uncertainty squarely on the table. At the same time, it did not resolve the tensions within Britain's empiricist culture between some who saw the absence of evidence as evidence of the absence of risk, and others who took the same absence as sheer ignorance, pointing to as yet unknown and unimagined threats. An exchange between Robin Grove-White, professor of environment and society at Lancaster University and chairman of UK Greenpeace, and the chairman of ACRE, who gave evidence for the AEBC report *Crops on Trial*, [60] illustrates the clash of perspectives and the emergence of a new line of conflict:

> RGW: Do you think people are *reasonable* to have concerns about possible "unknown unknowns" where GM plants are concerned?
>
> Advisory scientist: *Which* unknowns?
>
> RGW: That's precisely the point. They aren't possible to specify in advance. Possibly they could be surprises arising from unforeseen synergistic effects, or from unanticipated *social* interventions. All people have to go on is analogous historical experience with *other* technologies . . . .
>
> Advisory scientist: I'm afraid it's impossible for me to respond unless you can give a clearer indication of the unknowns you're speaking about.[61]

As the exchange makes clear, the question about the safety of organisms in Britain's GM advisory circles had subtly shifted from "How safe is safe enough?" to "When, with what evidence, and on whose assertion, is it reasonable to raise a safety concern?" These questions were anything but academic. AEBC took office in an atmosphere of crisis about the future of GM crops in Britain. Protests by environmental groups, ranging from destroying field trial sites to lawsuits, had brought field trials by firms such as AgrEvo (later Aventis) to a virtual halt. *Crops on Trial*, one of AEBC's first work products, recommended a systematic program for proceeding with farm-scale trials in Britain. In its response, the Department of Environment, Food, and Rural Affairs again stated that there would be no authorization of commercial planting of GM crops until the likely conclusion of farm scale trials and full regulatory evaluation of the results in 2003.

### Germany: Containing the Process

And what of Germany, where Green mobilization had put programmatic issues of transparency, participation, and institutional reform on the biotechnology policy agenda in the 1980s, several years ahead of comparable moves in Britain? Just as science and expertise proved insufficient to hold the line against skepticism in the United States and Britain, respectively, so Germany's process-based approach also gave way under strain. Indeed, one major procedural innovation of the 1990 GenTG, the public hearing on deliberate release, was repealed just three years later, leaving the ZKBS with greater freedom to establish risk categories and assess new releases without significant public oversight or intervention. How can we explain this retreat?

An official stocktaking of experiences in the first two years under the GenTG points to some of the reasons. The occasion was a hearing of the parliamentary Committee for Research, Technology and Technology Assessment held in February 1992. A list of questions circulated before the meeting asked specifically for reactions to the public consultation requirements of the new law. Several respondents addressed the question, and opinion was sharply divided. The submission from Hoechst, one of the first German companies to be targeted by antibiotechnology forces,[62] noted that no hearings pursuant to the GenTG had been held in its home state of Hessen but nonetheless criticized even hearings held under prior law as ill-suited for drawing the public into a constructive dialogue with science. Participants, Hoechst argued, had not posed pointed, factual questions but instead had raised general objections to genetic engineering. The company quoted from a leaflet of a women's group, Bürgerinnen beobachten Petunien (Citizenesses Observe Petunias), declaring that it would use every means at its disposal to damage the reputation of genetic engineering and prejudice the investment climate against biotechnology.[63] From within the government, the Robert Koch Institute (RKI) took a similar line, pointing out that the great majority of the 1,600 submissions to the petunia hearing had raised either general reservations about biotechnology or else purely formal complaints about the application materials, such as their incompleteness and the use of English. The hearing, moreover, had entailed costs on the order of DM 100,000 as compared with DM 1,000–1,500 for normal ZKBS evaluations. Under these circumstances, the RKI concluded, the benefits of citizen participation did not justify the expense.[64] On these utilitarian grounds, RKI experts in effect refuted the proposition that citizens *had* to be involved as watchdogs over the relations between science and government. It was the ultimate victory for the calculating administrative state.

It seems in retrospect that, for all these observers, it was not the petunia release but the public hearing that was the more unusual experiment. Hearings made sense, in their view, only if they served as a public sphere in the Habermasian sense. Participants were expected to engage in informed debate, conforming to industry's and government's preconceived notions of rational inquiry. A hearing was meant to corral public opinion within appropriate limits, much as physical containment devices corralled GMOs. Strategic use of the hearing by participants to advance a broader political agenda subverted this construction of the purpose of public consultation. Citizens had been given a chance to behave responsibly, as reasoning actors, and they had failed the test.

Gerd Winter, writing for the Center for European Law Politics (Zentrum für Europäische Rechtspolitik, ZERP) at the University of Bremen, offered a very different analysis. Citizens, he argued, had good reason to question the goals and benefits of biotechnological production. Hearing administrators had therefore erred in excluding queries on this score. An analysis of benefits was required, Winter suggested, not only to enable citizens to form a holistic picture of the risks they would be exposed to, but also pursuant to the precautionary principle. After all, how could state authorities assess the acceptability of a project's risks *without* considering its goals? If the purpose was of questionable social value, then the associated risks could not be socially tolerable. Winter also urged more openness in the operations of the ZKBS, citing the U.S. Government in the Sunshine Act as a model. The commission, he asserted, was not a secret service but represented a mechanism for activating social expertise.[65] Consistent with this role, information had to flow out of, as well as into, the ZKBS, and its members—especially those representing segments of civil society—had to be free to relay information to their constituencies outside the commission. In other words, Winter, too, stressed the informational and opinion-making functions of the public hearing and of ZKBS, but in his view the adequacy of these processes was to be judged from the standpoint of the "at risk" or "to be informed" citizen—not from that of the state's interest in scientific freedom, industrial productivity, or governmental efficiency.

The 1993 amendments to the GenTG paid more attention to complaints from science and industry, and to the government's own experts, than to arguments like Winter's or from environmentalists in favor of openness and participation. In particular, the research community's charge that excessive bureaucracy was stifling German science and destroying its competitiveness proved effective,[66] and the 1993 legislation streamlined many aspects of the approval process for both marketing and release. German authorities also cited the European directive, 90/220/EEC, as grounds for simplifying their own approvals process and for eliminating the public hearing on releases. Few noted that this victory for scientific experimentation entailed a defeat for the social experiment of involving the public more fully in the management and control of biotechnology.

Protests over field trials of GM crops continued in Germany as in Britain, although in Germany, too, commercial planting remained banned into the next decade. In 1999, for example, Rainder Steenblock, the Green Party environment minister of the northern Land of Schleswig Holstein objected to federal authorization of a trial of genetically modified oilseed rape.[67]

His objection raised renewed questions about federalism and the capacity of a central authority, the Robert Koch Institute, to resolve local environmental concerns regarding GM crops. As late as spring 2003, Greenpeace activists sabotaged the approved site for a planned trial of Syngenta's GM wheat, near the northern city of Hamburg, by sowing it with organic wheat seeds.[68] Such actions prompted German researchers, in turn, to keep secret the locations of fields sown with GM corn—a course that activists denounced as contrary to European law.[69]

## Conclusion

In the late 1980s and early 1990s, the United States, Britain, and Germany all put in place new procedures and principles for managing the risks of GMO releases into the environment. National strategies and policy discourses differed, reflecting the historical origins and forms of debate in each country, and the consequent framings of biotechnology as product, process, or program. More specifically, different institutional and discursive resources were mobilized to ward off criticism and reassure the public: in the United States "science" was said to have confirmed the safety of most releases; in Britain it was not so much science as expert judgment that formed the basis for assertions of safety; and in Germany, to begin with, bureaucratic procedures and public consultation were the instruments of choice for allaying the fears of a nervous citizenry.

Curiously, each mode of stabilization carried within it the seeds of its own vulnerability. U.S. science proved less monolithic and less quiescent than the most ardent biotechnology proponents would have wished. Ecologists' concerns about huge, unprecedented, and largely unmonitored environmental experimentation never fully faded, and findings like those of Losey and Chapela indicated, at the very least, how many aspects of environmental risk had never been fully tested. In Britain the regulatory and political upheavals of the early 1990s undermined the social role of the expert, forcing the creation of new advisory institutions that were both more diverse and more transparent. In these forums, traditional standards for evaluating evidence were challenged, revealing new fault lines between those demanding empirical evidence of risk to justify more stringent regulation and those urging greater precaution in the presence of unknown unknowns. The German case is in many ways most interesting, because here an experimental democratic settlement was undone in favor of a return to a more technocratic approach. The deeper point, however, is that the very idea of a universally acceptable process, creating an open deliberative space for science and technology, proved untenable in Germany, just as the facticity of science and the reliability of expertise had done in the United States and Britain.

The unraveling of the early regulatory settlements led, in turn, to new controversies and new attempts to forge consensus on biotechnology policy. The issue of deliberate release remained heavily contested in both Britain and Germany into the early years of the twenty-first century, though both countries had delayed commercialization in favor of continued, supervised scientific experimentation. In the next chapter, we turn to another contested site, the debates over the safety of foods derived through GM techniques.

# 5

# Food for Thought

On February 18, 1999, a pickup truck operated by Greenpeace UK dumped four tons of genetically modified soy beans outside the gates to No. 10 Downing Street, the London residence of Britain's Prime Minister Tony Blair and the famed nerve center of British politics. A wide yellow banner on the side of the truck carried the message, "Tony, don't swallow Bill's seed." It was a naughty, punning, and very British reference to President Bill Clinton's recently concluded impeachment hearing in the United States, on charges of high crimes and misdemeanors resulting from his fling with White House intern Monica Lewinsky. More important, it was part of the opening salvo in a trade war that would lead, by 2003, to the dispute resolution tribunals of the World Trade Organization (WTO). Why did such a dramatic standoff come into being, and what does it tell us about the democratic debate on biotechnology on the two sides of the Atlantic?

Bypassing the orthodox channels of deliberative politics, but playing directly to the media, the British arm of Greenpeace inserted itself into a process of technology assessment that seemed only partly within the control of official regulatory bodies and their expert advisers. Beans at the gates of Downing Street asserted, perhaps more loudly than words, that the British public would not lightly put up with an American technology that was muscling its way onto Britain's dining tables. A Greenpeace spokesperson said, "We are taking these GM soya beans to one of the few homes in the UK where they want to eat it."[1] The campaign vocally rejected the U.S. assertion that GM soy beans were only soy beans, with no consequences riding on their status as *genetically modified* food. Newspapers headlines and cartoons played on the specter of "Frankenfoods" invading the British diet. Giant supermarket chains saw the writing on the wall, and companies like Sainsbury's hastened to assure their customers that they would withdraw GM products from their shelves. The European Union declared a moratorium on the import of GM foods, lifting it in July 2003 only in the face of an incipient trade war, and then on condition that the products would be clearly labeled and traceable. Street theater and media reports, it appeared, had set the terms of debate on food safety far more effectively than government regulators.

Britain's, and more broadly Europe's, retreat from GM foods at the turn of the century caused much consternation in the United States, the would-be world exporter of these products. In the inevitable search for explanations, first on the list was the public's alleged misunderstanding of science and lack of trust in experts, particularly in Britain in the aftermath of the BSE crisis; many blamed irresponsible media hype for fanning the fears of an ignorant public. Careful commentators also cited the arrogance of U.S. multinationals such as Monsanto, and the British public's anxiety about anything that could harm their cherished countryside. But how persuasive are these explanations? Notably, several of these same factors had been invoked in the 1980s to explain—at that time—the *American* public's "irrational" fear of chemicals such as the plant growth regulator Alar.[2] *Europeans* at that time had seemed more complacent about risks related to food. How then did the shoe come to be on the other foot? Can factors such as scientific ignorance, irrationality, and distrust satisfactorily account for the apparent transatlantic switch in public attitudes with respect to GM crops and foods?

We will see in this chapter that concern about the safety and integrity of the food supply is not the prerogative of any single nation or particular segment of the consuming public. In fact, consumers in all three countries continually worry about the methods of food production and have asserted themselves at one time or another on the acceptability of GM foods. As with the issue of deliberate release, it is not so much the *fact* of protest in the GM case that was new or nation-specific, but rather the timing and occasion for expressing resistance, the forms and objectives of citizen action, and the shape of scientific, corporate, media, and governmental responses. In each country, the initial framings of biotechnology discussed in chapter 2 and their incorporation in legal and regulatory structures for consumer protection both constrained and enabled the modalities of citizen action. Thus, opposition to GM foods crystallized as direct action in Britain and as a fierce defense of the integrity of organic foods in the United States; in Germany, public protest was more muted, but national policy remained linked to skepticism about the need for green biotechnology and a strong commitment to organic farming.

GM foods, in other words, were never viewed in isolation from wider issues of agricultural practices, nature preservation, and the integrity of food. These products had to take—or, more accurately, *make*—a place for themselves on tables and supermarket shelves, as harbingers of a new mode of agricultural production, in competition with foods produced through more familiar, "natural" means. Within each country, the acceptability of GM crops and foods was evaluated in relation to dominant framings of the problem of biotechnology, and against the backdrop of systems of food production and consumption that had been selectively naturalized or taken for granted. In each nation, moreover, consumers had various means of opting out of the dominant eating traditions, critiquing them, or resisting perceived alterations to the food supply. Not surprisingly, the naturalness, or unnaturalness, of GM technology was discrepantly construed in these separate contexts. Although all three nations eventually moved toward acceptance of GM crops and foods, the legal and cultural meanings of these products changed in the process.

This chapter traces some of the connections between the national framings of biotechnology, the major expressions of resistance to GM foods in each country, and associated understandings about how to secure the safety and integrity of the food supply. Our object throughout is to identify, in each political system, how public concerns about food were deliberated, and how if at all reassurance was provided that GM foods would not cause harm to health, safety, or other social values. We begin with Britain, where the early focus on the process of genetic modification continued to play out in controversies over the government's handling of GM policy. In the United States, by contrast, the product framing initially muted the discussion of risks from GM foods, but it eventually gave rise to a market-driven backlash from producers and consumers of organic foods. In Germany, persistent concerns about agricultural biotechnology were refracted through the lens of national party politics and articulated as support for precaution at the European level. Finally, EU policy created the conditions for a larger global market in GM foods, but its role was driven by, rather than determinative of, member states' resolution of relevant questions of risk and uncertainty.

The politics of GM foods, I argue here, remained resolutely and importantly national, despite efforts at European harmonization. The central questions of citizens' rights to intervene in technological innovation, and the terms on which citizens might be entitled to do so, were addressed, tacitly or explicitly, in the framework of national politics.

## Britain: GM in the Shadow of BSE

It is hard to disentangle the story of opposition to GM foods in Britain in the late 1990s from the still unfolding story of BSE. Beginning quietly in 1986, BSE erupted into a full-blown national and international crisis a decade later when British authorities announced on March 20, 1996, that, contrary to previous governmental assurances, the disease had crossed the species barrier and was causing deaths among infected people. An illness hitherto observed only in cattle had now afflicted human beings in a new and fatal form known as variant Creutzfeld-Jakob Disease (CJD).[3] The announcement sent shock-waves through Britain and Europe and led to a worldwide ban on imports of British beef for two-and-a-half years, as well as a complete breakdown of confidence in the Ministry of Agriculture, Fisheries and Food.[4] The extremely public mishandling of the "mad cow" crisis by MAFF officials and experts eroded the Conservative Party's already tenuous hold on power and helped pave the way to Labour's landslide victory in 1997.[5]

The new Labour government kept its promise and ordered a massive public inquiry. MAFF was dismantled, its responsibilities were largely transferred to a new Department of the Environment, Food and Rural Affairs (DEFRA), and a new Food Standards Agency was formed to supply the government with reliable expert advice on food. The BSE inquiry, concluded in 2000 under the stewardship of Lord Phillips of Worth Matravers, found that British health and safety experts had acted as a narrow and secretive community: they were complacent about empirically unverifiable risks yet unwilling to commission new research to improve on available knowledge, and profoundly reluctant to display any uncertainty to a public they saw as irrational and prone to panic.[6] These revelations shook people's faith in a decision-making system founded on the premise that governmental experts know best—indeed they raised doubts whether, at moments of crisis, appropriate experts are available to cater to urgent public needs. BSE speeded a turn away from the "nanny state"; it was a spur to democratization.

When British environmental and consumer groups rose up to protest GM foods, it was therefore tempting to conclude that BSE was the root of their discontent. The issues, after all, were similar in striking respects. Both involved the industrialization (some said hyperindustrialization) of the food supply, with techniques that were seen as not natural. As far back as 1989, a committee chaired by Sir Richard Southwood, former head of the Royal Commission on Environmental Pollution and a member of Britain's scientific and policy elite, had called attention to the problematic practices that seemed to have caused BSE: "We note that this disease appears to have resulted from unnatural feeding practices as found in modern agriculture. We question the wisdom of methods which may expose susceptible species of animals to pathogens and ask for this general issue to be addressed."[7] The uneasy perception that "modern" and "unnatural" are somehow closely linked also surfaced in the debate over genetically modified foods.

Some hold the overwrought post-BSE environment directly responsible for a near-panic over GM foods in Britain in mid-1998. The triggering event was nothing more than a normal scientific controversy, involving arguably premature media disclosure of experimental data and charges of possible scientific misconduct. The story began on August 10, 1998, when Arpad Pusztai, a senior researcher at Scotland's Rowett Research Institute, announced on

Granada Television's *World in Action* that rats fed a pest-resistant transgenic potato in his lab suffered from stunted growth, depressed immune systems, and reduced body weight. Two days later the respected institute director Phil James announced that there had been a mixup over Pusztai's data, and he turned the results over to an independent audit committee for further investigation. Pusztai was briefly suspended and retired later in the year when his contract was not renewed. The committee exonerated Pusztai of wrongdoing but found his data too inconsistent to support his conclusions. His supporters meanwhile continued to insist that his findings of immune system effects were troublesome, and calls arose for a moratorium on the introduction of GM foods pending more rigorous regulation—a position the prime minister declined publicly to approve.[8]

From a political standpoint, however, treating the BSE episode as the "cause" of Britain's panic over GM leaves much unexplained. In particular, this easy linear attribution hardly touches on the discursive framing of the GM debate; nor does it satisfactorily account for the means by which the government and public responded to the crisis. In the U.K. case, it is especially important to address these specificities of public response, because they led to procedural innovations of a kind not seen in either Germany (which suffered its own BSE crisis) or the United States. Quite exceptionally, the British GM controversy operated as a site for an extraordinary reinvention of state-society relations in the management of science and technology. To understand this phenomenon, we need to take a closer look at the public's role in policy formation on GM food: who spoke for the public, in what forums and voices, who set the terms of the debate, and how were publics drawn into conversation with the state? In reverse, we also have to examine the state's response: how did it engage with disaffected publics, and how did it seek to build trust and acceptance on matters of contested scientific and technological expertise?

## The Prince and His Nature

The technology of GM food may be new, but the institutions it engages with are old, even archaic. This friction of old and new, producing recurrent crises of modernization, lies at the heart of Britain's political predicament at the turn of the century. It takes a novelist's imagination to capture the contradictions fully. In his wickedly comic send-up of contemporary Britain, *England, England*,[9] Julian Barnes imagined the Isle of Wight converted into a private, five-star tourist destination, in which everything the world thinks of as quintessentially English is re-created cheek-by-jowl, so that discriminating tourists can have a concentrated experience of the country without any of the inconveniences of travel. Preparing for this venture, Sir Jack Pitman, the tycoon behind the project, conducts a worldwide survey to determine the "Fifty Quintessences of Englishness." The Royal Family heads the list, immediately followed by Big Ben/Houses of Parliament, Manchester United Football Club, and the class system. In due course, Pitman persuades the royals, or their look-alikes, to take up residence in his luxurious theme park.

Irony is possibly the only register in which to record the peculiar status of the British monarchy at the turn of the twenty-first century, for this most unrepresentative and constitutionally powerless of institutions, lampooned as much as it is revered, still performs the task of representing the nation in bewilderingly complex ways. For most outsiders, the British monarchy quite simply symbolizes the nation, as in Julian Barnes's giddy confection; but within the nation, too, the royal family remains a kind of logoized emblem or embodiment of nationhood, a focal point for the national imagination, a family both like and unlike any other, and an abiding representation of the nation as nuclear family.[10] It is a role that Elizabeth I, one of Britain's most subtle and theatrically self-aware monarchs, well understood when, at the time of her accession to the throne in 1558, she told her nobles that "[t]his royal office made her a 'body politic to govern,' as well as a human 'body natural.'"[11] The historian Linda Colley writes that in the late eighteenth century an increase in the royalty's physical mobility helped spread the "deeply appealing myth that members of the Royal Family were just like everyone else, yet at the same time somehow different."[12] The ability to represent the nation by incorporation, encompassing the intangible national body within a common, physical, human body, remains even now a significant royal resource, in spite of all the foibles, follies, and scandals of the Windsor family. That power of representation played a nontrivial part in the GM controversy.

The threads that connect the British monarchy with Britain's voting public intersect with all their contradictions in the person of Prince Charles, the heir to the throne. Widely ridiculed for his romanticizing views on architecture and countryside, his reactionary support for aristocratic privilege, and his embarrassing extramarital entanglements,[13] Charles had reached a nadir in popular esteem at the time of his ex-wife Princess Diana's death on August 31, 1997. Yet, barely two years later, Charles significantly redeemed himself in the public eye not only as a considerate father to his two bereaved sons, but as an advocate for caution with GM foods. In June 1999 the Prince of Wales made news with an unusual public statement on the now heated GM controversy. He picked an unconventional medium: the *Daily Mail*, a tabloid that had not been particularly sympathetic to him during his matrimonial troubles. Here, Charles laid out ten questions (see table 5.1) that he felt had not been adequately answered by the advocates of genetic modification. His focus was on all the things that were not yet known, both about the products themselves and about the governance structures for dealing with them. In expressing these worries, Charles implicitly identified himself with the ordinary public, taking on their concerns as his own, the future king acting as sensible, empirically minded, if scientifically unschooled Everyman: "But what I believe

TABLE 5.1
Prince Charles's Ten Questions on GM Food

- Do we need GM foods in this country?

- Is GM food safe for us to eat?

- Why are the rules for approving GM foods so much less stringent than those for new medicines?

- How much do we really know about the environmental consequences of GM crops?

- Is it sensible to plant test crops without strict rules in place?

- How will consumers be able to have real choice?

- If something goes wrong with a GM crop, who will be held responsible?

- Are GM crops really the only way to feed the world's growing population?

- What effect will GM crops have on the world's poorest countries?

- What sort of world do we want to live in?

Source: The Daily Mail, June 1, 1999.

the public's reaction shows is that instinctively we are nervous about tampering with Nature when we can't be sure that we know enough about all the consequences."[14]

Reactions to the prince's forthrightness were mixed. Opinion surveys showed that a large majority of the British population shared Charles's concerns. Yet some saw the prince as being on a collision course with Downing Street's policy of favoring GM technologies and considered such active royal engagement with a live political issue to be seriously out of order. Derek Burke, a seasoned academic administrator and science adviser,[15] published answers to Charles's questions under the heading "comments from a puzzled subject." While the prince, along with many of his subjects, dwelt on the unknown, discontinuous, and unnatural dimensions of genetic modification, Burke, like many U.S. regulators and scientists, emphasized the continuities between GM and traditional agriculture, and by extension the similarities they present for public policy. More specifically, Burke called attention to the widespread cultivation of GM corn in the United States without any apparent risk or harm. In answer to the prince's final question ("What sort of world do we want to live in?"), Burke advocated "a world where we use new technology safely and constructively" and asserted that "we can do that if we keep our heads, which at the moment we are signally failing to do."[16] He thereby claimed, or reclaimed, the high ground of reason and enlightenment for thoughtful proponents of technological innovation.

A second intervention by Charles drew the lines of conflict still more sharply: science and rationalism on one side; god and nature on the other. Along with five other noted international figures, Charles delivered one of the BBC's prestigious Reith lectures in 2000, on the subject of sustainable development. His principal theme was the need to restore a sense of the sacred, a balance derived from harmony with nature, to a world he saw as too much in the grip of science and heedless technological progress:

> The idea that there is a sacred trust between mankind and our Creator, under which we accept a duty of stewardship for the earth, has been an important feature of most religious and spiritual thought throughout the ages. Even those whose beliefs have not included the existence of a Creator have, nevertheless, adopted a similar position on moral and ethical grounds. It is only recently that this guiding principle has become smothered by impenetrable layers of scientific rationalism.[17]

As a corrective, Charles endorsed the precautionary approach, urging his listeners to work "with the grain of nature and not apart from it," using "the wisdom of the heart" to guide future thought and action.

Whereas the ten questions had highlighted uncertainty and ignorance, the Reith lecture sounded a new note, with its frontal attack on science as a framework for understanding, let alone managing, nature. Not surprisingly, this royal pronouncement drew an impassioned response from the scientific community. Richard Dawkins, the famed Oxford geneticist known for his work on the "selfish gene," was an articulate if respectful critic. Professing to be "saddened" by Charles's lecture, Dawkins firmly rejected the heart as a viable source of wisdom in living with nature: "My own passionate concern for world stewardship is as emotional as yours. But where I allow feelings to influence my aims, when it comes to deciding the best method of achieving them, I'd rather think than feel. And thinking, here, means scientific thinking."[18] Other commentators, however, were more charitable, crediting Charles with having raised essential concerns about human consumption and the need for ethics in dealing with nature. In transition from a post at the *Guardian* to one with the BBC, the journalist Andrew Marr summed up Charles's contribution from this more benevolent standpoint: "The prince is a very rich and unelected man whose views have blighted good people's careers, but here he asks a hugely important question—is humanity as a species charging out of control, following one kind of thinking, lemming-like; or are we really *sapiens*, with the internal, democratic, philosophical levers to control the consequences of our own curiosity?"[19]

There can be no question that on the GM issue, as perhaps on none other, Charles succeeded—even as a "very rich and unelected man"—in expressing anxieties shared by a large cross-section of his people. He questioned the direction and pace of changes affecting nature and had the standing to pose ethical questions that few others could have so compellingly placed on the public agenda. In underlining the need to be aware of the limits of current knowledge, he urged a kind of reflexivity in the use of science that was consistent with certain native British notions of appropriate evidence and proof, as well as Europe's precautionary approach to technology policy. But the flurry over Charles's antiscience proclivities drew attention away from another note in his Reith lecture: a kind of fundamentalism about Nature with a capital "N," and the rights conferred by that Nature.[20] When Charles spoke of a "sacred trust between mankind and our Creator, under which we accept a duty of stewardship for the earth," one could almost hear a thousand years of inherited privilege talking. This was less the self-denying "we" of the populist politician than the solipsism of the royal "we," echoing the Stuart doctrine of the divine right of kings. The notion of a divinely ordained stewardship over a fixed state of nature coheres better with ideas of sovereign power than with the experimental dynamics of a working democracy, continually sorting out how to order, or reorder, its relations with nature.[21] Even as Everyman, then, Charles unconsciously spoke on behalf of his subjects. British *citizens* needed more active, participatory modes of representation, detached from the nation's historical embodiment in the monarchy. How to organize representation on more democratic lines became a preoccupation of the state as well as of civil society.

### Inventing Democracy

In 2003 the British government launched a remarkable exercise in constructing a new deliberative politics around GM foods. Proceeding with many hiccups and under much criticism, the effort nonetheless deserves attention as a novel experiment in democratic governance—a coproductionist experiment that required the simultaneous constitution of a process, an interested polity, and a body of reliable knowledge. It played out, moreover, in three strands on three separate stages, all overseen by Margaret Beckett, Labour's secretary of state for the environment, food and rural affairs. One strand was a study of the costs and benefits of GM crops conducted by the prime minister's Strategy Unit. The second was a Science Review headed by Sir David King, the government's chief science adviser. The third, a public debate called "GM Nation?," was overseen by a steering board with members drawn mainly from the Agriculture and Environment Biotechnology Commission, which had advised the government to organize such a debate in the first place.

Each strand led to its own forms of intellectual and social innovation, constituting in effect a micropolitics around each focus of deliberation. Yet, interestingly, the results converged. First to reach a conclusion was the Strategy Unit. Set up only in 2002, and reporting directly to the prime minister, the Unit is responsible for improving the government's capacity to respond to strategic, cross-cutting issues, develop innovative policy, and deliver on its objectives. The GM issue was thus a test case for the Unit's own agility in the face of unprecedented problems, and several participants gave it high marks for adapting well to the challenge. A key moment in the Unit's work was a seminar on "shocks and surprises" held in April 2003. The event partly responded to criticism of its earlier Scoping Note, which suggested that the Unit was thinking too conservatively about cost-benefit analysis and paying insufficient attention to uncertainty. On this April occasion, the Unit placed uncertainty front and center, asking participants to take account of possible disruptive events that were not captured in its own scenarios. Seminar participants were drawn from all interested sectors and constituted a virtual "who's who" of academics and stakeholders with credentials to comment on GM issues.[22] At least one participant saw this event as a major turning point in a potentially complacent analytic process that led instead to unexpectedly cautious findings about the prospects for GM crops in Britain.[23] In July 2003 the Strategy Unit published its report, *Field Work: Weighing Up the Costs and Benefits of GM Crops*,[24] which concluded that benefits from GM technology were likely to be limited in the short term by the lack of products specifically suited to U.K. conditions and by weak consumer demand. The report foresaw possible future gains, especially from crops delivering health benefits, but warned that these too could be compromised by negative public attitudes and by regulatory incapacity to manage uncertainties.

In contrast to the Strategy Unit, the Science Review was set up on the model of an orthodox expert advisory process, but with greater attention to transparency, breadth, and access than had been the norm in pre-BSE Britain. A twenty-four-member Science Review Panel, chaired by the science adviser David King, consisted mainly of natural scientists from relevant disciplines, but with two lay representatives and one social scientist to add balance.[25] Panel meetings were open, although interested persons were asked to indicate in advance by electronic mail their reasons for attending. In addition, the panel held several public meetings whose format (short presentations with questions and answers) was geared more toward allowing well-known experts to state their positions on key issues than to generating intensive debate. For example, at the first open meeting in January 2003, Derek Burke, Prince Charles's interlocutor on the ten questions, reiterated his point that the United States had served the rest of the world as a de facto natural laboratory for the safety of GM: "But three hundred million Americans have been eating GM food for 11 years, and in the most litigious country in the world there hasn't been a legal case yet, so, so far, okay."[26] The precarious logic of this argument passed unquestioned at the meeting (e.g., would adverse effects have been pronounced enough to be detected, and was the absence of litigation an appropriate indicator of safety?), as it was almost bound to do in such a tightly structured environment. It was, however, flagged in the review's July 2003 report, which (like the Strategy Unit) highlighted the many unanswered scientific questions still surrounding the issue of GM crop commercialization. The panel concluded that genetic modification was not a single homogeneous technology, and that U.K. science as well as regulation needed to keep pace with new developments.

The public debate initially left many dissatisfied, but it too led to a wide-ranging exploration of uncertainty. It was conducted, to start with, under severe resource and time constraints by the government's dubiously legitimate and competent public relations unit, the Central Office of Information.[27] As a result, many regional meetings drew those already knowledgeable about the issues, who were least likely to contribute fresh perspectives to the exchanges.[28] Coordination with the other two strands of the process proved difficult. Even the website, organized around bland questions and oversimplified answers, seemed ill suited to arousing the interest of persons not already involved in the debate. In sum, the effort underscored a dilemma confronting state efforts to democratize the politics of new and emerging technologies: on the one hand, interacting only with identifiable stakeholders may simply strengthen the traditionally cozy relations between business and government; on the other hand, the public that needs to be engaged in broader debates about the pros and cons of technology is elusive and, in the absence of reliable precedents, hard to engage in deliberations whose very authenticity and purpose are widely questioned.

By opening the door to novel forms of deliberation, however, the U.K. government created a fluid space in which many actors could take part in configuring the rules of the participatory game. There were, to begin with, more than six-hundred meetings (defined as involving thirty or more people) conducted as part of the public debate. Not only the state, it further emerged, but a variety of private actors had an interest in identifying and giving voice to the phantom GM-attentive public. These groups were not satisfied that the government's half-hearted and poorly resourced efforts would tap into the voice of the people, and they actively intervened in various ways. The strategy chosen by Genewatch, a small but technically competent nongovernmental organization (NGO), was to publicize the GM Nation? debate through its own informative website, thus partnering with the state's efforts at democratization. Greenpeace UK pursued a predictably more confrontational strategy, approaching the powerful Consumers' Association (CA) to organize a parallel consultative process in the form of a citizens' jury. CA was concerned about its own appearance of neutrality but agreed to the proposal when two additional sponsors, Unilever and the Co-op Group, offered support. In collaboration with researchers at the University of Newcastle, an eight-week process involving thirty jurors was launched in July 2003. According to one of the sponsors, "This is a chance for the Government to hear the real concerns of real people."[29] CA obtained agreement from the Steering Board that timely results from the jury process would be taken into account in its final report. A member of the Steering Board was appointed to the CA Oversight Panel to provide a link into the official debate. In this way, an ad hoc network of environmental activists, food suppliers, industry, consumers, and academics took on the task of putting *their* conception of the *demos* back into the democratic debate on GM foods.

Running through all three strands of the U.K. public debate were themes of uncertainty that Prince Charles's ten questions had also broached. Some of the concerns focused on what was scientifically unknown; others, however, centered on the economic, social, and moral risks associated with GM foods. In particular, public reluctance to accept as already demonstrated the benefits of GM technology resurfaced in these interlinked inquiries, which also reinforced Britain's original commitment to viewing biotechnology as a process. In view of these powerful displays of scientific and social unknowns, the government's announcement in March 2004 that it had decided to go ahead with the commercial growing of GM corn came not only as a surprise to many, but almost as a betrayal. Apparently guided by scientific considerations alone, this product-focused decision left out of account the message flowing in from many quarters: that the British public, when consulted through relatively open channels, did not want to adopt a technology with uncertain benefits, no matter how firmly science had established case-specific instances of its biological safety.

## United States: The Market Speaks

To many U.S. observers, the intensity of Europe's negative response to GM foods was incomprehensible and verged on the irrational. How could European publics seem so nonchalant about the well-documented health risks of smoking, some wondered, and yet be so intolerant of the speculative, scientifically ungrounded worries about GM foods? At the 2000 annual meeting of the American Association for the Advancement of Science (AAAS), President Clinton's Secretary of State Madeleine Albright expressed her frustration in no uncertain terms: "But science does not support the Frankenfood fears of some, particularly outside the United States, that biotech foods or other products will harm human health."[30] Speaking at the same meeting, Senator Christopher ("Kit") Bond, a Republican from Missouri, echoed Albright's disenchantment: "I am passionate because I believe the greatest risk associated with biotechnology is not to the monarch butterfly larvae, but from the naysayers, who may succeed in their goal to undermine biotech and condemn the world's population to unnecessary malnutrition, blindness, sickness, and environmental degradation."[31] For mainstream U.S. opinion from both political parties, the "unnatural" thing, the thing to fear most, was not genetic modification in agriculture, but the opposition to it. Certified by science as "safe," genetic modification of crops and foods was seen to be as natural as any other agricultural practices, indeed more so than treating growing things with "unnatural" chemical pesticides.

Naturalization, however, is not simply a matter of getting back to the true state of nature, but a process of social construction whose components can be analyzed and dissected.[32] The demarcation between the natural and the unnatural in any society is not given in advance but is crafted through situated, culturally specific forms of boundary work.[33] In the U.S. case, the notion that GM crops and foods are safe, and indeed the science that underwrites this conclusion, is embedded in a longer historical process of coming to terms with uncertainty about this mode of production. As we saw in chapter 2, an important milestone was the decision by scientists and policymakers in the 1980s that the process of genetic modification was not risky in and of itself: only products merited regulatory concern. Risk assessment techniques correspondingly focused on a relatively narrow range of possible effects—amounting, for GM foods, largely to questions about their toxic and allergenic properties. Virtually unthinkable within such a framing was the full-blown discussion of the technology's aims, costs, and benefits that engulfed British policy circles in 2002 and 2003. Opposition, when it came, arose from tensions within the U.S. product framework itself, not from challenges external to it.

The first GM food products were introduced into the U.S. market in the 1990s, their safety constructed partly through the regulatory process and partly through strategic channeling and silencing of opposition. It was, in all, as much a process of making the world safe for the introduction of GM as making GM safe for introduction to the world. As the agency most centrally in charge of protecting the wholesomeness of the nation's food, the Food and Drug Administration took the lead in developing safety assessment principles for GM foods. In the early 1990s the venerable agency[34] was struggling to regain its public standing, its reputation battered by huge regulatory backlogs and financial scandals. In late 1991 the journal *Science* reported that FDA was holding up approval of at least 21 biotech drugs that had completed clinical trials, and the outlook was not promising for 111 more in the pipeline.[35] Along with clearing the hurdles for these drugs, a vigorous response to the newer product stream emerging from agricultural biotechnology seemed an obvious way to revive the agency's flagging reputation. David Kessler, who became FDA commissioner in December 1991, recognized the problems as well as the opportunities the situation presented.

Trained in both medicine and law, and scarcely forty years old when he took on the job of FDA chief, David Kessler displayed a tenacious energy that served him well through a Republican and a Democratic presidential administration, under presidents George Bush and Bill Clinton, respectively.[36] By June of his first year in office, he announced a streamlined approach to approving biotech foods, consistent with the desire of the President's Council on

131

Competitiveness to maintain the U.S. lead in this economically vital techno-logical sector.[37] FDA's assessment strategy centered on exempting genetically modified food components, such as proteins and enzymes, that either had a history of safe use (that is, were "generally recognized as safe," GRAS) or were similar to already approved components. In an article in *Science* describing the agency's approach, Kessler and his coauthors repeatedly stressed that "substantial similarity" to safely used substances would lead to reduced regulatory oversight.[38] This was merely the logical consequence of adopting the product framework: to the extent that biotechnology simply replaced things already in commercial circulation, its products were not novel enough qua products to require the regulator's attention. When questioned, Kessler in-variably noted that changes in the nutritional content of traditional foods or the addition of substances that might trigger toxic and allergenic reactions would initiate FDA review. In the main, however, FDA's regulatory approach amounted to a laissez faire attitude toward most GM crops and foods. A sys-tem of voluntary consultation left the biotech industry largely free to decide when to seek agency approval.[39] Only in 2001, in the face of rising charges of underregulation, did FDA consider making consultation mandatory through required advance notification, but the rule was tabled when FDA's general counsel determined that it was not within the agency's jurisdiction.[40]

It is important to recognize that FDA's use of the similarity criterion (formally known as the criterion of "substantial equivalence") crucially rested on the prior determination that process was not a relevant factor in policy im-plementation: the fact that a new product, similar to an existing one, had been produced through the process of genetic modification was not an issue for U.S. regulators. By definition, this approach made no sense in a regulatory environment like Europe's where process *was* deemed a relevant, indeed a cen-tral, concern. As events showed, the attempt to rule process out-of-bounds for regulatory purposes was not entirely sustainable even in the United States.

### Early Tests: Of Milk, Tomatoes, and Corn

The first significant product of food biotechnology to enter the U.S. market won few friends for the new technology. It was Monsanto's recombinant bovine growth hormone (rBGH) or bovine somatotropin (rBST), approved for marketing by FDA in November 1993. Promoted as an aid to more effi-cient milk production, rBST was roundly criticized on economic, social, and ethical grounds: economically, it was of questionable utility in the United States, increasing production in the heavily subsidized, surplus-producing dairy industry; socially, it hurt small family farmers, who were not in a posi-tion to benefit from the economies of scale offered by rBST use; ethically, it aroused the ire of animal welfare groups by measurably increasing the rate of diseases such as mastitis in rBST-treated cows. Yet, in a move later seen as a serious public relations failure, Monsanto pushed forward with tactics designed to build a market for the product.

An important component of this effort was to make sure that consumers could not easily opt out of consuming milk produced by means of recombinant BST. Monsanto therefore vigorously opposed the labeling of rBST-milk, winning partial FDA support for this position. As long as the milk itself was unaffected by the use of rBST, FDA saw no health-based justification for labeling. Indeed, the agency concluded that unqualified labeling could actually be misleading, since it might lead consumers to suspect—erroneously, the agency maintained—that there was something wrong with the product. FDA therefore insisted that labels, if any, should explicitly indicate that the agency had found no differences between milk from rBST-treated and untreated cows.[41] As a health and safety agency, FDA declared that it was powerless to respond to consumers who demanded labeling simply as a means of increasing consumer choice; the agency likewise disclaimed any authority to evaluate the socioeconomic impacts of marketing rBST.

Opposition to rBST manifested itself in characteristically American forms, particularly in disputes over the scientific studies designed to establish the amount of increase in milk yields, the extent of health effects in treated animals and in humans, and the sameness (or not) of the recombinant and nonrecombinant growth hormones.[42] Not surprisingly, proponents of the technology claimed to find higher yields, with fewer ill-effects on animal welfare or human health, than did their opponents. Opponents for their part noted that the genetically engineered version of BST had one or two additional amino acid subunits, known as linker proteins, attached to the end of the hormone molecule; hence the two versions of the hormone were not technically identical.[43] But a prescient observer would have noted, besides the normal scientific controversies, deeper challenges to the underlying regulatory framing. In effect, the opponents of rBST use argued that, in this case, process *did* matter, and legitimately so—for reasons that were not connected with health risks alone. Worries about the future of family dairy farms and concern for labeling and consumer choice, as well as cartoons and columns deploring the conversion of cows into yet more efficient instruments of production, were all symptoms of an instability at the core of the product-based approach. In challenging rBST use, critics questioned a fundamental premise of green biotechnology: that, in the absence of risks to human health and safety, there is nothing wrong with pushing nature to yet more intensive feats of production. The dominant U.S. policy frame, centered on the structural and functional similarity of GM and "natural" products, left no official space for debating such concerns. Indeed, official policy ruled these primal political and ethical concerns out-of-bounds.

The low levels of public deliberation did not mean that U.S. consumers were generally satisfied with the regulatory status quo on GM products. A brief episode on the Internet in early 1994 gave some indication of troubled waters. The triggering event was a posting by Pamela Andre, acting director of the National Agricultural Library, on SANET-mg, an e-mail discussion group initiated in 1991 to facilitate information exchange about sustainable agriculture.[44] Andre accused Lara Wiggert, a graduate research assistant at the library, of having posted biased and misleading statements about the safety of rBST and the professional integrity of Dale Bauman, a highly re-garded Cornell University scientist and an expert on the effects of rBST.[45] This accusation provoked an intense outpouring of support for Wiggert, as well as some countervailing comments deploring the misuse of the electronic network for inappropriately political purposes. One colorful posting noted the Internet's potential for democratizing debates on science and technology:

> Finally it frosts my shorts when the empowered mouth self-serving myths about a class of people ill-equipped to choose between the subtleties of shiny, supported, warranted Scientific Opinions, and drecky icky rabbly ones. To me, the rBGH dia-logue both on and off the Net shows citizens questioning the very fabric of the so-cial production of knowledge, of policy, and of technology, often in terms more complex than I hear coming from some of the individuals who are replying. Citizens may not be doing this in Edgicated fora bearing somebody's stamp of approval, but, hell—neither did the hosts of the Boston Tea Party.[46]

Both the Internet and the popular technology assessment advocated by this author would come into play, with considerably more dramatic repercussions, in the controversy over labeling organic foods that beset the U.S. Depart-ment of Agriculture (USDA) in the late 1990s.

For much of the 1990s, however, concern about GM foods remained below the threshold of public visibility. Occasional product breakthroughs aroused almost ritualized responses from biotech proponents and opponents. The first genetically modified whole food to attract such attention was Calgene Inc.'s genetically altered tomato, named the Flavr Savr, which had been engi-neered to ripen on the vine without becoming mushy inside. Calgene devel-oped a way of slowing down the production of an enzyme that breaks down the pectin in tomatoes, causing the cell walls to lose their rigidity and the fruit to rot. As a result, the Flavr Savr could remain firm a week longer than normal tomatoes and be picked at an allegedly riper, more flavorful stage. Recognizing the need to win consumer confidence, Calgene worked with FDA for some five years to win the agency's approval in 1994[47] and also promised to label its product. Nonetheless, organizations such as Jeremy Rifkin's Foundation on Economic Trends and the Union of Concerned Scientists expressed doubts about its safety, and Rifkin launched a campaign to get top chefs and restaura-teurs to boycott the Flavr Savr and other GM foods.[48]

Though discussions between FDA and Calgene remained largely be-hind the scenes, there were sporadic media reports on the Flavr Savr from

1992 to 1994. One exchange illustrates how the public framing of the risks continued to be shaped by molecular biology's narrative of control and its focus on human health. In August 1993 Maxine Singer, president of the Carnegie Institution and former guiding spirit behind the Asilomar rDNA conference, wrote an editorial on the Calgene tomato in the *Washington Post*.[49] After noting that "almost all the foods we eat are the product of previous genetic engineering by cross-breeding," Singer went on to provide a short primer on DNA and its manipulation in the tomato. She mentioned that the process included, in addition to a very precise modification of the target enzyme, the insertion of a marker gene that confers resistance to the antibiotic kanamycin. The common device of a marker allows GM plant breeders to identify the plants that have been successfully modified with the desired primary trait, in this case delayed ripening. A major concern about the use of antibiotic resistance genes as markers has been their potential transmission into the environment, producing resistant organisms and constituting a possible threat to public health. Singer, however, focused on a more immediate health risk, noting in terms familiar to any toxicologist that the tiny doses of this extra enzyme in GM tomatoes would easily be destroyed in the consumer's digestive tract. Interestingly, a letter responding to Singer's op-ed piece by two staff scientists from the National Wildlife Federation, Jane Rissler and Margaret Mellon, both of whom subsequently joined the Union of Concerned Scientists, also did not mention the environmental risk (which FDA, to be sure, had considered and deemed unimportant, but which loomed as much more significant in European debates).[50] In the end, consistent with a product-based policy framework, the market, not the regulators, determined the fate of the Flavr Savr. Following disappointing sales in California and the Midwest, attributed in part to its bland taste, the product was quietly withdrawn.[51]

A more public furor accompanied the discovery in September 2000 that a pesticidal protein, Cry9C, from the genetically engineered StarLink corn produced by Aventis, a French multinational, had been found in taco shells produced by Kraft Foods and sold by the Taco Bell fast-food chain.[52] The U.S. Environmental Protection Agency had approved StarLink for use as animal feed, but not for human consumption, because of unresolved concerns about the possibly allergenic properties of Cry9C. Random tests by a scientist working for Friends of the Earth showed that Cry9C had not stayed in place as the regulators intended, safely out of the human food chain. Subsequent to the initial discovery, Cry9C was found in other processed foods, leading to the recall of some three hundred food products, a buyback program for Cry9C-containing seed by USDA, large drops in U.S. corn shipments to Japan and South Korea, and a voluntary withdrawal of the registration for StarLink by Aventis. In a gesture reminiscent of protests in Britain, Greenpeace dumped a ton of StarLink corn at EPA headquarters on February 7, 2001, and cautioned the newly appointed administrator, Christie Todd Whitman, not to go along with "genetic experiments" by the biotech industry.[53]

In another worrying incident, the Prodigene Corporation was permitted to conduct a field test of genetically modified corn containing an insulin precursor, Trypsin.[54] It was an important trial because the plant represented a crossover between agricultural and pharmaceutical biotechnology, and its success had substantial economic implications. As agreed with USDA, the GM corn was planted in an unmarked field in rural Iowa, in the heart of the U.S. corn belt. The field was to be treated the following year so as to remove any adventitious reappearance of the GM plants. In fact, this effort failed, and an undetermined quantity of GM corn was harvested along with about 500,000 bushels of soybeans during the following season. USDA had the soybeans destroyed, thereby removing any risk of harm, but at considerable cost.[55]

Much publicized and economically burdensome, the StarLink and Prodigene episodes were seen by American commentators as massive regulatory failures, but they were interpreted more as incentives for better monitoring and enforcement than as warning signs about the manageability, or unmanageability, of commercial biotechnology. In a wealthy country with money to pay for mistakes, the value of innovation tacitly triumphed over the cost of error. A typical reaction was that of former Agriculture Secretary Dan Glickman, who characterized StarLink as a "high-profile regulatory misstep" and asserted that thoughtful conversation among diverse viewpoints would lead to an eventual meeting of the minds on the value of agricultural biotechnology.[56] EPA decided that "split" approvals (for animals but not humans) as for StarLink were a bad idea, and people even began asking whether it was appropriate to treat genetically engineered pesticidal traits within plants as equivalent to chemical pesticides. Nothing in these reactions, however, approached the systemic critique that had led the sociologist Charles Perrow to speak of "normal accidents" in tightly coupled, high-risk technologies in the 1980s.[57] Nor was there much public discussion of the destabilizing implications of this event for the idea of coexistence—or managed separation—of GM and non-GM products. In this respect, StarLink and Prodigene did little to upset official U.S. assumptions of regulatory control.

### The Boundaries of the Natural

What, though, was the public reaction? On a dreary day in November 2002, a forlorn-looking young woman sat at a small table outside the Shaw's Supermarket at Porter Square in Cambridge, Massachusetts, collecting signatures and handing out leaflets. Her object was to get passers-by to pressure the local Shaw's and Star markets to remove genetically engineered ingredients from their store brands. I asked her what the problem was with these foods. "They really haven't tested this stuff properly at all," she replied, "It probably causes cancer." Her leaflets told a different and more nuanced story. "Hundreds of Americans have reported allergic reactions to the FDA after eating Kraft and other brand name corn products containing genetically engineered (GE) ingredients," one leaflet warned, adding, "Lab tests and industry disclosures indicate that 60–75% of all non-organic supermarket foods now 'test positive' for the presence of GE ingredients." Another urged consumers to "speak out" and "take action" so as not to become part of an "unwanted and unpredictable experiment."[58] One recommendation was to buy certified organic foods and join the Organic Consumers Association. There is then an Eden that still holds promise for American consumers seeking refuge from the march of agricultural biotechnology, namely, the organic products market. But how was this safe haven created?

It took a decade. The 1990 Farm Bill contained the Organic Foods Production Act, which laid the foundation for a National Organic Program. One of the program's missions was to replace the patchwork of competing and conflicting state provisions with consistent national standards for the certification of organic foods, along with a workable set of practices for keeping organics separate from conventional foods. Disagreements about standards had created uncertainty for consumers and problems for handlers of multi-ingredient products.[59] On December 16, 1997, USDA released its massive proposed rule on the labeling of organic foods. In brief, the department intended to monopolize the designation "organic" and to permit under that heading several practices that growers and consumers of organic foods had traditionally rejected as unacceptable or unsafe. What particularly drew the ire of the organics community was the department's decision to allow under the organic label the so-called Big Three—biotechnology, irradiation, and the use of biosolids such as sewage sludge as fertilizer. In the case of biotechnology, USDA said it was following through on the logic of the product framing, just as FDA had done before. To deny bioengineered foods the organic label, USDA argued, would be to admit that there was something specially risky or unnatural about this process. Yet official U.S. policy had already determined that the manufacturing process was irrelevant to the evaluation of GM products. On a product-centered approach, it should make no difference whether a food had been prepared with or without genetic modification; this principle should hold for both organics and other types of food.

As word of the proposed rule circulated, the organics lobby mobilized as never before.[60] USDA was inundated by public comments, the great majority opposed to the rule, with an estimated 300,000 flowing in by the end of the four-month comment period. That response was some twenty times higher than for any rule previously proposed by USDA.[61] Secretary Glickman, a strong proponent of biotechnology like other high officials in the Clinton administration, admitted that the department had been caught by surprise at the intensity of the reaction and would bow to consumer demand.[62] On

December 20, 2000, USDA issued the final standards for organic foods containing the desired concessions with respect to the Big Three. It was a resounding political victory for opponents of agricultural biotechnology and other forms of intensively industrialized agriculture.

This was popular democracy at work, on a scale that dwarfed the earlier reaction to BSE, but the popular will was tied in this case to serious economic considerations that mattered to USDA. Domestically, organic agriculture had become a six-billion-dollar industry by 1999, accounting for around 2 percent of the nation's total food production and growing phenomenally at the rate of some 20 percent a year.[63] It was also an important export industry, and as long as other countries remained skeptical of American agricultural production methods, such as biotechnology, the export market for organics was expected to become even more important.

Economic interest also helps explain why pro-GM forces adopted a position in the labeling debate that was in one way contradictory to their stance on other issues.[64] Industry groups like the Biotechnology Industry Organization, who had stoutly fought labeling through the FDA, seemed happy to accept USDA labeling if their products could thereby gain a share of the lucrative organics market. Antibiotech groups for their part were also inconsistent at one level: skeptical of industry's claims regarding the safety, traceability, and segregation of GM products like StarLink corn, they nonetheless seemed prepared to accept that proper labeling and monitoring could maintain a workable separation between organic and conventional products. But the contradictions on each side were in a sense epiphenomenal to the deeper issue at stake for all participants: the meaning of the organic-conventional boundary. Breaching this boundary was important for an industry intent on showing that biotechnology was simply an extension of traditional agriculture by other means; maintaining its integrity was similarly crucial for the organics lobby. To take the category of "organic" as given, and only needing specification through objective criteria, is to miss the fact that the designation itself was an instrument of sense-making—of ordering the world in ways that people wanted it to be ordered.

Organic food production evolved during the 1970s and 1980s as the sector of U.S. agribusiness dedicated to the needs of a consuming public committed to such postmaterial values as the environment, animal welfare, freshness and taste, food in season, small-scale farming, and closeness to the sources of production. Consumers showed that they were prepared to pay a premium for food produced through processes that reliably reflected these values. Advertising for organic or "whole food" products began to play on these sensibilities by establishing a personal connection from grower to consumer, often through first-person stories told in first-name terms on packages or on company websites.[65] (The predilection for first-person connections between producer and consumer was not limited to the United States.[66])

Unlike Maxine Singer, who professed to be eager to bite into a raw Flavr Savr from her supermarket shelf, many "alternative" consumers found the idea of genetic modification profoundly alien to their sense of the natural order of things. For these affluent and increasingly more numerous consumers, the *process* of food production was itself a valuable product. Buying organics was buying into a mode of production that reinforces basic ethical commitments toward community and nature. It used economic power to sustain a moral and political order, much as Prince Charles sought to do in Britain with his royal prerogative on inherited lands. From this perspective, the label organic is best seen as an ontological placeholder for intense sociopolitical convictions: it is a naming device that maintains the viability of a way of eating, and of life. No wonder, then, that the placement of GM on the allowed versus the disallowed side of the label proved to be a matter of deepest contention.

The point to ponder for democratic theory, however, is that so fundamental a metaphysical boundary was drawn in the United States through an unreflexive process of national market integration. Other ironies ensued. The labeling rule standardized and rendered more economically and geographically tractable a product category whose earliest claims on consumer loyalty had included closeness to its places of production. In achieving commercial purity, the organics label restored attention to the process of GM, but organic foods themselves became a globally mobile commodity. Fittingly enough within America's product-based framing of biotechnology, organics gained national certification and market share as the idea of nature became commercially valuable to rich consumers. But the commitment to natural food as a way of life that animated the lonely leaflet distributor outside a Cambridge supermarket was arguably lost in translation.

## Germany: An Eloquent Silence

On the issue of GM foods, politics in Germany was notably more tranquil than in the other two countries. There were no public displays to match the emotionalism of demonstrations in Britain and the United States, no media frenzy or mass popular mobilization around GM products, no discursive rallying points like "Frankenfoods" or "genetic pollution," and little in the way of visible regulatory or institutional upheavals tied specifically to GM. These absences are all the more notable because Germany, too, had its own BSE scandal as noted below—nowhere near in scale to Britain's, but equally interpreted as a symptom of governmental incompetence. As in Britain, it led to a thoroughgoing shake-up at the cabinet level, but without similar calls for reforming the mechanisms of public participation.

To understand this state of affairs, we have to look at two sites of containment that matter equally for the politics of biotechnology in Germany: the

European theater and the theater of domestic politics. In both contexts, the German solution was to use state power to steer clear of borderline cases,[67] thereby preempting some of the possibilities for messy political emergence seen in the U.S. and U.K. cases.

### Containment Abroad: The Mantle of Europe

As an interested and conscientious European citizen, Germany has been a major player in the design of European policy. One effect is to export a substantial fraction of potentially divisive regulatory politics to the relatively technocratic political environment of Brussels. In most important particulars, Germany simply took the EU GM directives on board as national law but went further in some areas that proved much more divisive in other countries.

German GM food policy is nested within the legal framework supplied by the European Novel Foods Regulation (NFR) of 1997 and associated labeling regulations of 2000.[68] Administered by DG Enterprise, the NFR defines the regulatory process for all foods that have not previously been consumed to a significant degree in the EU, including foods derived from GM crops. The regulation describes two alternative decision-making procedures for GM foods. The more complex authorization procedure parallels the process for placing on the market live genetically modified organisms (GMOs) that are not substantially equivalent to existing organisms. It permits a product to be approved at EU level based on a positive report of the rapporteur member state (the state of first introduction) on the substantial equivalence of the "novel food" to an existing product.[69] The second simplified procedure only requires notification by the rapporteur member state, but this has triggered some controversy from a number of states who insist that all members should be able to contribute to decision making at the EU level. These discussions are still in progress and may lead to the replacement of notification by the more elaborate authorization.

### Internal Containment: Lines of Control

Responsibility for implementing the NFR in Germany has lodged since January 2001 with the newly created Federal Ministry for Consumer Protection, Food and Agriculture (Bundesministerium für Verbraucherschutz, Ernährung und Landwirtschaft, BMVEL), formed through a merger of the former Agriculture Ministry and a division of the Federal Ministry of Health (Bundesministerium für Gesundheit und Soziale Sicherung, BMGS). This restructuring followed the shocking disclosure in late 2000 that several cases of BSE had been identified in Germany after years of insistence that German cattle were BSE-free; with the number of infections expected to climb as high as five hundred, officials saw no alternative but the slaughter of herds, a painful decision in a country where animal welfare was a perennial topic of political concern. The crisis led to the forced resignations of Andrea Fischer, the Green minister of health, and Karl-Heinz Funke, the SPD agriculture minister. Renate Künast, a tough, city-bred lawyer and member of the Greens, was appointed to head the new ministry and restore shattered confidence in the food regulatory system.

Testing and inspection, along with expert advice, were the principal instruments Germany relied on for this purpose. Two expert bodies perform the technical assessment of GM foods: the Federal Institute for Risk Assessment (Bundesinstitut für Risikobewertung, BfR),[70] a part of the new BMVEL, and the Robert Koch Institute of the Health Ministry. BfR, advised by a thirteen-member committee of food scientists, is responsible for food produced with GM techniques but not containing GMOs, whereas the RKI assesses foods containing or consisting of GMOs. According to an EU mission of inquiry that visited Germany in 2001, no applicant had yet selected that country for initiating a novel foods authorization, although German expert bodies were active in evaluating and commenting on proposals submitted elsewhere.[71] Under German law, inspection and quality control are the responsibility of the individual Länder. The 2001 EU mission found these state-level controls to be both technically proficient and well coordinated. This conforms to Germany's historical commitment to strict legal implementation.[72] The public plays a negligible role in the process. Though notice is given of authorization requests, the government's legal advisers were of the opinion that "substantial equivalence" determinations could not be made public, as these were done under contract for applicants.

Domestically, the strategy of heading off possible areas of conflict manifested itself most notably in Germany's early commitment to labeling. Along with only two other EU member states (Austria and the Netherlands), Germany introduced a "GM-free" label for food in 1998, as a precursor to and possible model for eventual EU regulation of this issue.[73] To be designated GM-free, products may not contain more than 1 percent (subsequently reduced to 0.9 percent) of the genetically modified ingredient or of modified DNA. This compromise was the result of a curious domestic battle, in which the "normal" political alignments around the purity and danger of GM appeared temporarily to have switched sides. The powerful state of Bavaria, allied with several major GM-friendly food producers, insisted to start with that any level of GM contamination, however small, would deceive consumers, and that absolute purity was the only basis on which a GM-free label made sense. By contrast, organic farmers and consumer groups were willing to tolerate the realistic prospect of some contamination in order to have a meaningful label in place. The 1 percent limit initially written into the law represented a victory for the segregationists, who ironically were the ones prepared to admit that contamination was a fact of life in an already GM-laden agricultural environment.[74]

On becoming agriculture minister in 2001, Künast made clear her support for Germany's organic farmers, though she also signaled by 2003 that she wished to move the GM debate away from the unproductive yes-no positions most strongly associated with her own party, the Greens. Künast announced a goal of boosting organic agriculture from 2.5 percent of current production to 20 percent in ten years—a policy that the opposing Christian Democrats claimed in the 2002 general election would cost the government more than five billion euros in subsidies. Others questioned the Greens' assertion that organic production guarantees higher-quality food, a doubt also expressed in both the United States and Britain. In April 2002 the skeptics were handed a ready-made weapon with the discovery that poultry feed distributed by the organic processor GS agri had been contaminated by the herbicide nitrofen. The respected conservative newspaper *Frankfurter Allgemeine Zeitung* (FAZ) took obvious pleasure at Künast's discomfiture. A FAZ reporter challenged the very concept of an organic "Eden" as economically and administratively untenable:

> Like the walled-in Gardens of Eden depicted in medieval miniatures: people, chickens, and pigs living together in harmony—animal factories and cracked and creviced fields would be a thing of the past. Protected and cared for by the new "organic seal" brand, the system is nourished from the manna of state subsidies. The role of the angel with flaming sword driving away the sinners is played by a network of government authorities, and the private control organizations representing them. These keep an eye on things all year round.[75]

For this writer, it was organic agriculture that was the unnatural order of things, unable to sustain itself without undesirable and unworkable state supports, and without constant policing This was the artificially confected nature of the modern "gardening state,"[76] bearing no relation to "real nature." Implicitly, then, conventional agriculture was the truly natural alternative, a product of free enterprise, regulated by market forces, and untrammeled by state planning. In naturalizing the market, its proponents, such as this FAZ writer, also naturalized the intensive, industrialized forms of agriculture that the market had rendered familiar and economically desirable.

## European Solutions: Terms of Integration

The wheels of European policy continued to turn as EU member states conducted their national debates on what to do about the fruits of GM agriculture. What emerged from deliberations at the EU level, however, was a formal restatement that the process of genetic modification had to be recognized as integral to its resulting products. GM foods could move about in the European market, but only if they bore clear indications of their GM origins. Instrumentally, the devices through which the EU asserted the continued relevance of process-based regulation of GM foods were the 2003 regulations concerning the authorization, traceability, and labeling of GMOs and GMO-derived products.[77] These provisions require full traceability even when the final products derived through GM lack any foreign DNA. In other words, the new EU legal regime asserts, if anything more explicitly than before, that process matters in the marketing of foods. If GM agriculture is to win its place in the sun of European commerce, it will be for the foreseeable future only by acknowledging its roots in the laboratories of genetic manipulation.

The EU labeling and traceability regulations can be seen at one level as a case of regulatory convergence, contradicting the predictions of permanent polarization and regulatory standoff put forward by the political scientist Thomas Bernauer.[78] Indeed, once EU rules were in place, some national governments felt compelled to follow, overriding in some cases deeply felt countertendencies at home. Germany was a case in point. In February 2004 Schroeder's cabinet approved a bill to allow the regulated cultivation of GM crops in Germany.[79] For our purposes, however, these developments confirm the persistence of frames as ways of ordering the politics of knowledge societies. Both at the European and at the national level, there was no turning back from the assumption that consumers have a right to know whether the products they consume come from GM or non-GM sources. At the same time, it is important to see that the thick social dynamics of how to ascertain the safety and acceptability of GM foods continued to be played out in national forums rather than in EU institutions. These controversies, which this book argues are sites for the reenactment of political culture, remained attached to national traditions of generating and debating public knowledge. We will return to this point in the discussion of civic epistemology in chapter 10.

## Conclusion

A comparison of the GM food controversies of the 1990s rules out easy explanations for the gulf that opened up between the United States and several European nations in the last decade of the twentieth century. The notion that GM suffered in consequence of public hysteria occasioned by the BSE crisis fails to capture important similarities and differences in the politics of GM in all three countries. I have argued in this chapter that many of the concerns that crystallized around the advent of GM food production were already simmering beneath the surface in each country. The particular form in which these concerns achieved political expression reflected different understandings of the role of the state, of science and agribusiness, and of farmers and consumers as actors in food policy.

In Germany, even the shock of the BSE scandal failed to rock the commitment to conducting politics through political parties and the institutions of the state; confidence in statist solutions was reinforced by the long national experience of expert agencies enforcing bright-line regulations by means of technical expertise and attentiveness to the law. In Britain, by contrast, BSE helped open up the very fabric of expert decision making, converting official ignorance and uncertainty into salient political questions. Both government and other major players realized that this was a time to reopen institutional features of decision making to take account of the limits of official knowledge; one consequence in Britain was an unprecedented experimentation with the forms of public engagement, whose effect in part was to constitute an issue-specific public around GM crops. In the United States, the initial product versus process line remained largely intact, although somewhat weakened by the StarLink and Prodigene scandals. As in Germany, more attention was given to beefing up the regulatory structure, evidenced by FDA's attempted move away from purely voluntary consultation and USDA's eventual promulgation of complex organics standards.

An exclusive focus on transatlantic quarrels over the status of GM foods also obscures the extent to which the politics of GM was tied to the politics of alternative agricultural practices. In the United States, where the GM horse had largely fled the regulatory barn before the public took notice, once-inchoate concerns about process values in food production gathered political force around the question of organics labeling. As so commonly in American politics, it was the creation of a market around these popular concerns that enabled the organics lobby to wield influence; labeling simply became the instrument with which this lobby created a space for alternative forms of food production seen to be natural. In Germany, the SPD-Green coalition tied its rhetorical support for GM agriculture to an absolute commitment to organic production and indicated that it was prepared to commit very substantial state resources to preserving, and substantially enlarging, the zone of GM-free food production. In Britain, where the public outcry against GM was both sharpest and most explicitly tied to the values of tradition and countryside, it seemed certain that commercial GM agriculture would henceforth have to answer public questions about need, benefit, and level of certainty that it could fulfill its promises of health and environmental safety.

By 2004 European resistance to GM foods appeared to be weakening as the EU adopted traceability and labeling regulations to allow such foods entry into the European market, Germany took official steps in the same direction, and U.K. policy on GM crop commercialization followed scientific recommendations of safety rather than public expressions of uncertainty. For economists, these actions may seem to be predictable convergence in a world committed to technological innovation and free trade. It had to be only a matter of time before market principles, underwritten by science, would overcome softer cultural resistance. And yet the GM foods that Europe seemed ready to take into its territorial jurisdiction were not the same GM foods that

U.S. authorities and producers had originally hoped to distribute in global trade. They were now clearly marked as GM, the products of an agricultural process that U.S. policy had firmly sought to naturalize and render invisible. The trade barrier against GM foods was lifting, but only on condition that U.S. manufacturers accepted Europe's framing of GM as a process that has social and legal meaning. Economic victory was premised, in other words, on ontological capitulation.

# 6

## Natural Mothers and Other Kinds

The industrialization of agricultural biotechnology, as we saw in the last chapter, ran ahead of its political uptake in many ways, causing severe crises of public confidence. The same cannot be said of biotechnologies aimed at human beings. In this case, political debate (though not necessarily political *action*) kept more nearly in step with invention. As scientific research opened up new frontiers for intervening in human biology—from the birth of Louise Brown, the first "test-tube baby," in England in 1978 through debates on human cloning and stem cell research in the 1990s and beyond—discussion of how far to proceed with these techniques, and under what kinds of supervision, gathered worldwide momentum. Debate swirled in the mass media, as well as in traditional decisionmaking forums such as courts and parliamentary committees; countless new deliberative bodies also took up the challenge, nationally and internationally, including ethics commissions, patients' organizations, and industrial pressure groups. It seemed that every political community with the technological capacity to manipulate human life felt compelled to participate in these debates; all across the industrial world, the extraordinary moral dilemmas presented by the possibility of such intervention captured, and partly transformed, the processes of politics as usual.

It is hardly surprising, then, that these questions gripped the attention of high-level policymakers in Britain, Germany, and the United States. All three states had selected biotechnology as an instrument for advancing economic competitiveness and human health, and all had made substantial public investments in biomedical research.[1] In each, a vigorous pharmaceutical sector stood ready to translate scientific discoveries into lucrative treatments for infertility, cancer, and various inherited illnesses. In all three countries, dominant religious traditions favored the remediation of perceived mental and physical problems and endorsed the use of technology to achieve human betterment. At the same time, sectarian differences had given rise to sharp divergences in each nation about the appropriateness of specific types of intervention. Far from inchoate or powerless, these contrasting moral sensibilities were associated with deep political cleavages that raised questions about manipulating human life to exceptional salience in law and politics. If the growth of agricultural biotechnologies was marked by too little deliberation, then human biotechnologies were burdened by almost a surfeit of public soul-searching. In the former case, an arguably premature policy consensus impeded the articulation of basic issues of governance; in the latter, there was no dearth of explicit deliberation on the issues, but the very intensity of opinion made closure often seem remote.

The debates over human biological intervention in our three study countries are the subject of this chapter and the next. We begin here with the issues surrounding technologically assisted human reproduction, which led by the mid-1980s to major regulatory initiatives in all three countries. The focus in this chapter is on the development of legal and administrative controls on reproductive science and technology, and the ways in which they demarcated permissible from impermissible interventions. In the next chapter, we will look at the rise of bioethics as a discourse for setting limits on human biological research, particularly in debates surrounding cloning and stem cells. Together, these histories illustrate very different national strategies of normalization, that is, for making human biotechnology seem mundane and governable in the face of moral uncertainty and conflict.

The issues posed by the new reproductive technologies in each national context were virtually identical to start with: when and on what basis can a pregnancy be intentionally terminated; under what conditions is it appropriate to manipulate the course of human embryonic development; and, in an era of growing technological sophistication, how if at all should legal concepts of "natural" family relationships be extended or modified? Political and policy responses to these questions, however, diverged considerably. Britain and Germany adopted comprehensive legislation governing assisted reproduction and embryo research, but these laws codified strikingly different assumptions about the continuity of human development, as well as the role of the state in preserving human dignity. By contrast, U.S. policy remained a patchwork of competing and overlapping rules, both state and federal; these set some limits on reproductive science and technology but offered no principled justification for either sanctioning or prohibiting embryo research. In comparing these trajectories cross-nationally, we will identify not only what led to particular forms of closure in each country, but also what accounts for key differences in the substantive terms of each national settlement.

The vexed politics of abortion forms an essential node in the wider politics of reproductive technologies in each country. Reviewing the three national abortion debates, as I briefly do in this chapter, is not intended to add to the voluminous literature on this topic but to highlight the ways in which the abortion issue helped to frame and constrain other conflicts arising around the new human biotechnologies. The question when life begins and its ethical correlate, what is the moral status of the fetus, have been at the heart of attempts to define workable limits on abortion; they are also implicated in many controversies about technologically assisted reproduction. Issues about women's rights and the sanctity of the traditional family—consisting of a heterosexual couple and their biological children—have similarly helped to define the battle lines over both abortion and reproductive technologies. Put differently, the abortion struggles created some of the conceptual and political resources that actors later drew on as they sought to expand or contract the range of application of the new human biotechnologies.

Abortion, however, is only part of the story told in this chapter. The new technologies of reproduction enabled taboo-breaking moves, generating threats of social disorder in each society beyond scientific overreaching and harm to the fetus. These technologies not only further decoupled intercourse from child-bearing—as contraception and abortion already had done in varying ways—but split conception, now capable of occurring outside the womb, from implantation and gestation. A hitherto seamless biological process from conception to birth, entailing relatively little social or scientific ambiguity, became in this way fragmented, distributed among multiple agents, and subject to purposive human intervention. Additional participants made their way into what had once been, biologically, the exclusive preserve of one woman and one man—and had been cautiously expanded only to include artificial insemination by third-party donors. The resulting novel linkages made close relations of strangers, recombined generations in unexpected ways, and even cut across the line of life and death. Through in vitro fertilization, surrogate motherhood, and the storage of frozen sperm and eggs, children could now be born of postmenopausal mothers, be carried to term by aunts or grandmothers, and claim kinship with long-dead parents. How, if at all, to render these unexpected relationships natural or unnatural, and how to adjust existing legal and moral dispensations so as to accommodate or disallow particular practices provoked sharp clashes. The resolution of these conflicts followed radically different pathways in the three countries and exemplified their political cultures in action.

## Britain: A Consensus for Science

The birth of the world's first test-tube baby in 1978 stood for many things in the popular mind—among them, the perennial themes of science out of control, playing god, and tampering with nature—but it also stood as a miracle cure for infertility and a triumph for British science, which once again had rolled back a frontier of impossibility. Starting in the late 1960s, at about the time that Britain opened the door to legal abortions,[2] the embryologist Robert Edwards and the gynecologist Patrick Steptoe, an expert on laparoscopy, began the experiments with in vitro fertilization that led, with Louise Brown's birth, to the first successful use of IVF. Not everyone was equally thrilled at Louise's advent. The spectacle of the embryo as an object of research aroused intense opposition from anti-abortion forces, who saw IVF as a further assault on the sanctity of human life and a symptom of moral decline. Yet, despite mobilization and vociferous lobbying by conservative religious and political groups, British policymakers, unlike their American and German counterparts, were able to define a relatively unburdened, authorized space for embryo research, as well as to keep abortion rights from being seriously rolled back. Over several years of parliamentary deliberations, IVF came to be seen as a legitimate offshoot of scientific research—arousing worries about misuse, to be sure, but benefiting, too, from science's claim to autonomy. Key to maintaining this framing were certain entrenched features of British policymaking that we have seen in earlier chapters: leadership by trusted individuals and institutions; faith in empirical observation and demonstration; and confidence in the state's ability, through law, to apply the brakes to runaway social forces, including in this instance the power of medical science.

### An Enactment of Trust

The Human Fertilisation and Embryology Act of 1990 (hereafter Embryology Act), which established the regulatory framework for embryo research, was unexpectedly long in the making. Its history says much about the means by which seemingly irreducible differences of opinion can be made tractable within the British political system. As Michael Mulkay relates in his careful study of the embryo research debates, the work done by Edwards, Steptoe, and their colleagues to produce embryos outside the human body prompted sporadic criticism during the 1970s, largely on the ground that these procedures might produce fetal abnormalities or be used to transfer embryos from one woman to another.[3] Public debate began in earnest in 1982, when the Department of Health and Social Security (DHSS) appointed a blue-ribbon commission of inquiry, chaired by the Oxford moral philosopher Mary Warnock, to examine the social, ethical, and legal implications of developments in the field of assisted reproduction. The commission's report was to provide guidance for future policymaking, including legislative action if need be.

The Warnock Report,[4] presented to Parliament in 1984, recommended the formation of a statutory body to license embryo research within strict guidelines; perhaps more important, and certainly more controversial, was the report's conclusion that research should be permitted only on embryos aged less than fourteen days. The biological rationale for this cutoff point was the appearance of the so-called primitive streak at about fourteen days; this marks the end of the period when embryonic cells stop simply dividing and instead begin to differentiate. The committee split on whether such a distinction made biological or moral sense, with a minority report authored by its one Roman Catholic member and two others; but Warnock felt that this disagreement only strengthened the report's overall balance and "authenticity."[5] The research community found the fourteen-day barrier scientifically unsupported and worried that it would become overly restrictive as biological science made further strides in understanding human development.[6] Some noted the contradiction between a research regime according near-personhood to the embryo after fourteen days and a permissive abortion law that, at the time, permitted termination as late as twenty-eight weeks into a pregnancy.[7] Anti-abortion groups for their part saw the bifurcation of the embryo into an early, prehuman object-state and a later state entitled to full respect as a potential human as an attack on their core belief that conception is the starting point in a single, continuous process of human development.

Although the Warnock Report was supposed to inform legislation, formal action on its recommendations took another six years, during which a voluntary Interim Licensing Authority was formed to provide regulatory oversight. To arrive at a stable legal arrangement, many disputed issues had to be resolved, and the eventual shape of the legislation was often in doubt. Almost before the ink on the report was dry, anti-abortion groups led by the Society for Protection of Unborn Children (SPUC) began mobilizing parliamentary support against it, swinging many legislators to the view that embryo research meant, in effect, the sacrifice of "unborn children" at the altars of science.[8] In 1985 these objections crystallized in the form of the Unborn Children (Protection) Bill, introduced on behalf of the prolife lobby by the Conservative former Minister Enoch Powell, and prohibiting embryo research. To the scientific community's dismay, Powell's bill won large majorities in two votes in the House of Commons, although it was eventually eliminated from the legislative agenda through a series of parliamentary maneuvers.[9] Clearly, countermobilization was called for if proresearch interests and their government backers were to prevail; both sets of actors responded creatively to the challenge.

All-important in the government's arsenal was the procedural device of consultation, which could be used not just to build consensus but also to stave off unwanted parliamentary action. In 1987 the government undertook a year of consultation with recognized professional bodies to review the provisions of the Warnock Report. Under normal circumstances, such an exercise might have culminated in prompt legislative action, but—wary of fights ahead and unwilling to be seen as too forceful—the government instead adopted a more velvet-gloved approach: it issued a White Paper in November 1987, presenting a framework for legislation.[10] The document ratified the fourteen-day limit on the use of embryos for research but went beyond the Warnock proposals in supporting bans on some genetic manipulations (e.g., cloning, long-term storage) that scientists still regarded as speculative, and some (e.g., artificial creation of human beings with predetermined genetic characteristics) that the drafters of the White Paper concluded were "extremely remote."[11] The White Paper also took a firm stand against the interspecies transfer of embryos and the fusion of human and nonhuman cells to produce a chimera.

As debate on these proposals began in the House of Lords in January 1988, the strategic significance of the White Paper became very clear. Lord Skelmersdale, the Conservative undersecretary of state at DHSS, introduced the document, setting aside the charge that "the Government are making consultation a substitute for action." Rather, he insisted: "We felt it important to set out the details of our proposals in a White Paper, including an explanation of the alternative possibilities for dealing with embryo research, and to allow Parliament to debate these. Your Lordships' debate today, and that in another place shortly [i.e., House of Commons], will help frame the Bill to be introduced during this Parliament."[12] In other words, the ultimate decision-maker in this case was not to be the government of the day, but Parliament, voting independent of party lines, as was customary in so-called matters of conscience such as abortion and capital punishment. By informing and facilitating this process, the government represented itself as a trustworthy servant of the democratic will, offering its services to Parliament in accordance with the rule of law instead of pursuing its own predetermined political agenda.

Strategic too was the launching of this round of debate in the House of Lords. One experienced observer, the Scottish Labour MP and science columnist Tam Dalyell, noted that this move "created a different atmosphere from the ethos it would have had, had it come to the Commons first."[13] An activist, Christine Lavery, representative of the proresearch Genetics Interest Group, commented that the peers were more informed than most members of Parliament, and hence presumably more likely to favor research.[14] Mulkay offers a more sociological view,[15] noting that the peers had relatively unquestioningly taken instruction on the pros and cons of research from leading members of the scientific community, such as Robert Winston, founder of the proresearch group PROGRESS, and Anne McLaren, a distinguished developmental biologist and member of the Warnock committee.[16] Without inquiring into the basis for the scientists' assertions about the facts of human biology, the peers assumed that theirs was the enlightened, nonfundamentalist position that reasonable people ought to adopt; in this way the Lords forged a powerful alliance between the voices of science and the politically elevated common sense represented by the Lords themselves.

The stage-setting debate on the White Paper in the House of Lords showcased a number of individuals whose authority to address this issue in the public interest was above question. Speakers in favor of the government's position included John Habgood, the Anglican Archbishop of York, and Earl Jellicoe, president of the Medical Research Council; their credentials mattered, even though in the spirit of free debate each professed to express only his personal views. At the center of the stage was Baroness Warnock herself: no more than plain Mrs. Warnock, a senior research fellow of Lady Margaret Hall when the state first called upon her services, she was named Dame of the British Empire in 1984 and elevated to a life peerage in 1985, shortly after publication of the Warnock Report. In the same year, she became Mistress of Girton College at Cambridge. From that point on, with her name indelibly linked to the work of the commission she had chaired, Warnock became virtually synonymous with British bioethics. Like many of Britain's most prominent public servants, she personified in her own self her nation's commitment to the ethical conduct of embryo research at home and abroad; as we saw in chapter 3, she was the U.K. representative for the first bioethics body convened by European Commission president Jacques Delors. To the extent that successful politics demands the public production of trust,[17] the government could not have asked for a safer circle of players to represent its point of view.

### A Distinction with a Difference

The success of the Embryo Bill depended crucially on another figure—a figure frequently referred to in the scientific and popular press, and constantly invoked during the parliamentary debates, though absent from both the Warnock Report and the text of the law, and curiously no longer much in evidence after the law was enacted. This was the pre-embryo, the name given to the pre-fourteen-day human conceptus before the appearance of the primitive streak. As we have seen, intensely aware that it was expected to provide legislators with a definite, enforceable demarcation line, the Warnock committee built a consensus around the fourteen-day limit. But for research to be permitted, it was essential to repudiate the countervailing notion that the developing embryo was from the moment of fertilization an "unborn child." By giving a name to the pre-fourteen-day entity, research scientists and their political allies met and parried the nominalist tactics of the prolife lobby. Thus, in a neat display of coproduction,[18] the adoption of new rules for the ethical conduct of science resulted, at one and the same time, in the coming into being of a new natural entity. Characterized as a mass of undifferentiated cells, the pre-embryo was safely bounded off from personhood, and hence could be an object of research, as opposed to the embryo proper, the authentic precursor of human life.[19] Seen this way, the pre-embryo could not, by definition, be an unborn child. But how was consensus achieved on this vital ontological distinction?

Looking back at the pre-1990 British debates, one is struck by the confidence of an empiricist culture persuaded that observable distinctions exist in the real world, that they can be witnessed in common,[20] and that they accordingly form a stable basis for public policy. In a 1993 interview with me, Warnock conceded that "seeing" distinctions is a complex affair.[21] Observation may be important to making distinctions that will stick, but so are justification and persuasion. It required careful management to produce an agreement around the fourteen-day cutoff, both in the Warnock committee and later in the House of Lords. No one, Warnock indicated, came to her committee believing in advance that the appropriate place to hold the line on embryo research was at fourteen days. Rather, once the need for a clear criterion was acknowledged, that limit served as a plausible, not wholly arbitrary solution to the problem; the committee then undertook to play an educative role, explaining its conclusions to the public in understandable language. The term "pre-embryo" was ready to hand from the field of animal development familiar to Anne McLaren and others. Warnock, however, steered the committee away from this and other potentially confusing terms, preferring instead an approach directly tied to observable physical changes in the embryo.

The wisdom of that move became clear during the debate in the House of Lords, where the reactions of some peers showed that the term pre-embryo could promote dissension as well as consensus. Lord Kennet decried the use of "purpose-serving new language" and noted: "Pre-embryo tells us nothing. A pre-embryo goes on to be an embryo, which is a pre-foetus, and a foetus, which is a pre-baby, goes on to be a baby: and a baby, of course, is a pre-adult. We do not call any other stage of human life story something beginning with 'pre.' It is a negative definition which merely says that it is not an embryo in order to avoid the stigma of destroying embryos.[22] More than a claim of empirical observation was needed to counter this kind of skepticism and to give the pre-embryo the ontological robustness it needed to support scientific research. That added assurance came from uniting mere observation with cultural authority, both religious and secular.

Several speeches from the 1988 Lords debate illustrate the interplay of cognitive, cultural, and social authority that shored up the pre-embryo's reality at this politically uncertain juncture. One was a statement by the Archbishop of York that carried substantial weight with his peers. The archbishop placed himself firmly among those who see "personhood in terms of the historical continuity of the individual subject." From that vantage point, he noted that "there is a certain fluidity about the identity of the organism [the embryo] at this early stage, and that uncertainty is not resolved until the cells, instead of just going through the process of multiplying as happens in the very early stages, begin differentiating. It is only at that stage that one can begin to talk in any meaningful way about the embryo." In other words, biological criteria mattered to the primate, though it was their confluence with moral ones that proved to be clinching. The archbishop went on to observe that, on a sliding scale of human development, "it makes no sense . . . to ascribe full personal value to human matter which possesses none of the attributes which normally belong to persons."[23] Represented in this way as being both morally *and* biologically discontinuous with full-blown individuality, the pre-embryo was categorized as a non-person—as mere matter.

Another instructive statement came from Lord Henderson of Brompton, who claimed that he had originally found the subject "misty" but had gained enlightenment through the consultative process, and through "the schism of the word embryo into two parts":

> There is the pre-embryo from the fertilised ovum to the primitive streak and implantation.
>
> Secondly, there is the post-primitive streak period which can be described as an embryo, later as a foetus, and later still as a child. But in no sense can the 14-day period be regarded as an embryo. That has been made very clear by today's speeches. It has been made very clear by the experts to whom I have listened over the years. I instance particularly the very humane doctors, Dr. Winston of Hammersmith and Dr. Anne McLaren and Professor Shaw of the Medical Research Council. These humane people made the distinction so clear to me that I find that the moral problem of dealing with the first 14 days after fertilisation of the embryo has disappeared. It is a precision I have found helpful and in no way misleading.[24]

Lord Henderson had no qualms about telling his peers how *he* had attained his state of intellectual and moral enlightenment. He was led there by trusted experts, whose humaneness had given them the power to translate biological distinctions into terms that could be understood and made legally binding by lay politicians such as Lord Henderson.

Most interesting of all was Mary Warnock's own statement to the Lords, which included the following passage: "there is a very noble prayer frequently uttered in my college chapel that we may be given the grace to distinguish things that differ. I, as a member of my college chapel, pray this most fervently because I believe that it is our moral duty to make a distinction in this case between the pre-embryo and the embryo which is to become the individual and to act on this distinction."[25] The prayer Warnock referred to is attributed to the fifteenth-century Christian mystic Thomas a Kempis. A common English version reads as follows, and it speaks to the insufficiency of unguided human sight:

> Grant me, O Lord, to know what is worth knowing,
> to love what is worth loving,
> to praise what delights you most,
> to value what is precious in your sight,
> to hate what is offensive to you.
> Do not let me judge by what I see,
> nor pass sentence according to what I hear,

but to judge rightly between things that differ
and above all to search out and to do what pleases you,
through Jesus Christ our Lord. Amen.[26]

In quoting Thomas a Kempis, Warnock herself distinguished judging, perhaps wrongly, by "what I see" from judging "rightly between things that differ."[27] The former requires mere human competence; the latter happens only through God's grace, but transmitted on Earth through persons who—like Warnock—can be trusted to concentrate in their own person the combined authority of scholarship (Cambridge University), state power (the House of Lords), and established religion.

The pre-embryo, we may justly conclude from these interventions, became an effective figure in the British IVF debate, even if not one formally recognized under law. Its ontological and political reality was not a product of biological knowledge alone but was created (or coproduced) out of a complex mix of pragmatism, empiricism, and trust in experts. One sentence about the fourteen-day rule offered by Earl Jellicoe underscores the hybrid foundation of the rule's success: "It is workable, it is observable, it has a quality of certainty about it—there is an important change at fourteen days."[28] One must note, too, that other concurrent attempts to make the embryo politically visible did not work so well, presumably because they lacked comparably persuasive social and cultural certification. In particular, the anti-abortion lobby was widely seen as having made a tactical error in its attempt to reduce the time limit on abortions from twenty-four to twenty weeks. SPUC sent every MP a life-sized model of a twenty-week-old fetus, but undecided MPs found this move distasteful and rejected SPUC's position on research as well as abortion.[29] The model violated unwritten canons of good taste, and SPUC succeeded in attracting rather than mobilizing collective disgust. The pre-embryo, which Earl Jellicoe described as "equivalent in size to the full stop" at the end of a sentence in *Hansard*,[30] proved to have more action-forcing power—though it was only a name—than SPUC's disturbingly physical representation of an unborn child.

Not even the insertion of an abortion provision served to stall the final passage of the Embryo Bill. An amendment lowering the time limit for legal abortions from twenty-eight to twenty-four weeks was added in the House of Lords, but the fear that this might arouse unmanageable passions proved to be unfounded.[31] Though abortion and embryo research were thus coupled in the same piece of legislation, there was no discussion of possible inconsistencies in the way these provisions conceptualized the origin of human life.

## Stopping on the Slippery Slope

The metaphor of the slippery slope, so beloved of common law judges in rationalizing conservative legal rulings, repeatedly surfaced in the public discourse on embryo research. Mulkay notes that in the eight major parliamentary

155

debates, the metaphor was explicitly used in 33 out of 189 speeches, and was accompanied by a cluster of related metaphors, such as "the thin edge of the wedge" and "the downward path."[32] The notable point for our purposes is not so much the prevalence of these metaphors as the means by which experts and policymakers sought to reassure their audiences that one did not have to slide pell-mell down the slippery slope: there were effective safeguards against moral degeneration. Faith in the rule of law was the paramount source of confidence. In the 1988 House of Lords debate, for example, Warnock told her colleagues, "I think your Lordships should not be frightened of the slippery slope argument. I am sure you will not be and that the general public may be aware of this fact. We can stop our descent down the slippery slope at any point when we wish to do so and the way of stopping ourselves descending into unknown horrors is by legislation."[33] Indeed, regardless of their opinions on the advisability of embryo research, all sides agreed on the need for a statutory (that is, legally appointed) licensing body.

Outside the parliamentary setting, too, proresearch advocates rejected the specter of unavoidable decline. Writing in *Nature* in 1989, Anne McLaren was convinced of British society's ability to control morally risky initiatives in *any* context, not only that of research: "Slippery slopes are not unique to science. They are a general characteristic of our society, and we have well-tried and long-established methods of making sure that we do not slide too far down them. Regulation, licensing authorities, inspections, surveillance—the trappings of slippery-slope control may be irksome, but they are remarkably effective".[34] A 1990 editorial in *Nature* noted that a statutory licensing body "should give the lie to those who also argue that the manipulation of embryos before implantation is the beginning of a dangerous slippery slope."[35] Questioned on the issue at about the same time, Robert Winston, founder of PROGRESS, rejected the notion of an inevitable slide on the ground that "scientists and doctors take their moral responsibilities 'extremely seriously.'"[36]

To what extent were these optimistic expectations borne out? The passage of the Embryo Bill in 1990 forbade certain abhorrent activities outright, such as placing an animal embryo in a woman or a human embryo in an animal. For the rest, the law charged the Human Fertilisation and Embryology Authority (HFEA) with carrying out two major functions: licensing clinics where infertility treatment is done, and approving research projects involving embryos. While some of HFEA's activities are precisely controlled by the governing statute, the authority retains considerable discretion to act on borderline cases not expressly provided for by law. In these situations, HFEA is guided by notions of natural justice.[37] Three examples indicate the range of pressures that HFEA has been subjected to since its creation, and how it has responded. These cases illustrate how this statutory body lived up to the expectation of McLaren and others that embryo research would not set Britain off down the slippery slope of unchecked human experimentation. They also delineate some important contrasts between the British regulatory regime and its counterparts in Germany and the United States.

In one of the first really public cases to come before HFEA, Diane Blood, a thirty-two-year-old public relations consultant from Nottinghamshire, unsuccessfully sought to be impregnated with sperm taken from her husband Stephen, who had died of bacterial meningitis in 1995. The sperm had been removed at his wife's request while Mr. Blood was in a coma, and he had thus given no prior written consent for its removal or use. Accordingly, under the terms of the British law, Mrs. Blood's wish to be inseminated could not be granted: the law specifies that gametes may only be used with the donor's explicit consent.[38] Mrs. Blood then asked to be allowed to take the sperm to a clinic in Belgium that was willing to perform the procedure she sought, but HFEA denied that request. In the final phase of the legal battle, the Court of Appeal overturned HFEA's decision and granted Mrs. Blood the right to carry out her plan elsewhere, on the ground that to do otherwise would violate European human rights law.[39] Diane Blood bore one son, Liam, in 1998 using her dead husband's sperm; another son, Joel, was born through the same procedure in 2002.

A second test of HFEA's effectiveness involved a large public consultation, in which the authority tried to ascertain people's attitudes regarding infertility treatment with ovarian tissue from dead persons and aborted fetuses. The authority distributed leaflets to twenty-five thousand people and received nine thousand responses, considered a very high rate of return. These results, however, reflected mobilization by prolife groups such as SPUC, who took advantage of the occasion to press their own moral agenda. The process introduced a new term, expressive of public repugnance, into the British discourse on embryo research: the "yuk factor."[40] Many respondents were instinctively repelled by the idea of telling a child that her mother was a dead person or an aborted fetus, and HFEA thought it appropriate to respect these instincts. It was in some ways a democratizing move, paralleling the public debate on GM crops, in which a U.K. state agency sought to identify and consult with an interested public. Some, however, criticized the consultation as antidemocratic in its effects because the choice of a public relations strategy simply opened the way for SPUC to pursue its own well-funded private campaign to influence public opinion.[41]

A third well-publicized and emotional issue concerned so-called designer babies: children with particular genetic characteristics selected after conception to provide tissue for treating their sick siblings. HFEA had refused to allow tissue typing in one case where the process carried no benefits for the yet unborn child but was designed solely to find a match for an existing child with a rare form of anemia. The child's parents, Michelle and Jayson Whitaker, traveled to Chicago as "procreative tourists" to receive the desired treatment, and Michelle eventually gave birth to a second son who was a perfect match for their older boy. In the case of Shahana and Raj

Hashimi, whose child Zain was suffering from the inherited blood disease beta thalassemia, HFEA did grant permission for preimplantation genetic diagnosis and screening. The difference in this case was that testing would help to produce a healthy child who was free of Zain's genetic defect and, secondarily, could also serve as a "savior sibling" for Zain.

HFEA was taken to court for this decision by Josephine Quintavalle of the prolife group Comment on Reproductive Ethics, who charged that the authority did not have the legal right to license tissue typing. Quintavalle said she was interested in removing these sorts of decisions from an unelected body acting behind closed doors. Parliament, she contended, was the appropriate institution to be dealing with such issues in a democratic manner. A lower court decision in her favor was overturned by the Court of Appeal in May 2003.[42] The appellate court reasoned that IVF accompanied by preimplantation genetic diagnosis in this case assisted a woman to carry a child within the meaning of the law, by assuring the mother that the new baby would not be born with an inherited disease. That the IVF child could then be selected for other characteristics (such as tissue compatibility with a sibling) raised questions that HFEA was legally competent to decide. At the same time, growing pressure on HFEA from such high-visibility cases strengthened calls for the reevaluation and revision of the 1990 law.

All three episodes (as well as the cloning case discussed in the next chapter) show a firm consensus among establishment actors, including the courts, that HFEA's authority in the field of embryo research was supreme and deserved to be protected. The principal opponents were disaffected but politically weaker prolifers, who disapproved of the whole legal regime that had created the HFEA. This meant that new issues arising around embryo research and assisted reproduction were promptly depoliticized and taken out of the legislative arena, where the prolifers wished to place them. HFEA at the same time was careful to operate with an eye on its public legitimacy, as in its decision not to use fetal tissue in fertility treatment. A sign of the government's continued sensitivity to public concern about embryo research came with the appointment of Suzi Leather as HFEA chair in March 2002. Leather, a respected figure in British consumer and health policy, had most recently served in another high-profile post, as deputy director of the post-BSE Food Standards Agency. The British government clearly had no intention of risking in the context of embryo research the breakdown of trust that had engulfed agricultural biotechnology.

## Germany: Ruling out Monsters

The German and British debates on assisted reproduction and embryo research occurred almost in parallel and both led to national legislation, but the provisions of the German Embryo Protection Law (Embryonenschutzgesetz,

ESchG), which took effect on January 1, 1991, were sharply different from those of the comparable British statute. These divergent outcomes have to do with a number of factors, one of which undoubtedly is the memory of the Nazi period in Germany, and the associated extreme sensitivity to any possible state-sponsored inroads upon the sanctity of human life and human dignity. But the approaches followed in each setting also reflect the contingencies of national politics, the professional roles of science and medicine, and fundamental differences in the two nations' legal cultures. These contrasts become apparent if we examine the developments around embryo protection in Germany together with the ongoing conflicts over the law and politics of abortion, which raised similar questions about the origin of, and state responsibility for, human life.

## Abortion: Divisions under Law

Of all the issues that had to be settled during the reunification of East and West Germany in 1990, abortion turned out to be one of the most unexpectedly tricky[43]—although the trickiness should not perhaps have been altogether unexpected. Formally, the legal situation on the two sides of the divided nation had gone in different directions since the 1970s. In 1972 East Germany adopted a permissive abortion law allowing abortion on demand during the first trimester. In West Germany, by contrast, a socialist-feminist coalition had tried in the mid-1970s to liberalize section 218 of the Penal Code, the century-old law criminalizing abortions, but with only limited success.[44] The attempted law reform adopted a term-based approach (*Fristenlösung*) similar to East Germany's, which would have allowed relatively unrestricted abortions during the first trimester, but it was challenged by the minority Christian Democrats and struck down as unconstitutional by the Federal Constitutional Court in its first abortion decision of 1975.[45]

The starting point of judicial analysis was the provision of Article 2, Section 2, paragraph 1 of the Basic Law, which provides that "Everyone shall have the right to life." This right, the court held, encompasses the unborn as well as the born, since the development of human life is a continuous process, extending without sharp breaks from the fourteenth day after conception, even beyond the moment of birth, as the child attains consciousness and grows into an adult. This idea of the continuity of the individual has implications for the state, since—consistent with its duty to protect human life as well as its constitution as a state committed to the rule of law (*Rechtsstaat*)—the state cannot abandon its responsibility to the developing fetus by creating an essentially law-free zone, ruled only by private moral intuitions. The 1974 law, by providing free access to abortions with only a counseling requirement, had created precisely such an unprotected zone for developing human life in the court's view. Hence that law needed to be replaced by one that much more explicitly specified how to balance women's legitimate interest in self-determination against respect for the potential life of the developing fetus.

The next abortion law responded to the court's decision by specifying four "indications" (*Indikationslösung*) that could justify legal abortions: a medical indication of a threat to the mother's health; a so-called eugenic indication of danger to the fetus; an ethical indication associated with criminal acts such as rape or incest; and a social indication that the experience of childbearing would be too distressing for the mother. Some 70–80 percent of abortions in West Germany relied on the social indication, but how burdensome this was in practice varied from one Land to another, and also depended on the physician;[46] women in northern and urban regions were less constrained than women from Catholic Bavaria or rural Germany. Violations of the law carried potential criminal penalties for women as well as their physicians. In the notorious Memmingen trials of 1988–1989, the physician Horst Theissen was sentenced to a prison term, and his largely immigrant patients received stiff fines and unwanted publicity for failure to comply with the law. West German women often traveled abroad to the Netherlands or elsewhere in order to avoid the stresses of medical certification.

Feminists and liberals on both sides of the border initially saw reunification as an opportunity to loosen the restrictions of the West German law by approximating it more nearly to the East German provision. At this overwhelming "national moment," however, the effort to build a coherent German identity demanded the complete overthrow of communism and all it stood for. Some saw the process as an outright annexation of the East by the West, a cultural superimposition rather than a merger. The feminist political scientist Angela Wuerth has argued that the patriarchal "master narrative" of national unification marginalized women's issues as secondary.[47] The time, in any case, was scarcely propitious for celebrating any liberating achievements of the East German state, and the expansive abortion law was represented by the ruling Christian Democrats in the West as just another of the exploitative, dehumanizing practices of a failed and discredited regime. Statistical arguments called attention to the fact that the number of abortions was similar in the two states even though the West had four times the population, suggesting coercion in the East. Women from the two Germanies proved unable to build a winning coalition on abortion. West-East divisions surfaced as non-negotiable line of political cleavage, with women on each side blaming the other for political ineptitude or disloyalty to the agendas of either feminism or unity.

The Unity Treaty of 1990 did nothing to resolve the abortion issue but called for new legislation, ensuring more effective protection of unborn life, by the parliament of reunited Germany. In 1992 the Bundestag passed a law allowing abortions in the first trimester, following counseling by a state-licensed counselor. Immediately, the prime minister's Christian Democratic Party and the state of Bavaria took the issue to the Federal Constitutional Court, where it was heard by the heavily male and more conservative Second Senate.[48] The court declared that the law had erroneously failed to make abortions unlawful, consistent with the Basic Law's Article 2(2) guarantee of unmitigated respect for human life; abortions would have to be authorized on the basis of clear countervailing indications having to do with fundamental interests of the mother. In an action seen by some as antagonistic to the liberal organization Pro Familia, the court also struck down the law's provisions on counseling. Under a new 1995 compromise, the earlier social and eugenic indications were set aside in favor of abortions following authorized counseling, while the provisions for abortion on medical and criminal (ethical) indications remained virtually unchanged. The older "eugenic" indication was formally repudiated as coming uncomfortably close to the Nazi worldview that a deficient life is life not worth living (although Pro Familia's website indicates that the risk of severe health defects in the child may still be considered in connection with the medical indication for the mother[49]). Abortion law in this way affirmed the transcendent moral status of the embryo but detached this issue from the wider discourses of genetic stigmatization and discrimination against lives already in being that the SPD and the Greens had initiated around genetic engineering in the 1980s. Legal clarification satisfied the German penchant for clear ontologies, coupled to clear rules of governance, but in drawing workable lines it also repudiated the holistic, programmatic vision of biotechnology that the Greens had tried to keep alive on the political agenda.

### Assisted Reproduction and Embryo Research

The German law on assisted reproduction and embryo protection was publicly debated in the same charged atmosphere of nation-building, or rebuilding, although legislative deliberations on these issues had gone on for years in both public and private forums.[50] Early movers included the powerful professional society of German physicians (Bundesärztekammer, BÄK), which at its eithty-eight annual meeting in 1985 approved guidelines for IVF and embryo transfer for infertility treatment.[51] Several of the Länder, including Bavaria and Rheinland-Pfalz, also began considering possible legislative action on aspects of assisted reproduction in the mid-1980s. The Bundestag appointed its Enquiry Commission (Enquete Kommission) on gene technology in 1984 and (as discussed in chapter 2) received its final report in 1987. In 1984 the federal ministries of Justice and of Research and Technology appointed a working group on IVF, genome analysis, and gene therapy chaired by former Constitutional Court president Ernst Benda. The Benda commission's final report a year later provided the basis for a provocative discussion draft issued by the Justice Ministry in 1986. Particularly controversial were the draft's provisions regarding restrictions on research and the use of the penal code to forbid specified actions. Following much debate in academic and professional circles, and consideration by a Bundesrat-sponsored federal-state working party, a legislative draft was drawn up in 1988 and enacted into law in 1990. Although the Benda commission had allowed for the possibility of high-priority research on embryos, the 1990 law strictly banned all such research.

A hearing before the parliamentary judicial committee in March 1990 showed that opinion on the bill was divided along party, professional, and denominational lines.[52] The Social Democrats and Greens would have preferred to see embryo protection treated as part of a wide-ranging reform of the civil law related to issues of women and family raised by the new technologies. Parliamentarians and experts of the Left identified numerous concerns that the bill had not addressed: for example, the medicalization of childlessness and the psychological and material consequences of this move for women; the unequal treatment of the sexes in infertility treatment; the relationship between abortion and embryo research; and the basic framing of the law as a provision of the penal code. Representing the Max-Planck Institute for Criminal Law, Dr. Albin Eser noted that the proposed law's absolute prohibition on research was inconsistent with the balancing approach of Germany's abortion law, which permitted the termination of embryonic life under stated circumstances.[53] The German Research Foundation (Deutsche Forschungsgemeinschaft, DFG), represented by its vice-president for medicine, stood by its 1987 report, in which it had supported the Benda commission's conclusions that some research might be permitted under well-defined conditions.[54] The BÄK representatives spoke strongly in favor of self-regulation, asserting that five years of implementing their society's ethical guidelines had proved effective in regulating medical practice.[55] The sociologist Elisabeth Beck-Gernsheim, however, countered these optimistic assessments, observing that self-regulation could only lead to an ever-expanding list of research priorities and the doing of whatever was considered technically feasible.[56]

In the end conservative voices won the day, as the ruling Christian Democratic coalition carried its highly restrictive bill into law. A list of absolute prohibitions, punishable by fines and prison sentences up to three years, included the following: the use of egg cells from one woman to impregnate another; the implantation of more than three embryos in a woman at one time; the creation of more embryos than could be implanted in a single cycle; and the practice of surrogate motherhood using IVF. Just as the elusive figure of the pre-embryo had helped to focus the legislative imagination in Britain, so the German legislative imagination also fixed upon an absent figure—the supernumerary (überzählig) embryo. Only, whereas the pre-embryo had served as an instrument with which to advance the wishes of the U.K. research community, the supernumerary embryo was perceived and presented in Germany as a threat to order, a being whose existence might create an unregulated gray zone and so weaken the constitutional protection of human life and dignity, promoting dangerous scientific inquiry. The judicial committee's March 1990 hearing ended with an informal count of these unwished-for beings. Professor Wuermeling of Erlangen reported that one IVF center he had visited housed some thirty or so frozen embryos, all products of treatments still in process; some parliamentarians requested a more accurate count of these legally ambiguous entities.[57]

Whatever scientists may have thought in private, they did not challenge the embryo law as inconsistent with the constitutional guarantee of freedom of research. The attitude of the noted geneticist and historian Benno Müller-Hill of the University of Köln helps explain the scientific community's silence on this score. Author of an important history of Nazi eugenics,[58] Müller-Hill was at once a vociferous advocate of research freedom and a relentless opponent of anything touching on abuses of genetic research. In 1994, for example, he unsuccessfully campaigned against the appointment of Hans-Hilger Ropers as director of the Max-Planck Institute for Molecular Genetics in Berlin, on the ground that Ropers had conducted a behavioral genetic study of the basis of violence without properly considering its social implications.[59] In a 1993 interview with me, Müller-Hill was scathing in his denunciation of "that stupid Asilomar" and the subsequent German regulations governing rDNA research, which he saw as stifling and "totally neurotic."[60] Yet he had little critical to say about the Embryo Protection Law. Research on humans, he insisted, was "totally different" from research on bacteria, plants, or even transgenic mice.

Müller-Hill's sense of absolute distinctions accords well with the spirit of a law that was remarkable for its intolerance of ambiguity and of biosocial experimentation. It not only severely restricted research but also criminalized most technologically assisted deviations from the ideal of the biologically related nuclear family: in other words—and this fact deserves emphasis—Germany's 1990 law was as negatively disposed to social as to scientific experimentation. A Diane Blood case was virtually unthinkable under that law, as were cases involving designer babies; the contrast between this regime and the one that developed in the United States around assisted reproduction could hardly have been more stark.

## United States: A Regime of Rights

Abortion politics played a part in the social response to assisted reproduction in the United States as in the other two countries, but it did so with little of the explicit attempt to link these domains that we saw in Britain and in Germany. For more than a generation, since the Supreme Court's 1973 decision in *Roe v. Wade*,[61] the arena of national politics was dominated by fierce debates over the morality of abortion. The battle was fought in the courts, in elections, in Congress, and in the media. It polarized American society as almost no issue other than the Vietnam War had in the last third of the twentieth century. It gave rise to social movements and spawned whole industries of scholarship. The sharp divisions on abortion meant that deliberative rapprochement and legislative compromise became almost unimaginable, and opposing parties kept returning to the winner-take-all environment of the courts. Between 1973 and 1992 the Supreme Court handed down several major decisions on abortion; one of these, *Planned Parenthood of Southeastern Pennsylvania v. Casey*,[62] appeared for a time to bring about a measure of legal stability. Yet, as late as 2003, bitter conflict still raged around the abortion views of several Bush administration judicial nominees, and especially after the 2004 presidential election many worried that the next liberal retirement from the Supreme Court could lead to a further retreat from *Roe v. Wade*. Legislative attempts to restrict abortion rights and enhance the rights of the fetus also continued, with a federal ban on partial birth abortions in November 2003 and the passage in March 2004 of a law making it a crime to injure the unborn.

In Britain and Germany, questions about the embryo's status crossed over between the debates on abortion and assisted reproduction. By contrast, in the United States, the supercharged topic of abortion remained largely decoupled from developments around IVF. The root of this difference lies in the framing of both abortion and technologically assisted reproduction, through litigation, in terms of individual rights. As a result, the question of reproductive rights remained bounded off from issues of research, and the three-cornered interplay among women's rights, the status of the embryo, and freedom of scientific inquiry that had developed in Britain and Germany failed to materialize in the U.S. context; this situation changed materially only with the stem cell debate, as we will see in the next chapter.

### Abortion and the Limits of Law

The storm unleashed by *Roe v. Wade* had two centers: the extension of the right to privacy, previously recognized mainly in cases on contraception, that gave women an absolute right to terminate a pregnancy; and the mapping of that right onto a trimester framework, in which state regulation was banned in the first trimester, allowed only to protect the mother's health in the second, and authorized in the interests of the developing fetus, from the point of viability, in the third. The 1992 *Planned Parenthood v. Casey* ruling reaffirmed women's right to an abortion, but, departing from *Roe v. Wade*, it acknowledged that the state's interest in potential life permits regulation even in the first trimester, so long as it does not place "undue burdens" on women's fundamental right to terminate a pregnancy. Following *Casey*, numerous states enacted counseling, parental notification, and waiting period rules designed especially to discourage younger women from undergoing abortions.

For purposes of comparison, two other aspects of *Casey* deserve to be specially noted. First, the Court claimed to uphold *Roe*'s essential holding— including a "recognition of the right of the woman to choose to have an abortion before viability and to obtain it without undue interference from the State"[63]—but the majority grounded the woman's right in the liberty guarantee of the Fourteenth Amendment. Privacy, deemed a controlling right in *Roe*, was relegated to secondary functions in the *Casey* opinion, mainly applied to communications between women and their physicians and to relations between women and their marital partners. Thus, the Court ruled unconstitutional a requirement of mandatory disclosure to prospective fathers, saying that undue state interference with individuals could not be justified, even if the infringement was for their spouse's benefit. In other words, relations between the state and the pregnant woman were framed in terms of the individual's right to be free from undue, and unjustified, state intrusion within a zone of protected rights. Doctrinally, this was a very different resolution from the German Constitutional Court's balancing approach, which took as its starting point the state's affirmative responsibility to respect all human life, even that of the unborn.

The second point to note is that the majority in *Casey*, like the German legal authorities but unlike British lawmakers, did not find it necessary to decide when life begins. The Court recognized that the state has a legitimate interest in protecting developing life from the moment of conception and may regulate in order to do so. This, however, was not equivalent to addressing the ontological status of the fetus head-on, as did the debate on the U.K. Embryo Bill. Instead, the constitutionality of state regulations was left to be judged case by case according to the reasonableness of the burden placed on women seeking to exercise their right to an abortion.

### In the Marketplace of Reproduction

While the national trauma over abortion prevented any systematic federal regulation of reproductive technologies, their use and control developed in piecemeal fashion, through private enterprise, social experimentation, lawsuits, professional standard-setting, and state-by-state regulation. Compared with Germany and even Britain, assisted reproduction in the United States remains largely free of formal state controls. As a result, media stories about the developing field convey the carnival air of a fairground where almost any imaginable approach to baby-making is on offer: IVF, surrogate mothers, gestational mothers, artificial insemination for lesbian couples, frozen gametes from deceased donors, embryos for "adoption," Jewish surrogacy and egg donation, and sperm and eggs from Ivy League donors, who are handsomely paid if they meet the buyers' exacting selection criteria.[64] Behind the colorful stories lie human experiences of tragedy, desperation, and often deeply felt need. A wealth of private companies have sprung up to offer or broker services to meet the growing demand from infertile couples, with names like OPTIONS, IntegraMed, Tiny Treasures, Fertility Alternatives, Genetics and IVF Institute, and Center for Surrogate Parenting and Egg Donation.

Together with the commercial services goes a willingness on the part of many couples and individuals to try anything if the result might be the child of their dreams. How genetics and gestation are combined to produce the desired outcome seem almost irrelevant. A limiting case of sorts was reached in a California case known as In re Marriage of Buzzanca.[65] Here, a couple decided to employ a married surrogate to carry to term an embryo created from the ovum and sperm of unknown donors. Six days before the baby's birth, the couple's marriage dissolved and the male member refused to accept legal or custodial responsibility for the child. A trial court concluded that the child—who could potentially have claimed kinship to any of six different adults—lacked any legal parents at all. The Appeals Court reversed this bizarre judgment, ruling that the Buzzanca couple's intent to produce the child in the first place, by conceiving a conception as it were, made them the legal parents.

Though extreme in its display of reproductive libertarianism, Buzzanca was typical in one respect: it required the courts to make law in a place where law was otherwise silent. As I have discussed elsewhere,[66] U.S. state courts generally took the lead in shaping the legal order surrounding assisted reproduction. In the process, they also laid down the rules governing what would count as natural or unnatural uses of the new reproductive technologies. Biological considerations often played second fiddle to social ones in such determinations, as in the California case of Johnson v. Calvert.

In September 1990 Anna Johnson, an African American single mother of a three-year-old daughter, gave birth to a baby boy. Johnson had been impregnated with a zygote formed from the egg and sperm of a racially mixed couple, Crispina Calvert, a Filipino woman who had undergone a hysterectomy and hence could not bear a child, and her white husband, Mark Calvert. Relations between the parties broke down six months into the pregnancy, so that by the time the boy was born both sides were in court demanding to be recognized as his legitimate parents. The California courts were forced to consider for the first time whether to give priority to the "genetic" motherhood of Crispina, the ovum donor, or the "gestational" motherhood of Anna, the woman who had carried the child to term. Under state law, each had an arguable claim to being the mother, one based on blood tests and the other on having given birth. The word that proved to be controlling in sorting out these competing claims was "natural."

The judges did not doubt that the "natural" mother in this case was also the legal mother. But what, they asked themselves, should count as natural in the context of gestational surrogacy: the donation of gametes, which determine aspects of appearance, temperament, and behavior, or the equally essential contribution of carrying the child to term? Dominant narratives in law and culture had recognized the special claims of "birth mothers," acknowledging thereby women's special role in reproduction.[67] The California Supreme Court, however, concluded that the common law's emphasis on gestation was equivalent to demanding irrefutable evidence of the "more fundamental genetic relationship" between mother and child. The court concluded that when genetic consanguinity and giving birth "do not coincide in one woman, she who intended to procreate the child—that is, she who intended to bring about the birth of the child that she intended to raise as her own—is the natural mother under California law."[68] In this case, that intending mother was Crispina Calvert.

The decision to favor Calvert's maternal claim stabilized not only the boundaries of the contested family unit but, in a coproductionist move, also the underlying category of the "natural." The court solved the problem by interpreting the decision of Mark and Crispina to have a child "of their own genes" as an extension of the ordinary human urge to procreate; by contrast, Anna's provision of a womb was interpreted as only instrumental, designed to "facilitate" the Calvert's intentions. By normalizing the Calverts' actions in this way, the court set aside any deeper analysis of the relationship between the intending parents and the gestational surrogate. Indeed, Anna's contribution to the reproductive process was construed simply as an economic transaction, which the court then defended on libertarian grounds: "The argument that a woman cannot knowingly and intelligently agree to gestate and deliver a baby for intending parents carries overtones of the reasoning that for centuries prevented women from attaining equal economic rights and professional status under the law. To resurrect this view is both to foreclose a personal and economic choice on the part of the surrogate mother and to deny intending parents what may be their only means of procreating a child of their own genes."[69]

The phrase "of their own genes" calls attention to the normalizing work going on in this passage. The "natural" order as understood by the court resides in people who wish to perpetuate their genes, even if that requires the instrumental use of another's body for carrying the child—a biological function associated from time immemorial with the very essence of motherhood. It is worth noting that the construction of motherhood in this case as *excluding* gestation assigned parental rights to the racially and economically more advantaged parties, thereby (as years earlier in the New Jersey case of *Baby M*) rendering the court's conclusions more "natural" from a social as well as a biological standpoint. In a market-driven society, the superiority of the wealthier, service-purchasing, conception-conceiving parties' claims to parenthood was reinforced by the courts on the basis of tacit biological and social understandings of what is natural.

The market in the meantime played along with the courts by continuing to develop the logic of commodification around assisted reproduction. The promises made by the Genetics and IVF Institute (GIVF), a particularly aggressive advertiser, offer one striking example. In a March 2004 advertisement, GIVF offered a 100 percent money-back guarantee, as for any other defective product in the American marketplace, if a woman did not get pregnant and have a baby using the company's techniques.[70] Another GIVF advertisement featured egg, sperm, and embryo programs "with a large selection of screened, ethnically diverse donors" and offered gender selection services for "family balancing as well as disease prevention."[71]

## Conclusion

Assisted reproduction opened up what one German commentator called a "white field" in the law—a blank area needing to be filled in with ordering principles and practices. In some important respects the three national responses to this challenge followed similar paths. The timing of public action was roughly the same in all three countries. All treated issues of assisted reproduction against the background of prior and concurrent decisions involving abortion. In each country basic questions about the status of the embryo and the moral and social acceptability of new forms of kinship had to be resolved. Each confronted dilemmas about how far the state could or should go in regulating reproductive behavior, and with which sorts of legal instruments.

Upon this framework of parallels, however, were inscribed numerous deep divisions. There are two ways to represent these differences. One is to compare permitted and prohibited reproductive actions and relationships across the three countries. Table 6.1 provides a graphic illustration of the key cross-national divergences. They show, as discussed above, that Germany was significantly more rigid about disallowing novel social practices that could have emerged around technological change. Britain was more permissive but handled novelty case-by-case under centralized regulatory control. By contrast, fluidity and experimentation—technological, social, legal, and moral— were the order of the day in the United States.

This sort of representation, however, does not get to the issue at the center of this book, namely, the ways in which political culture shaped and was shaped by public deliberation around biotechnology. To address that issue, we need a different way of displaying the similarities and differences observed in the three countries. Figure 6.1 conceptualizes assisted reproduction as a four-cornered discursive field occupied by four possible sets of actors: the state, science, mothers, and the unborn. Rather strikingly, we see that political debates did not activate all four corners equally in each country. In Britain, the role and rights of women were hardly discussed in the public

TABLE 6.1
Reproductive Choices—Permitted and Prohibited Actions

|  | United States | United Kingdom | Germany |
|---|---|---|---|
| Abortion | Woman's right, strongest in first trimester, but subject to regulation | Permitted under 24 weeks | Illegal, but permitted with medical consultation up to 22 weeks |
| IVF | Permitted, largely private services | Permitted, under HFEA supervision | Permitted, subject to many prohibitions |
| Surrogacy | State-by-state regulation | Not legally permitted | Prohibited |
| Gestational surrogacy | State-by-state regulation | Not legally permitted | Prohibited |
| Nonimplanted (stored, frozen) embryos | No national controls | Regulated by law and HFEA | Prohibited (no "supernumerary embryos") |
| Embryo research | Not regulated by federal law | Permitted under 14 days (only on "pre-embryos") | Prohibited |

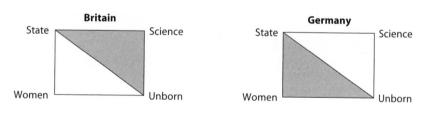

Figure 6.1. The Spaces of Political Deliberation

domain, even though the Embryology Act actually contained a provision somewhat restricting abortion rights. In Germany, the research community advised the state on pending legislation but was not politically mobilized or effective; the U.S. scientific community played hardly any role in framing debates around assisted reproduction. In other words, the practical zone of politics in each country was more a triangular space defined by at most three of the four possible actors, and these were not in each case the same ones.

Debates and lawmaking, moreover, drew upon the agency of participating actors in different ways. In Britain, a dominant coalition of scientists and the state succeeded in speaking for the embryo and, using the ontological device of the pre-embryo, created a space of political absence in which the interests of the developing fetus did not have to be considered by researchers. An administrative body, HFEA, guarded against the slide down the slippery slope into a state of moral anarchy or disorder. In Germany, the state remained the most active agent, entrusted with effectuating the interests of women as well as the unborn, but under strict supervision meted out by the constitutional court in Karlsruhe. In the United States, the agonistic field was again dominated by the courts, which extended the discourse of rights to cover both abortion and assisted reproduction. Abortion law gave preeminence to relations between women and the state, casting the battle as one of individual rights underpinned by the constitutional guarantee of individual liberty; prolife forces spoke powerfully for the unborn, but their actions played out largely in extralegal forums.

By contrast, in state-by-state adjudication of cases involving assisted reproduction, it was the implicit logic of the market that drove many U.S. legal outcomes. Even when courts decided according to the best interests of the child, it was the interests of the service-purchasing, and usually economically and socially better situated, parties that they wrote into the law as the "natural" order of things. In so doing, the courts echoed or reaffirmed the market logic of reproductive service providers, who increasingly stressed the contractual nature of their relations with their clients. We will return to these points in synthesizing the book's comparative conclusions in chapters 10 and 11.

# 7

---

## Ethical Sense and Sensibility

Although the debates around embryo research raised both moral and metaphysical questions, we saw in previous chapters that the intertwining of these domains occasioned relatively little explicit reflection in most cases. Even moral issues were raised at first only incidentally to more immediate policy concerns. Habituated to the formal discourses of economics, science, and law, political actors initially couched their analyses and proposals only in these familiar frameworks, with arguments phrased in terms of risks, rights, costs, benefits, property, human dignity, and the like. Advocates of biotechnology did not seem to realize at first how profoundly the new scientific and technological developments had destabilized the very terms in which those developments were being described and debated. How, for example, could the law speak persuasively of human dignity or of protecting nature when the meanings of "human" and "nature" were themselves under siege? Even when new discursive entities, such as pre-embryos or supernumerary embryos, were put forward to sharpen or foreclose debate, they appeared almost as stage props, incidental to the classic regulatory questions that preoccupied national policymakers—questions about the adequacy of existing laws and institutions in the face of problems created by human biotechnology. There was no self-conscious acknowledgment that mobilizing new entities like these in public discourse was to engage in what the philosopher Ian Hacking has called "world-making by kind-making."[1] Uncertainty, in the official language of public policy, focused not so much on the *nature* or *purposes* of the new things that biotechnology was bringing into being as on their possible *impacts* on settled expectations of safety and order.

Recognition nonetheless dawned in all three political systems that some of the risks and promises engendered by the multifaceted advances in genetics and biotechnology—particularly those associated with fears of "playing god"—called for a new language of deliberation, geared to the analysis of human values rather than the benefits of the market, the facts of science, or the norms of law. The language that actors seized on for this purpose was a branch of moral philosophy, specifically, *bioethics*. Combining life (*bios*) and moral custom (*ethos*) in a single portmanteau word, bioethics offered the promise of bringing order and principle to domains previously governed by irrational, emotive, and unanalyzed reactions like the "yuk factor," associated with the work of the U.S. physician and ethics expert Leon Kass,[2] and encountered by Britain's Human Fertilisation and Embryology Authority over the use of tissue from aborted fetuses in IVF.[3] Bioethics was, as well, an available instrument, having established its utility in connection with earlier, less encompassing developments in biomedicine. Unlike the centuries-old traditions of science and law, however, bioethics was itself in flux as politicians and policymakers reached to it for legitimation: the effort to extend bioethics into the new fields of genetics and genomics proved to be in salient ways constitutive of the discourse itself. In this chapter, we trace some of the cross-national consequences of this fluidity by looking at the rise of bioethics as an institutionalized policy discourse, and its specific uses in deliberation on the hazards of cloning and stem cells.

Comparing the role of bioethics across three political systems poses difficult conceptual and methodological challenges that we cannot hope to lay to rest in this work. To begin with, the definition of the term remains problematic. What, after all, is bioethics, and how is it understood in different political and cultural environments? If it is not the same thing everywhere, then how can we meaningfully compare bioethical issues and debates across countries? An additional difficulty has to do with the role of the European Union, as opposed to individual member states, in the development of bioethics. Since the early 1990s, both the European Parliament and the Commission have been active in bioethical matters, engaging staff, convening expert groups, commissioning reports, and more generally placing bioethics on the European political agenda.[4] The resulting diffusion of bioethical discourse across national boundaries makes it hard to show to what extent national debates matter in defining national policy, or to claim autonomy for national approaches to bioethics.

Without fully resolving these quandaries, and without undertaking a full-scale cross-national ethnography of bioethics, I note that some of the difficulties may be managed, or reduced in importance, by adopting an actor-centered comparative strategy. The task then becomes to show how each state sought to institutionalize bioethical discourse for particular functions, and how civil society actors in each country understood and strategically intervened in national bioethics debates. Using this approach, I develop three sets of comparisons.

First, I review the growth of "official bioethics" in each country. This part of the chapter focuses on bioethics as an instrument of public policy—that is, on bioethics as defined and employed by deliberative bodies devoted to public ethical inquiry. Of particular interest here is the interplay between institutional mechanisms and the development of bioethics itself as a more or less formal and disciplined analytic discourse. This section therefore also looks at the backdrop of older bioethical debates against which the new official bodies constituted their identity and legitimacy—and it examines how the composition and remit of these institutions helped frame the discussion of ethical and moral issues. Consistent with the book's focus on science and democracy, we ask as well who represents bioethics in each country: put differently, to whom does official bioethics give voice, and with respect to which sets of issues?

Equally important is to look at the uptake of bioethics as a strategic discourse by nonstate actors. The second part of the chapter accordingly looks at "unofficial bioethics." By this, I mean the array of efforts by civil society groups and actors, including industry and other nongovernmental organizations, to seize the discursive lead on bioethics so as to advance their normative goals with respect to biotechnology. Cross-national comparison of these initiatives shows variations in the values assigned by different national actors to different ethical questions surrounding genetics and biotechnology. This analysis underlines the decentralized character of bioethics as a discursive medium, as yet lacking the kinds of disciplining institutions that operate to produce shared ways of knowing in both science and law. The entry threshold for "speaking bioethics" is therefore lower than for other established policy discourses: interested actors are still in a position to invent and reinvent bioethics to some degree, in keeping with their desired ends.

The last part of the chapter examines how bioethics fared in addressing the related questions about cloning and stem cell research that erupted across Western societies at the turn of the twenty-first century. For each country, we will look most closely at the relationship between bioethical and legislative deliberations and ask how the boundaries between the moral, political, and scientific domains were negotiated. We will focus on the role of bioethics in simplifying ambiguity around these challenging developments in the life sciences. Our object in these comparisons is to assess the power of bioethical discourse as an instrument for addressing difficult political choices, as well as its role in coproducing a cognitive and moral order around particular products and applications of biotechnology.

## Official Bioethics: A Disciplined Vision of Morality

Public attempts to define the term "bioethics" usually stress its indeterminacy. Thus, a 1992 report of the European Parliament took as its point of departure a 1978 encyclopedia definition of the term: "Bioethics is the systematic study of human conduct towards life examined under the light of ethical values and principles."[5] For its own purposes, however, the report considered bioethics not as a discipline in itself, but as a forum of communication and reflection, open to laypeople as well as experts, on hard choices in the life sciences and technologies that affect human existence.[6] This tension between bioethics conceived as a narrow disciplinary discourse, dominated by experts in moral analysis, and bioethics as a medium of democratic deliberation permeates attempts to develop an ethics of the life sciences. For example, a 1995 report of the U.S. Institute of Medicine noted, "It must be clear that 'ethical analysis' is not a single, straightforward method, like algebra or geometry."[7] The purpose of ethical deliberation accordingly should be to clarify alternative views, not to provide clear answers.

Nonetheless, most analysts agree that two reservoirs of historical experience were central to the development of modern bioethics as conventionally understood. The first was the ancient tradition of the physician's duty to patients codified in the Hippocratic oath, a text attributed to the fifth-century B.C. Greek physician Hippocrates, which promises to hold the interests of patients paramount in the practice of medicine. The oath's simplest and most widely known injunction is to "do no harm." Many medical schools administer the oath as part of the graduation ceremony for physicians, although wordings of the text vary. In September 1948, the Second General Assembly of the World Medical Association, seeking to rebuild confidence in the medical profession after the atrocities perpetrated by the Nazi regime, adopted an updated version of the Hippocratic oath as the Declaration of Geneva.[8]

The second strand, also rooted in worldwide revulsion against the horrors of Nazi biomedicine, focuses on human experimentation. Its first formal expression was the Nuremberg Code, a set of principles drafted by an American doctor advising the prosecution in the Nazi doctors' trial at Nuremberg, and incorporated into the verdict as a ten-point code on "permissible medical experiments." The code's first and most salient principle is that the voluntary consent of the human subject is absolutely essential; "voluntary" is further interpreted as meaning free, uncoerced, informed, and based on a full understanding of the experiment's purposes and risks. In short, this provision forms the basis for the doctrine of informed consent. Like the Hippocratic oath, the Nuremberg code, too, was internationalized by the World Medical Association, through the Declaration of Helsinki adopted by its Eighteenth General Assembly in June 1964.

Together, the Hippocratic and Nuremberg ethical traditions have influenced Western discourses of bioethics in significant ways. Both traditions center on the relations between individual doctors and patients; they demand beneficence by medical experts toward lay patients and research subjects, call for self-regulation through peer control, and are most concerned with physical risks and impacts on the human body. These principles are useful for defining the permissible limits of experimentation, but they do not take account of the growing role of the state as funder and developer of biomedical science and technology.

A more subtle legacy of the Nuremberg trials was to tie the concern for ethical deliberation to the deployment of novel devices and therapies. It became over time an article of faith, particularly in the United States, that public ethical analysis is required for new developments in biomedicine. Used in this way, ethics tends to become just another assessment technique in a modern market economy. It is, like regulatory standard-setting, a filter against controversial or harmful products, but not an upstream discourse that engages with the political economy of production or questions the purposes for which we do science and develop technology. We turn now to the evolution of these principles and presumptions as official bioethics took on the challenges of biotechnology in each country.

### United States: Between Professionalization and Politics

Making codes is one thing; transforming them into forms of life, quite another. In many ways, the United States led the other countries on both fronts—serving both as an industry leader in the professionalization of bioethical analysis and as a site where attempts to regiment bioethics periodically overflowed into the messiness of politics and power.

As so frequently in America, harm to individuals was the major driver of change, in this case, harm that arose in the shortfall between ethical norms and actual practices in medical research. The Harvard anesthesiology professor Henry K. Beecher, who with strange foresight had changed his name in his youth to that of a noted New England abolitionist family,[9] did more than any other individual to bring these abuses to light. In 1966 Beecher published a famous exposé in the *New England Journal of Medicine* of twenty-two cases in which U.S. medical research had failed to live up to the standards of the Nuremberg code.[10] In each case the physician-researcher had grossly violated the principles of beneficence and informed consent. These charges hit home. Almost at once, the National Institutes of Health and the Food and Drug Administration adopted guidelines requiring the establishment of Institutional Review Boards (IRBs) to ensure peer review and informed consent in all publicly funded research.

Nonetheless, slippage between norms and practices continued to occur. The problem resurfaced again painfully in 1972, when the *New York Times* disclosed that the U.S. government had for the past forty years been conducting one of the longest-running experiments in history on unconsenting human subjects—the notorious Tuskegee Syphilis Study. Doctors from the U.S. Public Health Service had knowingly withheld treatment from hundreds

175

of poor African American sharecroppers suffering from the disease. Racial discrimination added its noxious stain to egregious violations of medical ethics. The subjects were not told the aims of the study or their role in it and were deprived of access to therapies even when they became generally available. In 1997 President Clinton formally apologized to the surviving participants of the study and pledged federal funds for a new bioethics center at Tuskegee University.

Partly in consequence of these revelations, and partly in response to other unsettling advances in new medical technologies (particularly life support systems), the turn from the 1960s to the 1970s proved to be an epochal moment for bioethics in the United States. Academically trained philosophers for the first time took up the problem of biomedical ethics, entering a domain previously dominated by their colleagues in medical research and clinical practice. Centers like the Hastings Institute of Society, Ethics, and Life Sciences (later the Hastings Center) and the Kennedy Institute of Ethics at Georgetown University, founded in 1969 and 1971, respectively, generated capacity for systematic interdisciplinary and professional reflection on ethical problems at biomedicine's moving frontiers. One achievement of these institutions was to begin to produce a pedagogical language for bioethics. In particular, the distinction between *utilitarian*, or interest balancing, and *deontological*, or based on obligation or necessity, became part of every card-carrying bioethicist's conceptual toolkit—even though these distinctions only highlighted the lines of possible conflict without showing how to traverse them.

The 1970s were also the period when official bioethics burst on the scene.[11] At the national level, the congressionally mandated National Commission for the Protection of Human Subjects of Biomedical and Behavioral Research functioned from 1974 to 1978. Its influential 1978 Belmont Report set out three basic ethical principles—codified as respect for persons, beneficence, and justice—for the conduct of research with human subjects. Slower off the ground, but enormously prolific in its thirty-nine months of existence from 1980 to 1983, was the President's Commission for the Study of Ethical Problems in Medicine and Biomedical and Behavioral Research. One of its reports, *Splicing Life*,[12] strongly endorsed gene therapy to cure inherited disease through manipulation of an individual's bodily (somatic) cells, but it took a more cautious position with respect to therapies that might alter the gametes (germline cells) and thus be passed on from one generation to the next. Supplementing the work of the two national commissions, a number of prominent ethics bodies also formed in executive branch agencies and in the states of New York and New Jersey. These included the Recombinant DNA Advisory Committee, formed in 1975 in response to the Asilomar conference (see chapter 2), and, more than a decade later, the Ethical, Legal, and Social Implications (ELSI) Working Group of the National Center for Human Genome Research (NCHGR).

The activities of bodies like these helped to frame and institutionalize certain canonical features of U.S. scientific and bioethical practice. First, the regulatory thrust of bioethics came at the point of delivery of the experimental medical product or service, where risk was the primary concern, so that patients were perforce given little say about broader questions of technological design or production. Second, and not unrelated, translating abstract ethical principles into increasingly routinized practices of informed consent reinforced the focus on the individual patient as the chief beneficiary of innovation and, by extension, of bioethical analysis; patients became in effect front-line consumers of bioethics. At the same time, this approach constructed the "good patient" as one prepared to engage knowledgeably with the therapies on offer, and to refuse them, if need be, on utilitarian calculations of risk and benefit. Third, the principle of informed consent underscored lay-expert divisions even while it sought to empower patients: the expert's duty of care formally complemented the patient's right to information, but the very idea of beneficence set up a seemingly unbreachable hierarchy between the caring expert and the cared-for patient. Finally, by producing a standard terminology for ethical analysis (as, for instance, in the Belmont Report), these processes did much to authorize a new class of professional bioethicists, both academics and practitioners, who threatened to take away from lay patients and consumers the capacity to deliberate on the values at stake in biomedical research and development—let alone in wider domains of agricultural and environmental biotechnology. On the whole, these tendencies pushed official bioethics in the United States toward utilitarian considerations—based on balancing risks and benefits—in ways that were compatible with the nation's pluralist political traditions, as well as with the market-oriented, product framing of the regulatory issues around biotechnology.

These constraining aspects of bioethics, and associated pragmatic problems of implementation and enforcement, generated a substantial critical literature, which deserves study in its own right.[13] One powerful strain came from anthropologists who found Western bioethics worse than useless in confronting the ills and troubles of medicine in other cultures. Even in its cultural birthplaces, however, efforts to discipline value debates through institutionalized ethical discourse proved to be less predictable than its most ardent proponents perhaps had wished.

The ELSI component of the Human Genome Project (HGP), the largest publicly funded bioethics program ever created,[14] provides a case in point. By most accounts, James Watson, the codiscoverer of the structure of DNA and charismatic first director of NCHGR, casually conceived the idea of ELSI during a press conference.[15] In answer to a question about how HGP would approach ethical issues, Watson promised that 3 percent (subsequently 5 percent) of the program's budget would be set aside for research on its ethical, legal, and social implications, and he subsequently sold the idea to Congress as a fait accompli. The resulting initiative had two components: an extramural research program administered by NCHGR's ELSI office, and an ELSI Working Group with broad but poorly defined oversight responsibilities.

In the eyes of close observers, this origin story (which deserves a more critical reading than I can give it here) protected ELSI in the short run because of Watson's extraordinary personal mystique, but it also contained within it the causes of ELSI's future vulnerability. In the eyes of a chronically hungry biomedical research community,[16] funding for the HGP was a zero-sum game: the 5 percent set-aside for ELSI was money taken away from research.[17] During Watson's tenure as director of the National Center, ELSI nonetheless attracted little adverse notice. The Working Group, consisting initially of just eight members active in the field from the earliest genetics and ethics debates, had no formal mandate and little opportunity to make political waves. Watson himself did not seem to take ELSI very seriously. For him ELSI was a sideshow compared to the "real" HGP constituted by the 95 percent of funds devoted to research.

Things changed with the appointment of Francis Collins to head NCHGR in 1993, after Watson's precipitate departure on what many saw as exaggerated charges of financial conflict of interest. Collins's appointment coincided with a changing of the guard in the Working Group, as most of the original members left and were replaced by a new cohort of twelve, including several prominent analysts of genetics and society. A micromanager by temperament, Collins was not convinced that the ELSI program had served the policy needs of the HGP: for him a major flaw of the program was its failure to deliver a genetic privacy bill that would have resolved, among other matters, researchers' worries about access to the billions of tissue samples stored away in existing data banks. His solution was to draw ELSI into the Center's general management and to appoint a new policy adviser, Kathy Hudson, who quickly got into trouble with the Working Group over her opposition to ELSI's involvement in the hot topic of behavioral genetics. Attempts to assert its independence led to open confrontation between Collins and Working Group, and the resignation of its chair, the well-known lawyer and commentator Lori Andrews.[18]

To calm the storm, Collins appointed a committee to review the ELSI program's mission and structure. A report submitted in December 1996 strongly endorsed the program's continuation as a multifaceted, policy-relevant research enterprise, but it also handed Collins the mandate he wanted to dissolve the ELSI Working Group.[19] The reviewers recommended distributing the group's activities among different levels of governance within the relevant agencies, the National Institutes of Health and the Department of Energy. Oversight of grants, coordination of ELSI activities throughout NIH, and policy advice to the secretary of health and human services were henceforth to be split up among different committees. The first two parts of this recommendation were acted upon in 1997; perhaps predictably, the more political third recommendation was not implemented in the proposed form.[20] Instead, the DHHS secretary appointed a narrower advisory committee on genetic testing, aptly abbreviated ACGT for the four bases of the DNA double helix.

At about the same time, NIH Director Harold Varmus also floated a plan to dismantle the twenty-year-old Recombinant DNA Advisory Committee, but he retreated in response to widespread criticism. While some saw RAC as needlessly duplicating peer review of individual studies, others regarded it as providing essential oversight in controversial areas of biomedical research, such as germline gene therapy and behavioral genetics.[21] RAC remained, but for our purposes this outcome was less important than the reasons for it. RAC was saved because its services were deemed invaluable in managing the risks of gene therapy. The stakes for biomedicine were not negligible. In 1999 the tragic death of eighteen-year-old Jesse Gelsinger, a patient at the University of Pennsylvania, sent a discouraging message about the future of gene therapy.[22] Substandard ethical practices were partly blamed for Gelsinger's death. Not surprisingly, the role of ethics in establishing acceptable levels of risk to individuals or groups remained the strongest justification for institutionalized bioethics at NIH.[23]

The push to create a national discourse on bioethics ran into political complications on other fronts as well, with grave consequences for biomedical research. Thus, a congressional move in the late 1980s to establish a Biomedical Ethics Advisory Committee to succeed the President's Commission foundered because of sharp divisions over abortion; since then, Congress has chartered no other national bioethics bodies.[24] An unintended consequence of the committee's demise was to put all federal funding of embryo research on hold, since the body that was to have approved such projects did not exist. In 1995 Congress formally banned such research through an appropriations bill for Health and Human Services; the Dickey-Wicker amendment to the bill provided that no federal funds may be used for research in which an embryo is "destroyed, discarded or knowingly subjected to risk of injury or death."

The initiative to centralize bioethics passed to the White House during the Clinton administration. In October 1995 President Clinton issued Executive Order 12975 to establish the National Bioethics Advisory Commission (NBAC), chaired by Harold T. Shapiro, president of Princeton University. Legally constituted as a federal advisory committee, the fifteen-member body was asked to identify broad principles governing the ethical conduct of research. The cloning of Dolly in February 1997, and the realization that human cloning might not be far behind, placed a more urgent issue on NBAC's agenda and raised the council's public visibility. The June 1997 report, *Cloning Human Beings*, was easily NBAC's most widely watched undertaking, although a 1999 report on ethical issues in stem cell research also attracted attention.[25]

NBAC's charter expired in October 2001. Two months later, by Executive Order 13237, President George W. Bush appointed his own ethics body, the President's Council on Bioethics (PCBE), "to undertake fundamental inquiry into the human and moral significance" of developments in biomedical and behavioral science and technology. The seventeen-member body, headed by Leon Kass, a physician from the University of Chicago, was generally seen as leaning toward the conservative side of the political spectrum; Kass himself was known for his conviction that modern science had lost its moral bearings and for his cautious views about research on reproduction and longevity. It was a controversial appointment because any reservations abut the desirability of research, let alone conservative-tending visions like Kass's, were unpalatable to the biomedical research community, as well as to feminists and political liberals. In appointing the committee, the president stressed the panel's educational mission, which he defined in expansive terms: "It'll help people like me understand what the terms mean and how to come to grips with how medicine and science interface with the dignity of the issue of life and the dignity of life, and the notion that life is—you know, that there is a Creator." The council, he iterated, would perform these services for him personally, for the country, and for the world, because "we're leading on a lot of fronts. And this is another front in which this country can lead."[26] PCBE's views would become clearer later that year through its first report, *Human Cloning and Human Dignity*.

### Germany: Foundational Questions

The history of bioethics, as we have seen, was tightly woven into Germany's history in the twentieth century. The topic stands as a barely healed scar on the national conscience. Not surprisingly, all attempts to incorporate bioethical discourse into official decision processes have been marked by that troubled and tragic relationship. Two features of Germany's political wrestling with bioethics emerge as particularly notable in comparison with the United States: a relatively greater emphasis on questions of social and collective good as opposed to individual risk, and a preference for deontological rather than utilitarian arguments in developing official policy options.[27] These orientations, however, are not set in stone. The German experience shows how, in the continual performance and reperformance of political culture, actors can either give renewed voice and expression to certain widely held cultural commitments or, under appropriate circumstances, find opportunities for reimagination and release from rigid fundamentalisms.

The early 1990s were a difficult period for bioethics in Germany. At a time when other countries were grappling with the first commercial successes of biotechnology, German politicians and citizens were preoccupied with the overwhelming question of national identity, centering on the terms on which to bring together the fractured pieces of the country's postwar settlement. As we saw in chapter 6, a discourse of Western triumphalism swept other things in its wake, inducing an absolutism in defining the status of the embryo that feminists and Social Democrats had to accept as the price of national unity. A similar absolutism marked ethical discussion on other biomedical issues as well. The controversies surrounding the Australian moral philosopher Peter Singer in and immediately after 1989 provide a telling illustration.

Singer, employed at Princeton since 1999, is best known for his advocacy of animal rights, but also for his outspoken support of the view that deeply compromised human lives are not worth living. One consequence of this conviction is Singer's tolerance for euthanasia and infanticide under some circumstances. For instance, he has argued (by no means alone) that parents of severely defective newborns should be free to decide, with the aid of their physicians and without state interference, whether such babies should live or die. These views are contested everywhere, but they have also been widely debated. Indeed, in the month when he took up his Princeton appointment an article in the New Yorker described Singer as possibly "the most controversial philosopher alive" and "certainly among the most influential."[28] Even ten years before, with his 1979 book Practical Ethics[29] translated into several European languages, Singer was already a well-known figure on the international lecture circuit. But nothing presaged the commotion that greeted his visit to the German-speaking world in 1989.

As Singer recounted in an article in the New York Review of Books,[30] news of his arrival caused such an outcry that all but one of his planned German appearances, a seminar at Saarland University in Saarbrücken, were called off; even the Saarbrücken event was disrupted by whistles and catcalls from the floor until the organizers asked members of the audience to come up to the microphones and explain their objections. Other philosophers representing positions similar to Singer's were also barred from various academic gatherings around that time, and, following student protest, an advertised position in practical ethics at the University of Hamburg was offered to a specialist in aesthetics. Singer's most unpleasant personal experience occurred in Zurich in 1991, where cries of "Singer raus" ("Singer out") reminded him of the earlier chant "Juden raus" and resulted in the lecture being canceled.

Singer was astonished at the violence of these reactions and ruefully concluded that German ethical discourse was too underdeveloped to support the nuanced moral deliberation that had gone on for some years in the English-speaking world on questions of life, death, and biomedicine. Yet a conversation he had with his German hosts on the evening of the Saarbrücken seminar hints at a more complex reality, in which two philosophical traditions failed to connect.[31] At one point, the interviewers asked Singer whether he did not feel he had raised a red flag by championing the hateful concept of "lebensunwertes Leben"—used by the Nazis to mean "life not worth living," and hence life the state could terminate at will. Singer replied that the translation was unfortunate, because he wished to distinguish a "life not worth living" from a "life unworthy of being lived." The former, he said, was to judge the quality of life from the viewpoint of the subjective individual living that particular life, whereas the latter implied the existence of a seemingly objective standard by which a fascist state might seek to end lives it saw as not worthy of being lived.

Singer wanted to preserve the individual's right to engage in the first calculus, and on that basis to permit the termination of lives subjectively experienced as unrewarding, without remotely sanctioning the state's embrace of the second. In so doing, however, he sidestepped a number of questions that his German interlocutors clearly thought important. One illuminating exchange centered on breaking taboos, such as that against euthanasia in Singer's work. Asked whether such taboo-breaking might not lead to unknown adverse consequences, Singer energetically asserted that it is always a good idea to shed light on official taboos lest unquestioned dogmas foster abuse; philosophy's task, he claimed, was to tear away the veil from rhetorical obfuscations like "letting nature take its course," which only hide a refusal to assume personal responsibility for one's actions. This espousal of reason against custom, accompanied by confidence in the power of skeptical inquiry to make correct distinctions and to unmask false argument, amounted to a recitation of a philosopher's credo.

Singer's belief in philosophy's ability to identify the right categories of thought reminds us of Baroness Warnock's invincible faith in distinguishing things that differ—and it was a product of the same analytic philosophical training favored in the Anglophone world. Not asked in Singer's scheme of things were nagging questions about the historical origins and social underpinnings of what he saw as absolute moral mandates. Where, for instance, does a person's subjective awareness of life's unworthiness originate, and how does it relate to wider social values concerning worthy and unworthy forms of life? How should we interpret the fact that people offered a legal right to end their lives sometimes choose not to, and how should law or morality respond to the fact that survivors of suicidal falls from bridges report they wished at the last moment they had not jumped? More historically minded analytic philosophers, such as Ian Hacking, Alasdair MacIntyre, and Richard Rorty, let alone Jürgen Habermas or other continental philosophers, might have dealt with the distinction between subjectivity and objectivity in a less absolutist way than Singer did in Germany; possibly, these philosophers would also have confronted Singer's German interlocutors with something closer to mutual understanding.

Native German philosophizing produced its own ethical controversies as well. The most dramatic episode in the postreunification period grew out of a lecture entitled "Rules for the Human Zoo" (*Regeln für den Menschenpark*) given in July 1999 by Peter Sloterdijk, considered by some to be Germany's greatest living philosopher. In this talk at a closed conference in a Bavarian castle, Sloterdijk bemoaned the growing barbarism and bestiality of human civilization and looked ahead to a bio-utopia in which genetic engineering might be used to purge human beings of their worst characteristics. Breeding and selection (*Selektion*) might lead to a better class of human, he suggested, somewhat on the model of animal husbandry. In an appearance at Harvard the following spring, Sloterdijk noted that the Nazi period had caused a rupture in an earlier tradition of progressive eugenics; that dark history, he suggested, should not be allowed permanently to cloud a more positive vision of human genetic manipulation.

Sloterdijk's apparent support for planned alterations of human genetic identity, his taboo-defying language of breeding and selection, and his approving references to such politically tainted philosophers as Martin Heidegger provoked a firestorm of criticism in leading German newspapers and magazines. Sloterdijk added fuel to the fire by penning an open letter to Germany's liberal philosophical icon Habermas, accusing him of having rounded up third parties to attack Sloterdijk on his behalf, thereby betraying Habermas's own cherished ideals of democratic discourse, and even of having seeded the debate with a pirated, error-laden text of Sloterdijk's talk. Though titillated by this clash of intellectual titans, many commentators saw the dispute as having more to do with German history than with the future of biotechnology.[32] They noted that, like the so-called *Historikerstreit* that broke out in the mid-1980s,[33] this controversy too revolved around the question of how to commemorate the twelve years of Nazi rule—whether to see it as an aberration to be set aside or as continuous with the rest of German history and culture.[34] Was Nazism an exception from which one should not generalize, or was it rooted in everyday features of German social life, in which case—and this was the most fearful thought—could something like it once again rear its monstrous head unless people maintained a steadfast watch on Germany's new democratic institutions and stamped out any attempts to transgress the stark commitment to human dignity that underpins them?

In short, German public intellectuals were intensely aware that bioethics and genetic engineering were but two fields among many in which the supreme question of German identity was still working itself out. Bioethics was just one more context, albeit a particularly sensitive one, in which there was an urgent need to find answers to such profound questions as the appropriateness of public language, the status of discursive taboos, the role of academic disciplines (including philosophy and the life sciences) in governance, the relationship of Nazi practices to earlier and later social traditions, and, above all, the meaning of being German in the momentous period of reunification and the end of the millennium. On all these issues, Germany's vibrant but deeply wounded intellectual culture displayed a degree of painful self-reflection that was unmatched in the more robust, but also more complacent, political environments of the United States and Britain.

Official bioethics had to construct its identity on this field of contention. The task of ethical deliberation in Germany as in the other two countries was to make the way safe for regulated advances in the life sciences and biomedicine while respecting core commitments to human dignity and integrity. German politicians of all major parties were as firmly convinced as their counterparts in other countries of the need to invest in and harness biological knowledge for competitive and productive ends. The challenge was to persuade the German public that state policy could accountably cope with the risks and opportunities, and moral ambiguities, presented by these novel sciences and technologies. An important step toward achieving credibility was through consultation with appropriate advisory bodies, and in Germany's parliamentary system the Bundestag controlled an important instrument for soliciting advice: the Enquete Commission. As seen in chapter 2, one such commission, on the "prospects and risks of modern medicine," had laid the groundwork for Germany's 1990 Genetic Engineering Law. In March 2000 the Bundestag created another body, the twenty-six-member Enquete Commission on the Law and Ethics of Modern Medicine, comprising thirteen elected representatives and thirteen independent experts.[35] The commission was charged with three tasks: to evaluate the ethical, constitutional, legal, and political implications of developments in medical research and therapy; to identify unregulated areas of research practice; and to establish criteria for the use and applications of biomedical science consistent with the protection of human dignity.

One sees in the commission's charge the anxiety about unregulated spaces that is a persistent feature of Germany's legal and political culture. In such unprotected corners, disorder may thrive: citizens may be exposed to risk, health needs remain unmet, human dignity be violated, and the state itself prove derelict in its constitutional obligations to serve the public good. Ethical and legal concerns go hand in hand in such an environment, for the identification of situations posing ethical problems almost necessarily calls up the power of the state to remedy the problem. The laissez-faire approach of the U.S. ELSI Working Group, which allowed a research program to thrive for years with little if any connection to legislative policy, would have been difficult to imagine in Germany. The Enquete Commission's constitutional duty was to inform the lawmaking process.

But just as Francis Collins evaluated ELSI's effectiveness from the standpoint of the highly competitive U.S. biomedical research establishment, so the German scientific community and its supporters in government also understood that bioethics was too important to be left to the vagaries of parliamentary politics. The Enquete Commission process had not served the needs of the research community all that well in the late 1980s in the run-up to the 1990 law, and, especially in the light of the Embryo Protection Law's absolute prohibitions on research, there was no reason to expect a new parliamentary commission to be much more sympathetic to science's interests. Responding to these worries and mindful of Germany's lagging technological competitiveness, the Schröder government in May 2001 took the unusual step of appointing a second ethics body, the National Ethics Council (Nationaler Ethikrat), "as a national forum for dialogue on ethical issues in the life sciences." The government declared that this council should serve as the central organ for promoting interdisciplinary discourse in this field and should evaluate the ethical aspects of new developments in biomedicine for individuals and society. The council's twenty-five members were drawn from the natural sciences, medicine, theology, philosophy, sociology, law, and economics; with the exception of one member of a patients' organization, no explicit representation was provided for laypeople or other political interests.

Creating such a body through cabinet action was seen as unusual by many and constitutionally problematic by some, although the Benda commission of the 1980s was cited as a precedent for executive action on bioethics. How the new body would define its relationship to the existing parliamentary commission remained a particular puzzle. A column in the liberal paper *Die Zeit*, for example, noted the peculiarity of the cabinet directing the Ethikrat to advise the Bundestag, and said that parliament should refuse the patronizing offer if it possessed even a spark of self-respect.[36] The stem cell debate, which we turn to in the last section of this chapter, would pose these questions in more pointed form.

### Britain: Privatized Power, Veiled Values

Official bioethics in Britain was notable for its far from official beginnings. By the late 1990s many British observers were deploring the absence of a national body for bioethics. The positive legacy of the Warnock committee and the perceived need for quick and integrated responses to ethical questions around new scientific developments provided part of the incentive for such a body. Important, too, was the desire for an effective counterweight to European developments in bioethics, which U.K. observers worried might unduly constrain British research unless a strong voice was mustered to represent Britain's position in Europe. The problem was where to locate such a body and how to support it. Margaret Thatcher, prime minister until 1990, declared her firm opposition to a governmental niche for such an entity, possibly fearing the influence of an official voice of conscience; a standing committee on the model of the Royal Commission on Environmental Pollution was therefore ruled out.[37]

Other actors stepped in to fill the void left by the state. In 1988 some senior scientists and research administrators approached the Nuffield Foundation, one of Britain's best known charitable trusts, to seek its help in creating a body for bioethics. In April 1990 the foundation sponsored a small conference of about thirty participants; in a familiar British consultative mode, not only scientists and professional society members were invited, but also representatives of law, philosophy, theology, the media, consumers, and public interest groups.[38] A consultation document issued at the end of the meeting reported the participants' conclusion that there was need for a new national bioethics council.[39] It proposed criteria for membership on such a body, including that members should be chosen for their expertise, on the model of the U.S. President's Commission, but that they should serve in person and not as institutional representatives. After several months of consultation, in which opinion was solicited from some 175 organizations and 60 individuals, a small steering group unanimously recommended to the foundation's trustees in December 1990 that the proposed body be formed. The Nuffield council on Bioethics was created in that very month. The government welcomed the move and asked the council to take on board not only ethical issues in biomedicine but also, more broadly, ethics in food, agricultural, and environmental biotechnology. This was a much broader agenda than was assigned to any of the U.S. national commissions and perhaps most comparable to that of the first German Enquiry Commission on the risks and prospects of genetic engineering.

The Nuffield Council's membership, remit, and source of authority reinforced one another. The original consultation document clearly intended to privilege nonscientific expertise among the council members. One sees in this another expression of the British urge to create authorized bodies who can be trusted to see correctly for the people. The proposed criteria stated that "a majority (however bare) of the members should be lay in the sense of being neither professional scientists nor clinically qualified," and that the "Chair should be lay in the same sense, possibly an eminent lawyer." The word "lay" did not appear in later documents, however, although it was accepted in principle that neither the chair nor a preponderance of the members should be scientists or clinicians. The steering group's report to the trustees suggests that the desire to constitute the council as a credible expert body, with the standing to commission studies, carried the day, taking precedence over worries that it might become too professional. The report did take note of the philosopher Onora O'Neill's "cogent" observation about the need for lay representation: "I sense a danger in some discussions of bioethics to underestimate many people's insecurity and vulnerability in the face of these problems. Lawyers, economists, people working in ethics are all accustomed to making quite generous assumptions about human rationality and independence. I don't know that anybody specialises in being perceptive about mundane vulnerability in the face of medical or genetic risk, to self or to others."[40] The steering group concluded that this was a valid concern, but that committee membership was not the right way to address it.

Instead, questions about membership fed into concern about the sources of the council's authority, since it would have no formal powers, regulatory or otherwise. Authority, its proponents concluded, would have to flow in the first instance from "the standing and quality of the individual members and staff."[41] Increased authority would come in time from "the value of the work done," its public impact, and the council's independence. In other words, building on the embodied virtue of its individual members, the collective body of the council would gradually acquire greater trust and acceptance: recognition would lead to more recognition and influence would breed influence.

Creating a trustworthy body was essential not only because the council would have no statutory powers, but also because of its wide-ranging mission. If U.S. policymakers saw the need for bioethics largely as an antidote to perceptions of risk by patients, and German politicians wanted bioethics to ensure the protection of human dignity, then Britain's scientific community saw bioethics first and foremost as a device for safeguarding a space for research—a space that was felt to be under threat from encroaching European anxieties. A background paper prepared for the April 1990 meeting noted that bioethics in the European context was subject to four influences, of which only one—the desire for competitiveness—was supportive of science; on the negative side were the views of the Catholic Church on embryology and German concerns about eugenics and genetic manipulation. Small wonder, then, that the council's tasks included, at the head of the list, responding to and anticipating public concern. The role of institutionalized bioethics in Britain was going to be to educate, to head off public anxiety before it materialized or migrated in from Europe, to promote public understanding (a phrase more typically used in conjunction with the public understanding of *science*), and to build confidence. This conception of the council's role was impossible to reconcile with the representation on the body itself of what Onora O'Neill called "people's insecurity and vulnerability." It was, at the same time, perfectly in keeping with a wider definition of ethics than in the other two countries, encompassing the process of genetic modification in all its applications.

Upon beginning work in 1991, the Nuffield Council took up genetic screening and human tissue as its first two topics. Since then, however, it has issued reports or position papers on a wide range of controversial subjects, including GM crops, DNA patents, genetics and behavior, xenotransplantation, and stem cell research, as well as the ethics of health care research and GM crops in developing countries. Its choices continue to be guided by the principles of anticipating public concern and not duplicating the work of other bodies. In the latter respect, however, the council's position changed dramatically with the addition of three new bodies to the British government's regulatory and advisory framework for biotechnology in 1999: the Human Genetics Commission, the Food Standards Agency, and the Agriculture and Environment Biotechnology Commission. All three have remits that substantially overlap with the Nuffield Council's terms of reference, most notably with respect to gaining public confidence and taking broad social and ethical issues into account. Only with this burst of institutional activity can we say that official bioethics in Britain came fully into its own. Perhaps the Nuffield Council will be seen in retrospect as an incubator or agent of *social* technology transfer, moving ethical analysis from the ambiguous format of private deliberation on the public good to a more visible and accountable position in the public domain. Seen in this way, the council exemplifies an empiricist culture's willingness to experiment with ways of governing itself. Put differently, the Nuffield Council instantiated the British national virtue of "muddling through."

## Speaking Ethics to Power: Civil Society's Role

As bioethics became a powerful instrument for framing policy choices around biotechnology, numerous social actors saw the advantages of participating in the development of the discourse, thereby challenging the state's (or, in Britain, the professional elite's) monopolistic control of the formal analysis of values. In some cases, groups formed to broaden or reshape the agenda of bioethics, pressing for ethical analysis of particular issues and developments; in others they fought for standing to insert their own ethical perspectives into policy debates. In each country, a major goal of nonstate actors seemed to be to use the rubric of ethics to create new deliberative spaces, and to some extent new languages, supplementing those offered by official policymakers. The politics of bioethics was a politics of diversification: bringing new issues on the agenda, new voices to speak for them, and new forums in which to engage with them. Bioethics, no matter how construed or where deployed, seemed in this way to overflow institutional attempts to contain it. Yet, the specific modalities of civil society intervention varied across the three countries, illustrating once again the continued influence of the early national framings of biotechnology as a matter of political and social concern.

### United States: Resurgent Interests

Official bioethics in the United States was, as we have seen, the product of a marriage between the needs and interests of the biomedical research and clinical communities, the rights of patients and research subjects, and the interests of the state—a state eager to sponsor biomedical advances but nervous about violating racial and gender equality, offending religious sensibilities, and incurring liability.[42] Like all formal policy discourses, bioethics constrained even as it enabled. We saw, in particular, that official bioethics focused on the delivery of novel medical products and services to individuals and homed in on the issue of informed consent. This relatively narrow definition of bioethics left wide open many areas, from GM foods to seed patenting, in which advances in biotechnology impinged on human values, but without formal institutional means of addressing them. Curiously, the biotech industry was among the first to perceive this deficit and seek to remedy it. Many in the biotech industry saw bioethics as a promising, even necessary, instrument for enrolling people more effectively into research involving human biological materials.[43]

This is not the place for an in-depth study of corporate bioethics, but one example may illuminate some of the points that are most important for our analysis. The case of the Biotechnology Industry Organization is especially instructive. BIO was formed in 1993 through the merger of two smaller trade associations representing some five hundred companies. It united under a single organizational umbrella a handful of multibillion-dollar companies involved in product development and hundreds of small to midsized firms active mainly in research and development. BIO's stated mission was to provide information to policymakers, the media, and its own members. But bioethics was on its agenda from its foundation, and a standing committee on bioethics was formed in 1995. According to BIO's own corporate history, this focus increased dramatically in importance after the cloning of the Dolly in 1997. BIO quickly issued a statement opposing human reproductive cloning. Similarly, after the terrorist attacks of September 11, 2001, BIO reaffirmed its policy of opposing the development of biological weapons. The organization has since made it a point to respond to questions from Congress and the media about other biotech breakthroughs. It has in this fashion become "the 'voice' to the public for hundreds of biotech companies, large and small."[44]

Looking at the range of activities that BIO lists under the heading of ethics, one is struck by a radical inconsistency. An industry that has historically pooh-poohed the idea of viewing genetic manipulation as a process of concern seems to take the process quite seriously when it comes to ethics. In BIO's Statement of Principles, for example, it is biotechnology—the process, not the products—that is the subject of attention. By adopting the discourse of bioethics, in other words, BIO in effect partitions the field of public deliberation into two spheres: a regulatory sphere focusing narrowly on products, and dominated by talk of risk assessment and sound science; and a sphere of public communication, encompassing virtually all other aspects of biotechnology, controlled by industry through what it calls bioethics—a self-conscious program of informing and educating society. BIO in this respect has adopted a role very similar to that of the Nuffield Council in Britain: to encourage informed debate by advancing public awareness and understanding. Only, where the council operated as a shadow governmental body, holding itself to well-understood British procedural forms like consultation, BIO uses the techniques of advertising and electronic mass communication to disseminate its interest-driven views of bioethics to national and global audiences.

Consciously or not, BIO's strategy is consistent with a signature development of late modernity: a marked technicization and moral hollowing out of public-sector regulatory discourse, and a shunting of questions of ethics and values into the sphere of private morality and "consumer choice." But in the era of the Internet it is not easy for anyone to monopolize the means of creating alternative deliberative spaces, as we saw in chapter 5 in connection with the outcry over organics labeling. Not only U.S. industry, then, but also citizen groups have turned to the electronic medium to express moral sentiments and mobilize public opinion regarding biotechnology. Two examples will illustrate this point.

The Cambridge-based Council for Responsible Genetics (CRG) was formed in 1983 by a group of liberal scientists, physicians, philosophers, and social scientists to foster public debate about genetic technologies. Like BIO, CRG does not tie itself to the product framing of biotechnology. Since the 1980s CRG has published a bimonthly newsletter, *GeneWatch*, which is described as America's only magazine dedicated to monitoring biotechnology's social, ethical, and environmental consequences. On each of these issues, CRG's mission is to promote an antireductionist view of genetics. In this respect CRG is intellectually allied with the Oakland-based Center for Genetics and Society (CGS), formed in 2001 in response to debates over human cloning. CGS, for example, posts on its website a critique of bioethics discourse, arguing that it systematically favors individual and utilitarian values at the expense of communitarian values such as social justice and human dignity. CGS is more exclusively focused than CRG on issues of human genetics. Consistent with its narrower self-definition and its mission to promote the use of genetics in the public interest, the California center also has sought to intervene more aggressively in national and international legislative processes than the older East Coast organization.

CRG's strategy by contrast is rather more conceptual. In the spring of 2000, for example, the CRG board of directors unveiled a Genetic Bill of Rights to prompt a global public dialogue on challenges posed by the life sciences to human integrity, individual liberty, and the health of the biosphere. The council aims, more broadly, to be a source of information and analysis to other actors rather than to pursue an active policy agenda of its own. Perhaps for this reason, CRG's successes have been either local or behind the scenes; its newsletter has never attracted a wide public following.[45]

### Germany: Networked Information

The German biotechnology industry, perhaps bespeaking its more troubled position in national policy and politics, has not aggressively sought to colonize the discourse of bioethics as BIO has done. On the other hand, an organization very similar in inspiration to the Council for Responsible Genetics arose in still-divided Germany in 1985. This was the Gen-Ethics Network (Genethisches Netzwerk, GeN), a politically independent charitable organization that had its roots in a fascination with the power of computers to undercut the mass media's control over the dissemination of news.[46] GeN was also concerned with the independence of information about biotechnology, since its founders claimed that universities, the traditional source of such information, had given up their critical role in society to become part of a larger industrial biotechnology complex. GeN's bimonthly newsletter sought to provide up-to-date reports on the legal, political, and social ramifications of genetic engineering, without incurring any debts to established social and political organizations.

Detachment, however, is a recipe not only for independence but potentially also for irrelevance. Though professionally produced and highly informative, it is unclear in how many cases GeN's newsletters successfully reached or mobilized large political audiences. In the earliest years, the publication seemed mainly a channel for reporting, with some critical bite, on biotechnology-related events in the distant United States. Later, especially after the German legislative process went into high gear at the end of the 1980s, the GeN newsletters published interviews with some of the major policy actors and critical commentaries by social activists. Beginning with its very first issue, a consistent thread in GeN's reportage was the substantive and symbolic elevation of a feminist perspective.[47] In 1990–1991 GeN undertook an intensive campaign on genetic patenting.[48] Working with a coalition of NGOs, and using its newsletter to publicize the campaign, GeN collected some thirty thousand signatures for a petition against the EU patent directive, which we consider in more detail in the next chapter. More limited interventions ranged from coordinating a campaign against GM foods to opposing the use of DNA fingerprinting in a 1991 criminal trial in Lower Saxony.[49]

### Britain: Deprofessionalizing Bioethics

"Official" bioethics in Britain was, as we have seen, only a semi-official response to concerns shared by scientists, science administrators, and sympathetic elements of the "great and the good," the nation's elite. It developed in part from anxieties about European religion and irrationality crossing into British shores unless a home-grown discursive barricade was raised against these inimical views. In part, too, it was driven by what the sociologist of science Brian Wynne has called the "deficit model" of the public—as a collective needing to be informed, enlightened, and guided so that it recognizes and acts in its own best interests.[50] It is not surprising that the response to official bioethics, whether by academics like Wynne or by other social actors, has been to resist the professionalization of this discourse and to reestablish ethical deliberation as a field of democratic engagement, accessible to ordinary people as well as experts.

A complex topic in its own right, these attempts to democratize bioethics can only be gestured at in our present discussion, but one or two examples are usefully illustrative. A powerful constellation of voices at Lancaster University has argued that ethical analysis should not be added on to the back end of regulatory evaluations or medical treatment as simply another technique of expert assessment. One reason for caution is that this approach to decisionmaking perpetuates a fact-value distinction that does grave disservice to the complexity of people's ethical judgments. As Wynne has persuasively shown in his own work, how people evaluate the ethics of technological developments, such as agricultural GMOs, inevitably also entails intellectual judgments about the limits of science, the fallibility of institutions, and the scope of societal ignorance.[51] Assessments of facts and values are thus thoroughly intermingled and inseparable. Others such as Lancaster sociologist Phil McNaghten have criticized both the dominant utilitarian and deontological traditions in bioethics for failing to grapple with new categories of harms, for example, in the case of GM animals—separating them from their "natures" through species-crossing or other disruptive techniques.[52] Put more generally, to the extent that biotechnology frequently involves kind-making (in Hacking's terms), ethical deliberation seldom seeks to plumb people's attitudes to these new ways of world-making—or remaking. The dialogue on this issue prompted by British academics is one of the few occasions when the tensions between expert bioethics and deliberative bioethics mentioned at the outset of this chapter have achieved explicit, public articulation.

If British academics have been at the forefront of questioning the meaning of ethics in the context of biotechnology, other organizations have sought more to open up space for lay participation in bioethical debates. One example is GeneWatch UK, a public interest group that uses web-based briefing and communication tools to "ensure that genetic technologies are developed and used in the public interest and in a way which promotes human health, protects the environment and respects human rights and the interests of animals." To this end, the organization produces briefing papers on key developments in biotechnology. In 2003 GeneWatch offered information on how to get involved in the public debate on GM foods (called GM *Nation?*) organized by the British government (see chapter 5); it also promised to monitor the consultation exercise, serving in this way almost as a consumer watchdog over biotechnology's unfolding political process.

## Bioethics on the Research Frontier: The Stem Cell Debate

As the old millennium wound to an end, biomedical science announced that it was on the threshold of a discovery that promised to unlock the secrets of youth.[53] The miracle-worker was the human embryonic stem cell. These pluripotent cells (unlike the more specialized adult stem cells in the fully grown human body) have the potential, when transplanted, to develop into virtually any cell type. They therefore hold enormous promise in treating diseases or injuries that require cell regeneration, such as Parkinson's, Alzheimer's, heart disease, and kidney failure. In late 1998 leaders of two different research teams in the United States announced that they had successfully kept alive cultures of human stem cells in their labs. They were John Gearhart of the Johns Hopkins Medical School and James Thomson of the University of Wisconsin, Madison. Their groups' work had been partly funded by Geron Corporation, a private company dedicated to aging research. It would not have been eligible for public funding under existing U.S. law.

Earlier research on the mouse had already shown the capacity of ES cells to develop into numerous types of tissues, including muscle, cartilage, bone, teeth, and hair.[54] The establishment of human ES cell lines opened the door to comparable advances in understanding and eventually aiding the process of cell generation in people. Scientific excitement at this news was tempered by serious ethical misgivings, linked to the sources from which the cells were derived. Four sources existed in principle: tissue from fetuses following elective abortion; "spare" IVF embryos not used in fertility treatment; embryos created from gametes donated for research; and (potentially) asexually created embryos through the process of cell nuclear replacement (CNR), in which the nucleus of an adult human somatic cell is transferred into an enucleated ovum.[55] Gearhart's lab had used embryonic germ cells (precursors of the egg and sperm) from therapeutically aborted fetuses; Thomson's had acquired theirs from unused embryos produced by IVF and donated for research by consenting couples.

Research with ES cells raises thorny ethical questions because it may involve the destruction of embryos, and therefore of what some see as potential human life; because embryos may be intentionally created for research, violating notions of human dignity for some; and because such research may encourage ancillary practices that some find morally repugnant, such as abortion or selective breeding of embryos for research and therapeutic purposes. In sum, the sources and uses of ES cells in research touch on some of the deepest moral and political divisions in Western societies, while speaking to some of their profoundest hopes and fears. It is therefore no surprise that the issue quickly made its way into national bioethical and political forums in all three countries of concern to us. In each country, however, it was not so much ethical deliberation that clarified the choices for politics as the dynamics of politics that shaped the discussion of bioethics.

### United States

Bill Clinton reacted to the stem cell reports with the same speed and energy he had displayed in 1997 on the cloning of Dolly. In November 1998 he asked NBAC to review the medical and ethical issues associated with human stem cell research. An optimist in technological matters as perhaps in all things, Clinton clearly intended the commission to help transfer scientific knowledge in this exciting research field as quickly as possible to the domain of medical therapy. When the commission reported back to him just ten months later, Clinton expressed the hope that "one day, stem cells will be used to replace cardiac muscle cells for people with heart disease, nerve cells for hundreds of thousands of Parkinson's patients, or insulin-producing cells for children who suffer from diabetes."[56] It was a typical pronouncement from a president whose vision of government, like his vision of stem cells, was as universal mother's helper, ready to reach in and rectify any and every day-to-day human misery.

NBAC did not disappoint its sponsor and prime cheerleader. The commission took the bull by the horns and declared that the congressional ban on the use of federal funds for embryo research was inconsistent with the bioethical principle of beneficence and medicine's commitment to healing, prevention, and research. The commission saw no ethical justification for distinguishing between the derivation and use of ES cell lines. It recommended that research using cadaveric fetal tissue and ES cells from spare IVF embryos should be eligible for federal funds under certain conditions, including oversight by a new stem cell review panel. Among the duties of such a panel would be to ensure that the tissues and cells had been obtained according to appropriate protocols, a role similar to that played by the HFEA in Britain. Accordingly, NBAC called on Congress to rescind in part its ban on federal funds for embryo research. Public funding and public review, NBAC concluded, were better for society than to continue this kind of research with only private funding and no public oversight. In short, NBAC favored giving a cautious and well-monitored green light to stem cell researchers, a position in line with twenty years of collaborative development between medical research and bioethics in the modern American administrative state.

At another time in U.S. history, NBAC's recommendations might have slipped into law and policy without great controversy. Indeed, in late 1999, under its popular and politically astute director Harold Varmus,[57] NIH issued cautious new rules for funding stem cell research. NIH would not support the derivation of cells lines from embryos (contrary to the position taken by NBAC), but it would fund research with privately derived cell lines, as well as both derivation of and research with cell lines from fetal tissue.[58] But normal politics was soon to go into remission in the United States. By December 2000 George W. Bush had been declared president over his rival Al Gore, who had won the popular vote. Keeping the Republican right happy was seen as key to the new president's success, and no issue so energized that segment of the party as abortion. The PCBE was appointed in the shadow of September 11, 2001, before the military victories that made the president politically all but invincible. Party political concerns were reflected in the choice of the chair, Leon Kass, and the members, most of whom were white men and many of whom were considered to be of conservative leanings. PCBE immediately took on the task of revisiting territory already covered by NBAC, choosing cloning as the topic of its first report.

As expected, PCBE unanimously opposed reproductive cloning. A 10 to 7 majority also recommended a four-year moratorium on therapeutic cloning—an outcome that neither pleased nor greatly surprised the medical research community. It was, however, at the discursive level that the council made its most original contribution. In the executive summary, the council indicated that it had chosen to avoid the standard terminology used in prior cloning debates. Instead, the report stated: "We have sought terminology that most accurately conveys the descriptive reality of the matter, in order that the moral arguments can then proceed on the merits. We have resisted the temptation to solve the moral questions by artful redefinition or by denying to some morally crucial element a name that makes clear that there is a moral question to be faced."[59] Acting on these principles, the council translated reproductive cloning into "cloning to produce children" and therapeutic cloning into "cloning-for-biomedical-research."

Several things can be said about this discursive shift, and all are important from the standpoint of understanding the framing power of bioethics. First, one is struck by the asymmetry of the council's analysis of language: its own simply relays the "descriptive reality of the matter," whereas others' is "artful" and distorting of reality. This representation of the speaker's position as empirically supported, hence objective, and the opponent's position as contingent, hence subjective, perfectly parallels observations about scientists' rhetorical strategies made by the sociologists Nigel Gilbert and Michael Mulkay.[60] Second, by separating "descriptive reality" from "moral arguments" on the merits, the council reinforced just the kind of boundary between facts and values that scholars like Brian Wynne find questionable in official ethical discourse. Third, in the guise of espousing a neopopulist, "just give me the facts," plain-language strategy, the council altered the framing of the issues at stake. For example, the term "reproductive cloning" connects the discourse of cloning to a wider discourse on "reproductive rights," making access to cloning for reproductive purposes seem more like an entitlement. The council's move sought to break that discursive connection. Similarly, by discarding the term "therapeutic cloning," the council set aside the positive connotations of therapy (including beneficence) in favor of a term—research—that carries greater connotations of uncertainty, experimentation, and possible lack of control.

The move toward plain language, in short, was not by any means morally neutral: it simply foregrounded and backgrounded different sets of social meanings and possibilities. In the guise of merely uncovering moral questions that are objectively *there*, the council strategically used language as a reframing device to transform the landscape of morality.

If President Bush had hoped that his ethics council would build a national moral consensus around stem cell research, then that hope soon proved illusory. The council's moral force eroded under internal as well as external pressures. Internally, the administration's failure to reappoint two of its more research-friendly members—Elizabeth Blackburn and William F. May—produced a firestorm of criticism. In March 2004 some two hundred bioethicists wrote a letter to the president charging him with narrowing the range of views on the council and thereby compromising its capacity to deliver effective advice. Externally, scientists who saw ES research as among the most promising on the biomedical horizon proved to be a powerful political force. In alliance with representatives of patients' groups, they mounted an effective attack on the president for mingling science with religious fundamentalism. Stem cells appeared on the agenda of presidential politics in the summer of 2004, producing the extraordinary spectacle of Ronald Reagan, Jr., the late Republican president's son, speaking about the promise of this line of research at the Democratic National Convention in Boston. In tones reminiscent of Mary Warnock, Reagan justified his support for ES research by claiming that one can *see* the difference between a mass of embryonic stem cells and an actual human being, who has a body and mind, fingers and toes. Only, in American fashion, he drew support for this distinction from the story of a thirteen-year-old girl's valiant fight to manage her juvenile diabetes (*she* was a real person) rather than from any appeals to divine grace.

## Germany

In Germany, too, the debate on stem cells led to a splitting of responsibility for ethical analysis, only the two German bodies that undertook the task were divided not chronologically nor by party politics, as in the case of the U.S. presidential commissions, but by their constitutional position. One was the Enquete Commission appointed by the Bundestag; the other was the national ethics council appointed by the Schröder government. From a political standpoint, the Ethikrat was an unprecedented and untested institutional form in competition with a body, the familiar parliamentary enquiry commission, whose form and function were well understood in Germany. How was this competition handled, and with what consequences for deliberation on stem cells and bioethics?

What triggered the stem cell debate was a request to the German Research Foundation for funds to conduct research on imported ES cells. The DFG turned to the government to seek clarification of the legal status of this request—whether, in short, ES cells could be legally imported from abroad. The DFG hoped perhaps for the kind of pragmatic approach NIH had countenanced in the United States, allowing research to continue with existing cell lines even though initial derivation was contrary to law. There was, of course, no possibility of producing embryonic stem cells for research purposes in Germany or deriving them from spare IVF embryos, whose very generation was forbidden by the embryo protection law. The immediate issue was the legality of importing cell lines from abroad, but as in the U.S. case the two bioethics commissions failed to agree on the right course of action. The parliamentary commission voted 17 to 7 against imports on November 12, 2001; on November 29 the Ethikrat voted 13 to 9 to allow the imports.

Both ethics bodies reported in the end to the Bundestag, for it was clear that parliament was the only body capable of definitively deciding the legal issue. On January 30, 2002, the Bundestag debated for five hours on the possible consequences of admitting ES cells into Germany. Three legislative options were on the table: first (sponsored by the CDU/CSU) to forbid imports, on the ground that protection of human dignity demands it; second (sponsored by the SPD and some Greens) to forbid embryo research resulting in the destruction of embryos, but to allow strictly supervised imports of existing cell lines on the ground that ES cells are not embryos proper and do not call for the same constitutional protection; and third (sponsored by the FDP) to permit imports on the grounds, among others, that responsible research on embryonic stem cells would benefit humanity and not harm potential humans.

The second option, favoring restricted imports, won handily in the Bundestag, by 360 to 190 votes. What draws our interest is the arguments offered by the three parties. For the Christian right, demanding a ban on imports was easy to defend and consistent with setting an absolute value on human life and human dignity from the moment of conception. CDU/CSU speakers repeatedly invoked the metaphors of dams, walls, and foundations, saying that these had to be built firmly enough to provide a clear expression of German values with regard to life. They feared a *Dammbruch*, a breach in the moral foundations of public policy, and unlike Britain's Anne McLaren, they were not uniformly optimistic about the legislature's capacity to contain the flood that might follow. In this respect, the rhetoric of disorder in the German political debate was not, as Habermas suggested, simply analogous to the slippery slope discourse of the U.K. embryo research debate.[61] The liberal FDP, who alone favored unrestricted import and even production of stem cell lines in Germany, were content with a utilitarian argument that placed a high value on medical advances to promote the public good.

Most interesting was the majority's justification for its seemingly inconsistent position: favoring imports, and thereby implicitly sanctioning embryo destruction, but insisting all the time that the law absolutely forbids the destructive use of embryos. With this, the SPD opened itself to charges of double standards (for Germany and other countries) and also an ethical

197

contradiction about the value of life. The solution paralleled in a way Germany's uneasy compromise on abortion. First, SPD representatives asserted a firm, apparently nonnegotiable principle—namely, no embryo-expending research (*keine vebrauchende Embryonenforschung*). Then, they distinguished the permitted import from the bounded-off, prohibited zone through an equally nonnegotiable rule: only admit existing cell lines, in part because the ethically reprehensible acts had already taken place, and in part because the resulting cell lines were not, after all, fully capable of becoming human. On this basis, the supporters of restrictive imports maintained that German research would create no incentives for future embryo sacrifice. The moral position was clear. Germany would not provide a home for any future ES cells produced after the controversial lowering of the barriers.

As earlier on the issue of embryo research, the debate on stem cells proved to be in fundamental ways a debate on the nature of the German nation-state.[62] Was this a state that would put the interests of researchers ahead of human dignity and allow biology rather than the Basic Law to frame the very meaning of being human? Was it a state that would follow the lead of other more scientifically advanced countries or one that would, by example, stand for a more reflective kind of compact between medicine and society? Was it a state capable of adhering to clear and consistent moral principles while making hard policy choices? Was it a state that could abide by a higher conception of law than other states seemed prepared to do? These issues were raised, often explicitly, during the stem cell debate, giving it a gravitas not normally encountered in dry discussions of science policy—but also a strange detachment from the prosaic concerns for gain and advancement that feed the dynamics of interest-group politics in more ontologically secure democratic societies.

The debate's resolution was almost anti climactic. In April 2002 the Federal Parliament passed a law requiring researchers wishing to import stem cells for research to obtain permission from the Robert Koch Institute. Following longstanding German custom, the actual work of approving applications was delegated to an advisory committee: the Central Ethics Commission on Stem Cells, whose eighteen members represented the fields of biology, medicine, ethics, and theology. With this step, the work of politics was safely concluded and a high dam was once again put in place, keeping disorder and moral confusion away from the domain of free scientific inquiry.

### Britain

And what finally of Britain, where embryo research was already permitted under strict regulatory supervision by HFEA and seen by many as a proud achievement of British science? Though HFEA's authority unquestionably extended to the derivation of stem cells from IVF embryos, new questions arose in connection with the fourth possible method for obtaining human ES cells: through the process of cell nuclear replacement, or CNR. The problem was one of statutory interpretation. The 1990 Human Fertilisation and Embryology Act defined the embryo as "a live human embryo whose fertilisation is complete." Technically, then, an embryo produced through CNR arguably did not fall within the meaning of the act because it was not the result of fertilization. Following this logic, research with CNR embryos, and stem cells derived from them, would not have been subject to HFEA's regulatory supervision, leaving a gap in the law requiring renewed parliamentary action and a reopening of the vagaries of the legislative process. A subsidiary question also arose about the government's authority to add further research uses of embryos to those already specified in the 1990 law.

To address these issues, the U.K. government appointed an expert panel to review the science and ethics of embryo research in the light of new technological developments.[63] The panel concluded, among other things, that the use of ES cells from IVF embryos that were five to six days old should be continued for research purposes, and it also supported limited use of stem cells derived through CNR. All this, the government held, could be done under the authority of the existing 1990 law, and it drafted new regulations to enable action; such incremental evolution had already been foreseen in the 1987 White Paper on Human Fertilisation and Embryology. A House of Lords Select Committee also considered the issues and reported back in early 2002 that medical research using human ES cells was highly promising and should be continued under existing law.[64] Parliament, too, stepped in to clarify the legal picture in one respect. The Human Reproductive Cloning Act of 2001 prohibited the implantation into a woman's body of any embryo created through a process other than fertilization—in other words, in the U.S. PCBE's terms, it forbade cloning to produce children.

While the legislative deliberations were taking place, the Pro-Life Alliance (represented by its director Bruno Quintavalle) challenged the government's proposed regulations in court, claiming that the 1990 law could not be extended to cover CNR embryos. In a High Court decision, Justice Crane held that the challengers were right and CNR embryos fell outside the remit of HFEA. That decision, however, was reversed on appeal, and in March 2003 the House of Lords, Britain's highest court, held that the 1990 law covered these embryos as well, thus reaffirming HFEA's authority to regulate research using them.[65]

The second Quintavalle case (see chapter 6 for an earlier one on tissue typing) raised a question of legal interpretation similar to that confronted by the U.S. Supreme Court in *Diamond v. Chakrabarty*[66] on the patenting of living organisms. When a legislature enacts a law against a backdrop of changing human circumstances, how can courts decide whether the law's original intent covers the altered conditions of life or whether things are sufficiently different to require new lawmaking? This question arises with particular urgency in the context of technological change, where it is often clear that the enactors of the original law could not possibly have had in mind the interventionist possibilities of a later day. In *Chakrabarty*, for example, Congress had to interpret an eighteenth-century law, drafted by Thomas Jefferson, in the light of capabilities made available by modern biotechnology in the twentieth century. For the U.K. House of Lords, the passage of time was much shorter, but the jurisprudential issue was equally challenging.

The Law Lords who judged the case all agreed on one interpretive point. As judges, they could not put on the mantle of the legislature and ask what Parliament would have done if the new facts had been placed before it. The judges' role was rather to discern whether the intent of the law *as written* covered the "genus of facts" that included the new ones under consideration.[67] Taking statutes as "always speaking,"[68] the court had to ask whether a given law could properly be construed so as to apply to the new case. The Lords showed little hesitation about their capacity, as individuals or as an institution, to make the necessary discriminations about whether new facts are like or unlike the ones that Parliament knew. The empirical judicial eye ruled triumphant.[69] Lord Bingham of Cornhill put the matter as follows: "If Parliament, however long ago, passed an Act applicable to dogs, it could not properly be applied to cats; but it could properly be applied to animals which were not regarded as dogs when the Act was passed but are so regarded now." By this reasoning, the Lords concluded that the 1990 law covered human embryos, by whatever means derived: CNR embryos belonged to the same genus as fertilized ones; it was not a matter of treating dogs as cats; it was dogs all the way. Lord Bingham concluded as follows: "[T]here is one important question which may permissibly be asked: it is whether Parliament, faced with the taxing task of enacting a legislative solution to the difficult religious, moral and scientific issues mentioned above, could rationally have intended to leave live human embryos created by CNR outside the scope of regulation had it known of them as a scientific possibility. There is only one possible answer to this question and it is negative."[70]

Since a lower court had ruled the opposite way, as urged by Bruno Quintavalle and his organization, we are forced to wonder what gave the Law Lords so strong a conviction of right reason, of there being "only one possible answer" to the question of parliamentary intent. The answer lies less in the appurtenances of formal reasoning than in the cultural production of authoritative voices, for whom reasoning is but one persuasive resource among others. One would perhaps have needed to be inside Lord Bingham's skin to share the full strength of his conviction, but that is precisely the sort of conviction that supreme judicial authorities are designed to project.

## Summing Up

The stem cell debate in three countries illustrates much about the nature and limits of bioethics as a discourse for policymaking in contemporary Western democracies. Accordingly, this chapter needs no lengthy further conclusion but only a brief résumé of the most salient features of the growth of bioethics in the three national contexts.

Originating in the relatively intimate environment of physician-patient relations, bioethics, the offspring of biomedicine, moved in the later twentieth century to occupy a central position in national political discourse. This evolution reflected the increasingly prominent role of the state as sponsor and user of the life sciences, and hence as a party responsible in the eyes of citizens for the ethical production and deployment of biological knowledge and power. The rise of bioethics illustrates, in short, the natural evolution of biopower as imagined by Michel Foucault. Despite its academic connections to philosophy, bioethics was seen in all three countries as a device for bridging potentially troublesome divides: among disciplines, professions, and institutions; and increasingly also among science, state, and society. Efforts to professionalize bioethics and to standardize its terms continually overflowed, as new actors and institutions saw advantages in speaking bioethics to power. A recurrent line of struggle was over whether bioethical analysis would function as simply another—usually the last—component in the analytic phases of policymaking or whether it would open up formal decision processes to wider questions about values and their implications for policy. Tied to this were questions about the proper boundaries of bioethical analysis: was it a discourse limited to concerns about the human body, or was it to be as encompassing as life itself?

On the whole, as we have seen, the agendas of politics shaped the use of bioethics more than the other way around. Bioethics provided some useful concepts, most notably that of beneficence, in efforts to promote the research agendas of biomedical science, but ethical debates were for the most part roped into the service of broader, more powerful national narratives: progress through medical innovation and opposition to abortion in the United States; building a principled *Rechtsstaat* in Germany; and maintaining a well-ordered space for research, in collaboration with the empiricist state, in Britain. The subservience of ethics to politics can be interestingly observed in the domain of language. The U.S. PCBE was prepared to invent new language to further its vision of the proper relationship between moral and factual inquiry. Britain's Law Lords interpreted Parliament's anything but definitive verbal injunction so as to produce moral certainty around a novel technological construct, the embryonic stem cell. And in Germany controversies around the two Peters, Singer and Sloterdijk, again gave evidence of the very high stakes involved in the choice of bioethical language.

Regardless of its specific national embeddings, official bioethics in all three countries functioned most often as a consequentialist discourse, reacting to novelties put forward in the first instance through science and technology. The ethics of changing the world through kind-making made only an occasional glimmering appearance on the discursive agenda, as in German discomfort over Sloterdijk's human zoo and Phil McNaghten's concern for animal natures in Britain. Arguments for a meaningful deliberative politics of kind-making did not emerge in any case from official bioethics in any of its initial close encounters with state policy.

# 8

## Making Something of Life

Things—hard, inanimate, material objects—are not normally seen to be in the mainstream of political life. By contrast, products of social interaction often are at the center of politics: laws, rules, standards, enforcement mechanisms, the design and shape of institutions. Biotechnology in some ways reverses this normal order. The things that biotechnology makes and deals in—genes, cells, plants, embryos, drugs, bacteria, DNA sequences, genetic tests, transgenic creatures—have often been the focus of controversy at one time or another in the past twenty years, and some still remain so. Yet, many social technologies that were vital to the birth of modern biotechnology have fallen curiously outside the center field of political analysis, for example, venture capital, start-up firms, technology transfer, and intellectual property law or, more simply for our purposes, patenting. In this chapter we look at the politics of patenting in the United States and Europe and ask why a device of such significance remained relatively insulated from democratic deliberation in America but generated controversy and long delays in European policymaking.

It is not too great an exaggeration to say that without patents the biotechnology industry as we know it, with all its revolutionary economic and social potential, could not have come into being. Especially in the United States, patents played a foundational role in the development of the biotechnology industry at several levels.[1] First, the extension of patents to the life sciences created new classes of property rights in things that were previously outside the realm of what could be owned, or even thought of as subject to ownership claims.[2] As a result, these objects became commodities that could have value, be exchanged, circulate in markets, and foster productivity.[3] Second, much of the early development of biotechnology occurred before there were any marketable products, and patents were the only evidence for eager venture capitalists that there might be something of future value to justify present investment. Third, patents provided some assurance to jittery investors that they would not be mired in endless legal wrangling if commercially useful products ever came on line. Fourth, patents proved to be a way of sorting out the competing claims of participants in an increasingly complex web of invention that linked together the disparate interests of patients, research subjects, farmers, academic researchers, universities, start-up firms, government, and industry.

Patents not only underwrite a scheme of property rights, but they order the process of invention in two ways that could be seen as intrinsically political. One is to designate classes of things that can be considered property. The extension of patents to new domains alters basic notions of what is a commodity and who can assert ownership over it. When a patent is awarded for a biological product, it has the effect of removing the thing being patented from the category of nature to the category of artifice—a profound metaphysical shift that, at least in theory, should invite public deliberation. The second political function is distributive. Patents assign ownership rights within production systems, rewarding some participants in the discovery process more than others. For instance, lab technicians' and research subjects' names are rarely written into patent applications; nor do these individuals normally share in the economic proceeds from specific inventions. The institutions in which inventive work is carried out do, by contrast, earn the lion's share of royalties. In this way, patents act as instruments of economic distribution.

Despite their distributive impacts, biotechnology patents have attracted relatively little public attention compared with GM foods or stem cells. Even in Europe, where there *was* prolonged debate, discussion stayed contained for the most part within the circles of governing institutions, expert bodies, and the specialist science media. There are some obvious reasons for the low degree of political mobilization. In the United States, judicial decisions removed some controversial issues from the sphere of political debate. In Europe, patent protection was internationalized under the European Patent Convention and later through the European Union's lawmaking process. This centralization left little room for a national politics of intellectual property, as there has been for other aspects of biotechnology. Intellectual property law, more generally, is regarded as being an extension of the technical work of invention rather than entangled with questions of justice, metaphysics, or people's immediate concerns about life, health, and the environment. Patent law, correspondingly, tends to be an esoteric field of practice, committed to specialized expert institutions in both Europe and the United States.

Nonetheless, to the extent that biotechnology patents became political, the issues raised in the U.S. and European debates were differently framed and led to different legal and policy outcomes. To put these comparative dimensions of patent politics in perspective, I begin by looking at the means by which social technologies such as intellectual property law channel and black-box political choices. I then turn to three moments at which patent issues either became political or else notably failed to do so, on each side of the Atlantic. In the U.S. case, the three moments are the U.S. Supreme Court's decision involving the patenting of living organisms, the California Supreme Court's decision on the ownership of human tissues and cells, and the dispute over the patenting of DNA sequences. In Europe, the episodes are the controversy over the patenting of the so-called Harvard oncomouse, the decade-long dispute over the 1998 EU biotechnology patent directive, and, paralleling the U.S. case, the debate over patenting DNA sequences. Both chronologically and analytically it makes sense to begin in the United States, where life patenting first appeared on the agenda of law and politics, and where it also remained more effectively insulated from political debate.

## The Politics of the Apolitical

To appreciate the role of intellectual property law in the politics of biotechnology, we can draw conceptual support from two lines of scholarship: an older tradition in critical technology studies and a newer body of work in the anthropology and sociology of technology. The former calls attention to the political economy of intellectual property law and its role in serving powerful social and political interests. The latter provides insights into the ways in which a device such as a patent helps to create and naturalize the very objects and rights that it claims to protect. Both areas of work remind us that things are not intrinsically political or apolitical by their very nature, but only as a result of conscious human decisions. Both also suggest reasons why the element of choice so often becomes invisible once a technology assumes its working form. It is an axiom of technology studies that we selectively attribute thing-like properties to some human inventions and institutions, thereby taking them out of the company of the social and political. Indeed, terms like reification or objectification or, in German, *verdinglichen* (to make into a thing) call attention to our awareness of the constructed aspects of objecthood. From this perspective, the relegation of things to the status of objects—devoid of political and social content—becomes a phenomenon to be explained, not to be taken for granted. A corollary is that technological objects and systems, though they may be regarded as inanimate or asocial, inevitably are repositories of human values, beliefs, imagination, power. Through their apparent firmness and immutability, certain human constructs may seem to stand outside the flows of politics, but how they achieve this appearance and with what consequences are not outside the realm of political inquiry.

The political scientist Langdon Winner forcefully made this point in a well-known article in which he asked whether artifacts have politics.[4] Winner

205

discussed two ways in which the question might be answered in the affirmative. Some technologies, he suggested, have political settlements built into their very design. To illustrate, he offered the example of the low height of bridges across the parkways of Long Island in New York State. Robert Moses, the great master-builder of New York's highway system, purposely designed the overpasses to be low, Winner said, to keep buses bearing poor people from the city from traveling into the rich outlying suburbs. Inequality between races and classes was in this way incorporated into the city's built environment, with long-lasting implications for social justice. Following Friedrich Engels and others, Winner designated a second class of technologies as "inherently political" because they entail particular social relationships and institutional arrangements for their operation. The example he gave was nuclear power, which in the eyes of critics could not have come into being or been sustained without a full-blown military-industrial complex to support it.

Later work than Winner's has tended to stress the multiplicity of interests that are incorporated into technological design, as well as the fact that neither social nor material structures have quite the obduracy that Winner assigned to them. Working in a framework called the social construction of technology,[5] some scholars have stressed the negotiations that go on among classes of potential users in order to arrive at particular solutions to technological problems. Design, in the constructivist view, is a product of compromise among those who have stakes in the outcome, not an ineluctable transcription of inflexible power into unchangeable material products. The school of actor-network theorists (ANT) places still greater emphasis on the contingencies built into complex technologies, noting that they are composed of heterogeneous elements that remain in active interplay with one another even after a technology acquires its "final" form. If the social constructivists look at design at its inception, from the standpoint of designers and users, the ANT framework looks at design equally after it begins to function, from the standpoint of those who operate a system and live with its day-to-day contradictions and idiosyncrasies. From this viewpoint, ANT scholars have argued, it does not make sense to establish hierarchical differences that privilege human actors over nonhuman *actants* in a technological system.[6] Each participant has a role to play in the functioning of the entire network; each acts upon, constrains, and influences the others, so that even the nonhuman elements of a technology partake in some sense of the social.

Together, these ideas provide a conceptual framework within which the on-again, off-again character of the politics of biotechnology patents can be better understood. The argument from political economy suggests how technologies can be made to seem apolitical. Artifacts—social no less than material ones—can become so hardened through design and use that the ways in which they incorporate political choice or economic power cease to be visible, like the highway overpasses whose form, once built, seems inevitable to the innocent driver motoring under them. Once a technology has been black-boxed and put to use, it takes unusual convulsions to make the underlying social choices apparent again: like a social movement of mothers against drunk-driving that deconstructs the design of cars and transportation systems, or a powerful lobby of organic food consumers that creates a market for things grown certifiably without the aid of agricultural biotechnology.

Approaches like ANT and the social construction of technology stress the fact that technological designs are never seamless, homogeneous, or entirely fixed but rather retain interpretive flexibility and hence room for maneuver. Even when a technological system appears to be completely stable and black-boxed, there are always nodes at which settlements can be reopened and renegotiated. When a technological system is working well, it may not be apparent, especially to outsiders, how much room exists for creative modification of the system's rules. Indeed, it may take breakdowns to make this flexibility apparent. A recurrent finding after technological disasters is that users or operators of the failed system behaved in ways that designers and regulators had not expected; users had discovered and exploited the system's soft spots and ambiguities, sometimes with tragic consequences.[7] More generally, though, this perspective on technology suggests that strategic users can find ways to question parts of any system's design, finding ways to make it work differently or, at the limit, to prompt wholesale dismantling and redesign.

Returning now to our puzzle about the selective politics around patent law, we begin to see reasons for quiescence as well as life. As a social technology, patents naturalize the idea of property, appearing to recognize property rights that are, quite simply, already present. Taken as merely ratifying an underlying status quo, patents are regarded as apolitical with respect to both the "what" questions (what can be owned?) and the "who" questions (who can own it?) that they regulate; patents operate, in this respect, not as norms but as tools. A counteranalysis similar to Winner's would point out, however, that patents are not simply declaratory instruments that affirm a prior order of ownership, but that they create and maintain property rights in specific forms that are anything but preordained. For example, scientific and technological patents, as enshrined in current U.S. law, presume the existence of identifiable inventors and a single moment of invention; the law also presumes that rewarding inventors and their institutions will lead to more socially beneficial invention. Neither presumption accords very well with the observed dynamics of biotechnological development, where discovery is more often polycentric and spasmodic in its evolution than progressing linearly from discrete bursts of creative genius.[8] Yet, though empirically unsupported, the presumptions built into patent law serve useful ordering functions. They locate ownership in ways that limit the potential for disputes, and they further capital accumulation by enhancing the ability of corporate actors and their inventive employees to reap the rewards of invention. In the scheme of liberal economies, the founding presumptions of intellectual property law are therefore not easily open to renegotiation.

Among social technologies, the law is unique in one respect that is critically important for our analysis of patents. Whereas most powerful institutions discourage self-reflection, the law makes progress precisely through case-by-case attempts to take apart its own settled meanings and reinterpret them in the light of altered social circumstances. Technological change potentially offers countless opportunities for such legal introspection, for it produces novelties that challenge existing classifications and the rules built upon them. Yet, the law's self-reflection is always partial and selective. In patent litigation it is never the entire regime of intellectual property law that is open to question; claims can be challenged only on relatively narrow grounds flowing from the text of the governing law. As in other areas of statutory interpretation, courts are committed to carrying out the legislature's intent, subordinating their own agency as interpreters of the law. Courts are institutionally mandated to apply the law as they find it—not to make new law or policy—and this is what judges say they are doing even in the most controversial cases, where the meaning of prior law is most in doubt.

The U.K. Law Lords dealt with such a problem of statutory interpretation when confronted by Bruno Quintavalle's objection to the regulation of embryos made through cell nuclear replacement under the 1990 Embryology Act (see chapter 7). In the context of intellectual property law, the U.S. Supreme Court faced a similar problem in deciding whether to extend patent protection to living organisms in *Diamond v. Chakrabarty*.[9] Both high courts determined that existing law covered the new facts and that no new legislation was required. There was nothing foreordained about these decisions; indeed, lower courts in both cases had held otherwise. We will return to the details of the reasoning in *Chakrabarty* below, but it is important to note that another decision would have led to other politics. If the Supreme Court had decided not to sanction life patenting, the issue would almost certainly have moved to Congress, where a less cut-and-dried decision about patentable subject matter might have followed. Perhaps the biggest difference between biotechnology patenting in Europe and the United States then is that, in Europe, the issue got caught up in the complexity of European lawmaking; in America, the framework of the Patent Act was not similarly breached by GMOs.

## Who Owns Life?—The U.S. Debate

The issue of patenting biological products and processes is often represented as being about "owning life." It is this most basic "what" of biotechnology patents that has tended to attract most attention in the media and in legislative debate, more so than the "who" questions—who owns, who benefits, who has incentives to innovate? Yet one of the more striking features of U.S. intellectual property disputes involving biotechnology is that actual patent conflicts rarely if ever get framed in these grand terms. The rhetoric of owning life circulates in public spaces, more in the language of opposition than of policy. Major legal disputes are disposed of as narrower questions of statutory interpretation, in accordance with technical criteria for granting patents, interpreted case-by-case by the courts. Only in a few instances—and *Chakrabarty* was one—have conflicts spilled over the retaining wall of the law into broader ethical and philosophical deliberation about the propriety of human dominion over life. Even then, the effects of such debates are hard to assess. When more than eighty religious groups mobilized in 1995 to oppose patenting human genes and genetically modified animals, Francis Collins, director of the National Center for Human Genome Research and a devout Christian, remained unmoved, saying, "Patenting is not a moral issue; it's a legal issue."[10] The successful containment of moral conflict within the legal process is the major theme of this section.

In the United States, disputes over intellectual property rights in biotechnology are litigated under authority of the Patent Act, which lays down four principal conditions for the award of a patent. First, the invention must be within the statutory subject matter definition; in addition, it must also be new, useful, and nonobvious.[11] On the whole, the terms new, useful, and nonobvious have given courts more trouble than the subject matter provision, which has been liberally interpreted. In the case of biotechnology patents, the line between what is natural, and hence not patentable as a human invention, and what is invented has shifted over the years to allow more and more claims on the invention side of the ledger. Only the most elemental physical phenomena such as electricity or gravity are today reliably regarded as definitely not patentable.

### Diamond v. Chakrabarty

*Chakrabarty*, as we saw in chapter 1, marked a seminal moment in the politics of biotechnology in the United States. The Supreme Court decision affirmed the legality of life patents at a formative period in the young industry's development and gave investors the incentives they needed to put money into genetic research and development. By brushing off Jeremy Rifkin's and the People's Business Commission's concerns about reductionism and the manufacture of life, the Court supported the emergent U.S. framing of biotechnology as being about useful, marketable commodities—not about the ethics or wisdom of genetic engineering writ large. At the same time, by making patentable "anything under the sun that is made by man," *Chakrabarty* construed the law's subject matter provision so broadly as to include almost all future products of genetic engineering. Indeed, though the invention in the case was a micro-organism, no effective legal dissent arose when, relying on *Chakrabarty*, the U.S. Patent and Trademark Office (PTO) awarded a patent for a genetically modified oyster in 1987 and a mammal in 1988, the famous Harvard oncomouse. The *New York Times*, the voice of the U.S. liberal establishment, editorialized in support of patenting higher-order life, saying it was the "wrong place" to address ethical concerns.[12] Following the oncomouse patent, a number of bills were introduced into Congress calling for a moratorium on the patenting of higher life forms, but these did not make headway. Legal challenges to the patent were dismissed as groundless.[13]

While interpreting the Patent Act's coverage broadly, the 5–4 majority in *Chakrabarty* interpreted its own powers narrowly, through a strategy of selective naturalization. The majority represented the process of invention as a natural phenomenon, and therefore unstoppable, while relegating the Court's role to that of a marginal spectator of inevitable scientific progress. "Judicial fiat as to patentability," the Court said, "will not deter the scientific mind from probing into the unknown any more than Canute could command the tides."[14] This institutional humility was more than a little incoherent.[15] The majority clearly presumed that the law has a part to play in encouraging technological progress and yet opined that judicial decisions are powerless to influence the tidal waves of research. Judicial self-abnegation, however, was consistent with the holding that, as to subject matter, the Patent Act had long since said all that needed to be said. This posture reduced judicial interpretation to a kind of automatic writing, or rewriting, of congressional intent. It was the Patent Act that spoke through the Court. The justices purported to be adding nothing new, certainly nothing remotely redistributive or political.

The basic argument confronting the justices in *Chakrabarty* was that the claimed subject matter in this case was sufficiently new and different from anything previously deemed patentable to require explicit congressional reauthorization. This issue posed a problem because it meant that novelty had to be treated in two different ways in the same case. On the one hand, Chakrabarty's patent could not be granted unless the invention was sufficiently new and nonobvious. Only by rewarding true novelty could the law claim to be advancing the legislature's purpose. On the other hand, the invention could not be regarded as so radically new that it fell outside the range of things that Congress had defined as patentable. The majority got around this dilemma by construing as normal the creation of things not found in nature. For example, the Court noted, "A rule that unanticipated inventions are without patent protection would conflict with the core concept of the patent law."[16] Invention thereby became a process without sharp qualitative breaks that might have merited renewed deliberation. Chakrabarty's invention was held to be "unanticipated" enough to merit protection; yet the process of making it was held to be continuous enough with normal processes of invention to require no new legislative authorization.

If nonobviousness justifies the validity of a patent, then the same cannot be said for judicial decisions. It is obviousness, or seeming obviousness, that confers legitimacy on court rulings, that is, the apparent derivation from applicable law through nothing more than reason and common sense. To see how this taken-for-granted quality is achieved, one must turn somewhere other than the majority opinion. A powerful dissent can be illuminating (though not all court systems tolerate these[17]), as can, in appropriate circumstances, the example of another jurisdiction in which the same issue was decided differently. The Canadian Supreme Court's 2002 decision to refuse a patent on the Harvard oncomouse is particularly helpful in this regard, for here a 5–4 majority, interpreting a law almost identical to that of the United States, reached a very different conclusion with regard to patenting higher organisms.[18] It is worth reviewing that case in some detail even though Canada is not part of our comparative study. The decision helps to pinpoint what was nonobvious in *Chakrabarty*'s reasoning.

Described as the most litigated mouse in history, the oncomouse was produced by researchers at Harvard University, with funding from DuPont Company. The technique used to modify the oncomouse's genetic makeup involves injecting a fertilized mouse egg, preferably while it is still at the single-cell stage, with an oncogene or gene that promotes cancer. The treated eggs are then implanted in a female mouse and carried to term, producing offspring that contain the oncogene in all their cells. The mice born with the oncogene are further bred to produce, by Mendelian inheritance, a strain of mice that is particularly susceptible to cancer. The resulting animals can be used in regulatory scientific studies to test the cancer-causing, or cancer-preventing, potential of various substances. The value of such a test system to the chemical and pharmaceutical industries is clear and considerable. Environmental and religious groups, however, have long objected on ethical grounds to the award of all life patents, including on animals used in research. Several such organizations filed a memorandum with the Canadian Supreme Court in May 2002 requesting it to deny the oncomouse patent.[19] Among their reasons were negative impacts on science, global inequity, and threats to biodiversity, human health, and the environment.

The oncomouse case had been brewing for years in the Canadian patent system. In 1995 the commissioner of patents granted Harvard's request for a patent on the process of inserting the cancer gene into mice, but not on the product, that is, not on the oncomouse itself. Harvard challenged the commissioner's decision and lost at the initial trial stage but won at the Federal Court of Appeal, where a majority of judges approved the product patent. By the time the case came to the Supreme Court, Harvard held patents on both the oncomouse and the process of modifying it in Europe and the United States, and similar mouse patents had been issued in Japan and New Zealand. The Canadian court was the first to deny the product patent on the oncomouse.

211

Countering a long and forcefully argued dissent, the five justices for the majority found differently from *Chakrabarty* on two important counts: should a higher organism such as a mouse be considered a "composition of matter" as understood in Canadian patent law; and was it deference to parliament or was it flouting parliamentary intent to refuse to extend patent protection to higher organisms? Key to the court's negative decision on the first issue was a conviction that "composition of matter" was too restrictive a term to encompass something so complex as a living animal, and that altering one gene in an animal's genome is not in any case equivalent to producing an entirely new "composition." Justice Michel Bastarache, writing for the majority, concluded this part of his analysis with a ringing statement against genetic reductionism:

> Higher life forms are generally regarded as possessing qualities and characteristics that transcend the particular genetic material of which they are composed. A person whose genetic make-up is modified by radiation does not cease to be him or herself. Likewise, the same mouse would exist absent the injection of the oncogene into the fertilized egg cell; it simply would not be predisposed to cancer. The fact that it has this predisposition to cancer that makes it valuable to humans does not mean that the mouse, along with other animal life forms, can be defined solely with reference to the genetic matter of which it is composed. The fact that animal life forms have numerous unique qualities that transcend the particular matter of which they are composed makes it difficult to conceptualize higher life forms as mere "composition[s] of matter." It is a phrase that seems inadequate as a description of a higher life form.[20]

Was Bastarache right in claiming that modification of one gene did not convert a mouse into a "composition of matter," an invented, patentable product? The answer, in turn, hinges partly on one's conception of the judicial role. In his vigorous dissent, Justice Binney wrote, "I do not think that a court is a forum that can properly debate the mystery of mouse life."[21] Binney seemed not to recognize that a refusal to debate this point in court would have amounted in effect to a judgment that the issue was *not* debatable at all—indeed, in the light of U.S. experience, not likely to be debated in any other forum. Binney's position would therefore have let stand a particular solution to "the mystery of mouse life," eliding the difference between higher animals and compositions of matter.

The comparative point worth emphasizing here is that an extension of ideas of mechanical invention to living creatures that had seemed quite natural in *Chakrabarty* in 1980 struck an equally thin majority of the Canadian justices as unnatural in 2002. There was, in short, nothing intrinsically natural about the U.S. Supreme Court's naturalization of the process of biological invention.

The Canadian oncomouse decision also approached the question of legislative intent differently from its U.S. counterpart. Both courts agreed as a matter of principle that broad social and environmental policy concerns were irrelevant to the implementation of patent law, but from that point their views diverged. For the Canadian Supreme Court, the extension of patents to higher organisms was a sufficiently large step that it raised concerns only parliament could appropriately address. Two issues that particularly exercised the majority were possible impacts on agriculture and the patentability of human life. With respect to the first, the court felt that extending patents to self-replicating plants and animals raised new issues that parliament clearly had not thought of in passing the original law and that only parliament had the institutional means to consider. On human life, similarly, the court concluded that there was need for forms of line-drawing that judges were ill equipped to undertake: "In my view, it is not an appropriate judicial function for the courts to create an exception from patentability for human life given that such an exception requires one to consider both what is human and which aspects of human life should be excluded."[22] Once again, the comparative point worth noting is the difference between the U.S. and Canadian courts on the law's interpretive flexibility. While both professed to defer to the legislature, the U.S. court expressed this deference by stretching the existing law to cover the new case, rendering new legislative action unnecessary; the Canadian court preferred to let the legislature have renewed say in an area it saw as raising questions that were morally and politically novel and perplexing.

## Moore v. Regents of the University of California

If *Chakrabarty* and the oncomouse cases raised questions about what can be patented, it fell to the California state courts to consider a landmark case dealing with the question of who can own biological materials. The case was *Moore v. Regents of the University of California*,[23] in which a patient in a university teaching hospital set the biomedical research establishment abuzz by claiming ownership of his own tissues and cells; the researchers, he charged, had taken them unlawfully in the course of treating him.

The claimant in the case was John Moore, an engineer from Seattle, Washington, who was diagnosed in 1976 with the potentially life-threatening disease hairy-cell leukemia, for which he sought treatment at the Medical Center of the University of California at Los Angeles (UCLA). On the recommendation of his treating physician David Golde and his assistant Shirley G. Quan, Moore underwent an operation to remove his diseased spleen. Unbeknownst to him, Golde and Quan had determined that Moore's T-lymphocytes, a type of white blood cell, overproduced certain lymphokines or proteins that regulate the immune system and that might be useful in the treatment of AIDS and other similar disorders. The UCLA researchers used sections of Moore's excised spleen to establish a cell line, named after its source the MO-cell line, on which they, together with UCLA, eventually filed for a patent. In 1984 a patent on the cell line was issued to Golde and Quan as inventors, with the university as patent assignee. Golde and the university in due course also negotiated a contract with Genetics Institute, a private company, for the development of products from the MO-cell line; possible derivatives were estimated at the time to be worth up to $3 billion, although the proceeds proved in practice to be more modest.

During this period, Moore was repeatedly summoned back to UCLA from Seattle, ostensibly for further treatment, but in fact to permit the research team to take more samples of his blood and tissues. He gradually became suspicious about the need for all these visits and finally sought legal advice when, in 1983, Golde insisted a little too strenuously on his signing an informed consent form permitting the use of his cells for research. Moore eventually sued UCLA and the research team, claiming, among other things, that his doctors had violated their fiduciary duty by not informing him of their research interests and that his cells had been unlawfully appropriated—in short, stolen—without his consent. The trial court ruled against Moore, but the court of appeal reversed this decision and the case moved to the California Supreme Court for final resolution.

In a widely watched decision,[24] the state's highest court ruled for Moore on the issue of the physicians' fiduciary duty, holding that research or economic interests unrelated to a patient's health must be disclosed to the patient before medical procedures are undertaken. On the issue of conversion (theft), however, the court ruled against Moore, stating that he had no legally recognized ownership rights in his cells. As in *Chakrabarty*, the outcome hinged importantly on what the court saw as new about Moore's case and its relationship with existing law.

Both the appeals court and the supreme court recognized that Moore's claim was unprecedented, but beyond this point the two courts reached strikingly different conclusions. For Judge Rothman, author of the majority opinion in the appellate court, the absence of law left the door open to a recognition of new rights. In language reminiscent of the PBC's *amicus* brief in *Chakrabarty*, Rothman argued that

> An inference of abandonment is particularly inappropriate when it comes to the use undertaken by defendants involving recombinant DNA technology. Almost from the beginning, this technology has incited intense moral, religious, and ethical concerns. There are many patients whose religious beliefs would be deeply violated by use of their cells in recombinant DNA experiments without their consent, and who, on being informed, would hardly be disinterested in the fate of their removed tissue.[25]

In other words, according to the appeals court, the novelty of the technology militated in favor of extending Moore's ownership rights over something so intimate and personal as tissues from his body. For Justice Panelli, who wrote the supreme court opinion, the absence of direct legal authority only provided a reason to reexamine the policy issues underlying the law of conversion.[26] This inquiry persuaded the high court that Moore's interest in the use of his cells should not be treated as a property right.

While denying ownership rights to the person whose cells were used for research, the California high court granted such rights to the researchers who made use of the cells. This was just one of several asymmetries that remain the most distinctive features of the *Moore* decision. Moore's conversion claim was viewed as novel, hence questionable; Golde's and Quan's patent claims, by contrast, were not held to be novel and hence were not open to further analysis. Moore's claim concerning the uniqueness of his genetic material likewise could not be supported, since the lymphokines his body produced "have the same molecular structure in every human being."[27] Yet the extraction of the very same proteins from a cell line based on Moore's cells was deemed sufficiently original to justify the award of a patent. Granting property rights to the human "sources" of biological specimens, the court opined, would introduce uncertainties that could be detrimental to academic research and industry. No such concerns, however, were voiced in connection with granting intellectual property rights to the UCLA research team; the patent was simply assumed to promote rather than hinder the unconstrained circulation of biological materials. Finally, in giving Moore a cause of action for lack of informed consent, the court recognized his sovereignty as a consumer of medical services; at the same time, by refusing him proprietary rights over his cells, it denied him autonomy as a research subject. In short, as the legal scholar James Boyle pungently observed, "As far as the majority was concerned, Moore was the author of his destiny but not of his spleen."[28]

Though one may criticize the reasoning in *Moore*, it is important to see that the case produced a powerful form of institutional closure. For more than a decade, no similar cases arose in other jurisdictions, suggesting that the California courts had successfully removed the question of cell and tissue ownership from the perimeter of political debate. How did it succeed? Pragmatism and compromise were important. The decision not only split the difference between the patient and the researchers—granting informational rights to the former and property rights to the latter—but also crafted a procedural solution that took advantage of existing administrative routines. Moore's remedy (and that of patients like him) was grafted onto the process of informed consent, the primary instrument used in controlling the ethics of medical therapy and research. Most important, though, was the decision's comfortable fit with the market logic that animated all of U.S. policy toward biotechnology. Under that logic, entrepreneurship leading to product innovation was always to be favored over other kinds of claims and values, however meritorious these might be. It is in the context of this "higher law" of the market that the contorted reasoning of *Moore*, and its curious splitting of the domain of rights, suddenly become transparent and comprehensible.

CHAPTER 8

### Controversial Sequences

Religious groups, animal rights activists, environmentalists, or research subjects—after *Chakrabarty* none succeeded in taking patents on living organisms out of the law courts and back into politics. When politics reentered the picture, it was over something much more humble than plants, animals, humans, or "life." The things at the center of the most intense biotech patent controversies at the turn of the century were small DNA sequences of uncertain biological function but great potential utility for research. They put life patenting back on a political agenda from which patents had all but disappeared for most of two decades, caused rifts to appear within the U.S. and international genomic research communities, and spurred bilateral U.S.-U.K. negotiations at the highest levels of government. Once again, controversy focused on the "what" of patenting—were these sequences patentable—but the reasons for the question's salience lay more with the "who" of the matter. For granting patents on sequences redistributed benefits within the network of research and development, favoring one powerful community (industrial developers) at the expense of another (basic researchers). It was this distributive effect within the marketplace of R&D that helped to reopen issues that moral arguments alone had not succeeded in making political.

The controversy over patenting DNA sequences was an offshoot of the publicly funded Human Genome Project conducted under the auspices of the National Institutes of Health and the Department of Energy. This massive effort to map and sequence the three billion or so base pairs of the human genome leapt to public notice when Craig Venter, an NIH scientist, broke with his employer over its extremely slow progress and announced that he was forming a private company, Celera (its very name connoting speed), that would get the job done in 2001, well ahead of the government's original target date of 2005. Key to Venter's boast were a technique that he had invented and a battery of instruments that he acquired with private funding to industrialize the sequencing process. The technique involved identifying stretches of DNA that provide a shortcut to identifying the whole genes in which they are embedded, even though their own biological function is unknown. Equipped with hundreds of sequencing machines, Celera began charting the genome at a fantastic rate with the aid of these identifier sequences; concurrently, Venter sought patents on the gene fragments that had proved to be useful in the gene mapping process.

The competition between Venter and NIH made newspapers happy and scientists nervous. Conflict began around a large number of so-called expressed sequence tags (ESTs) that become visible when a gene is being expressed, that is, producing a protein. To prevent what James Watson and others saw as a "land grab" by Venter, NIH itself sought patents on close to three thousand DNA sequences in order to use them for the public good. NIH's legal staff

216

backed the action, but leading scientists were appalled at this perversion of the patent process to reward "discoveries" that were, to them, scientifically meaningless and lacking obvious utility.[29] They also feared that ownership claims lodged around ESTs would greatly complicate any subsequent efforts to identify biologically meaningful structures and to develop them into commercially viable products; such attempts would have to work around preexisting patents held by enthusiastic sequencers like Celera, possibly necessitating the collection of licenses from multiple patent-holders.[30] Subsequently, conflict swirled around attempts to patent still another kind of fragment, single nucleotide polymorphisms or SNPs, which are single base pair mutations that occur in people at a rate of one per thousand nucleotides along the genome. These differences among individuals may be extremely useful for identifying a person's susceptibility to disease or response to drugs, but SNPs are mere markers of genetic variance among individuals and, like ESTs, are not obviously relevant to any bodily function. Holding patents on SNPs would once again enable holders to exclude others from exploiting them for drug development.

A political high-water mark of sorts for gene patenting came in March 2000, when U.S. President Bill Clinton and U.K. Prime Minister Tony Blair issued a joint statement declaring that the data from the Human Genome Project should be moved as quickly as possible into the public domain. The communiqué followed many months of intergovernmental negotiation, as if this were an issue of the greatest diplomatic significance. But scientists downplayed the event. At NIH, Francis Collins declared that the Clinton-Blair statement changed nothing in preexisting policy. Collins and others saw the statement as merely an injunction not to load up the human genome with thousands of disruptive patent claims. Nonetheless, Celera's and other biotech companies' stocks fell sharply in value immediately after the announcement, which some interpreted as a wholesale retreat from patenting DNA sequences.

A change of directorship at NIH and lobbying by the research community eventually pushed the U.S. Patent and Trademark Office to clarify its position on the patentability of gene fragments. Though a number of EST patents were granted, the PTO tightened up the utility requirement in January 2001, indicating that patents would be granted only for sequences used in identifying structures of predicted biological function.[31] As we will see, even this resolution did not completely satisfy the international genomic research community.

## A Place for Ethics: The European Debate

On the European side, the debate on biotechnology patents took a different turn, producing long delays in resolving the issue of animal patenting and in drafting new legislation for the European Union. The discourse, too, was markedly different. Patent controversies were linked from the start with questions of ethics—a term that rarely surfaced in U.S. patent deliberations even at their most heated. Not surprisingly, the settlement of some relevant issues was also different in significant respects. To put these differences in perspective, we begin with some key contrasts between the U.S. and European patent regimes.

European biotechnology patenting stands on two independent legal legs. One is the European Patent Convention (EPC), a 1973 international treaty in force since 1977, ratified by more than two dozen contracting states, including the members of the EU. The EPC provides patent protection equivalent to that available under national law within its members, and applicants may seek a patent in one or a number of participating states. EPC patents are awarded by the European Patent Office in Munich, which is not an organ of the European Union. The EPC differs from the American patent law in a few significant respects. Europe follows the more common first-to-file rule, under which the person who first files an application is entitled to the patent; by contrast, U.S. law since its inception has followed the first-to-invent rule, which awards the patent to the first inventor, regardless of who filed first. The first-to-invent rule is on its face more consistent with rewarding invention, but it is also less easy to administer and leads to more priority disputes than the rule that recognizes first filing.

More important from the standpoint of biotechnology patents is Article 53 of the EPC, which provides for two exceptions from the category of patentable inventions:

(a) inventions the commercial exploitation of which would be contrary to "ordre public" or morality, provided that the exception shall not be deemed to be so merely because it is prohibited by law or regulation in some or all of the Contracting states;

(b) plant or animal varieties or essentially biological processes for the production of plants and animals; this provision does not apply to microbiological processes or the products thereof.

Article 53 came into play in the debate over animal patenting and produced a more restrictive outcome than did *Chakrabarty*.

The second major legal provision governing European biotechnology patents is the EU directive on this subject, jointly approved by the European Commission and the European Parliament in 1998.[32] The product of some ten years of wrangling between Commission and Parliament, the 1998 directive too sets out a somewhat more restrictive framework of intellectual property protection than U.S. law. But for purposes of our analysis, it is the permeability of the boundary between law and ethics—a boundary that U.S. courts and interested parties like Francis Collins regard as sacrosanct and unbreachable—that sets the EPC and EU patent regimes most strikingly apart from the U.S. system.

## Oncomouse Meets EPC

The EPO received a patent application for the oncomouse from Harvard in 1985 and denied it in 1989 under Article 53(b), which the EPO construed as barring patents on transgenic animals. Harvard, along with DuPont, the sponsor of the research, appealed the decision, arguing that the term "varieties" in the convention refers only to products of breeding, not to newer techniques of genetic engineering. The appeals board agreed with this interpretation, thereby forcing challengers to resort to Article 53(a), the "ordre public" provision, which allows inventions to be excluded on ethical and moral grounds. Balancing risks and benefits, the EPO concluded that the oncomouse's purpose of facilitating cancer research was of paramount importance for the welfare of mankind, outweighing two other considerations— possible suffering to animals, and harm to the environment. After further legal proceedings, a patent on the mouse was issued in 1992, but new challenges again placed the patent in limbo pending a political resolution of animal patenting through EU legislation. The EPO decision established a precedent for case-by-case review of the ethical issues surrounding animal patenting, with no automatic guarantee of approval. For instance, the EPO later denied a patent on Upjohn Company's hairless mouse, which had been engineered to test products for stimulating hair and wool growth. These purposes, the EPO concluded, did not outweigh the possible risks of genetic modification to animals or the environment.

Though the oncomouse patent was eventually allowed, the applicability of Article 53 to other biotechnological inventions remained unclear. In 1995 the EPO reversed its own earlier position and refused a patent on a process for producing herbicide-tolerant plants.[33] Responding to a legal challenge by Greenpeace, the EPO decided that GM plants were, after all, varieties within the meaning of Article 53(b), and hence outside the class of patentable inventions. The uncertainty and unreliability of these decisions, coupled with growing numbers of unprocessed patent applications and worries about competitiveness vis-à-vis the U.S. and Japanese biotech industries, put enormous pressure on the EU to harmonize the disparate provisions of its member states concerning biotechnology patents. The European Commission took up the challenge of producing a Europe-wide directive, little knowing just how contentious the task would prove or how long it would take.

## Contentious Co-Decision

March 1995 witnessed a stunning defeat for the European Commission and a corresponding victory for the European Parliament, flexing its political muscle according to the new co-decision procedure under the Maastricht treaty. By a vote of 240 to 188, the Parliament voted down the patent directive that had been nearly seven years in the making. Most unusually, the defeat came on the bill's second reading, following intense lobbying by Green MEPs newly elected to the Parliament. Unlike the Supreme Court in *Chakrabarty*, the European Parliament provided a political forum for airing moral and ethical concerns and there were ears waiting to listen. Before the vote, Greenpeace sent to every MEP a postcard bearing the following message:

. . .

In the beginning, the Lord created heavens and the earth and all therein.
But scientists said let the seas be populated by fish containing human, mouse and rat genes.
And Hoechst and Monsanto said let the land produce plants and trees bearing fruit containing bacteria and virus genes.
And Ciba Geigy said let corn grow containing scorpion genes.
And Amoco said let there be tobacco containing hamster genes.
And Zeneca made all kinds of fruit and vegetables and they did not rot.
By the seventh day the work of the Lord had been undone and the companies saw they had recreated life for their own ends.
And the Parliament of Angels blessed the seventh day and made all life patentable.[34]

. . .

Uninspired as lyric, the text seems to have been more effective as propaganda. Its few short lines articulated many of the anxieties simmering just below the surface of debates on the governance of biotechnology: category violations, unnatural manipulation, playing God, corporate greed and ambition, and lack of effective regulation. Many in the Parliament also felt that insufficient safeguards had been included against the patenting of human bodies and body parts, or controversial processes such as cloning and germ line gene therapy. The draft patent law, in other words, had not made sufficient concessions to European ethical sensibilities.

The EP vote caught many patent experts by surprise, but anyone reading the tea leaves from the standpoint of European politics around biotech should not have been altogether astonished. European NGOs had worked hard from the early 1990s to link ethics to patenting. Germany's GenEthisches Netzwerk, for example, had collected thirty thousand signatures in its campaign to say "no to life patents," presenting them to the German justice minister in February 1991. The EP's own commissioned report on bioethics in Europe had called attention to the ethical dimensions of life patenting as far back as 1992.[35] Already in that document, the parliament's technology assessment staff had identified concerns about risks to animals and the environment, assaults on human dignity, and the impact of biotechnology patents on the conduct of scientific research.

The 1995 EP vote sent the directive back to the drawing boards for another three years of deliberation. By mid-1997 the elements of a compromise were visible as the EP agreed to the outline of a redrafted directive, but major environmental groups were still dissatisfied, and it took almost a year to iron out the wording of certain key provisions. The directive that was overwhelmingly approved in May 1998 incorporated numerous changes proposed by the EP.[36] Processes for human cloning, germ line gene therapy, and commercial use of embryos were expressly excluded. The directive also distinguished between "discovery" and "invention," disallowing patents for mere extensions in knowledge with no clearly identified technical application. Importantly, too, the directive provided for continuing ethical review through the Commission's European Group on Ethics in Science and New Technologies.

Mere adoption of the directive did not end the long agony over harmonizing European opinion with respect to patenting. In 2001 the directive survived a legal challenge by some member states at the European Court of Justice, which held that the law insufficiently respected human dignity and the integrity of the person. But by mid-2002 eight EU members, including France and Germany, still had not implemented a directive that should have been written into national law by July 2000. The European Commission threatened to take these foot-dragging members to court to resolve the legal and political impasse that still beset life patenting.

## Sequences of Trouble

Once again gene sequences were a major part of the trouble. The EU directive sought to use a strict concept of technical utility to resolve the conflict between those who categorically opposed the patenting of human biological materials and those who wanted to liberalize patent rules on these materials so as to encourage pharmaceutical innovation and compete with the liberal U.S. regime. In this spirit, Article 5(1) of the directive states, in wording provided verbatim by the European Parliament, that "The human body, at the various stages of its formation and development, and the simple discovery of one of its elements, including the sequence or partial sequence of a gene, cannot constitute patentable inventions." Article 5(2) and 5(3) relax this prohibition to the extent of allowing *isolated* gene sequences to be patented, even if they are identical to naturally occurring elements, but only if they are derived through a technical process and if they have an identifiable industrial application. Yet in 2003 the French Senate drafted a bioethics bill denying patentability to genes and gene sequences, contrary to EU law, while permitting research on stem cells derived from surplus embryos obtained through IVF.

Clearly, Article 5 of the biopatenting directive did not instantly establish a workable boundary between unpatentable discoveries of naturally occurring sequences of human DNA and invented, industrially applicable, and hence patentable, isolated copies of those same sequences. Ethical objections continued to arise in commentary on the directive. One line of critique took issue with the idea that something so fundamental to humanness as DNA sequences could be subjected to private ownership. International and European NGOs active on genetic issues also charged the EU with endorsing biopiracy and biocolonialism, thereby seeking to reframe an internal matter of EU policy in terms of global human rights. But even people committed to furthering genetic research and development expressed reservations about the extent of patent protection that should be granted for isolated DNA sequences.

A 2002 report of Britain's thoroughly mainstream Nuffield Council on Bioethics illustrates the intellectual difficulties that the EU directive failed to lay to rest. The distinguished panel of authors got into something of a metaphysical muddle in trying to restate the rationale for Article 5, as follows:

> As we have seen, scientific knowledge concerning a natural phenomenon is not eligible for patenting, since it is a mere "discovery." It follows from this that scientific knowledge about genetic information which is encoded in some naturally-occurring phenomenon is not eligible for patenting as such. But it does not follow that an artificial phenomenon that does not occur naturally (such as a molecule that has been isolated and cloned) that encodes genetic information may not be eligible.[37]

The distinction between scientific knowledge and genetic information, the report explained, is "a fine one" but crucial to the understanding of patent law.

A fine distinction indeed! Among the paradoxes it opens up is that, in the case of DNA sequences, the so-called artificial phenomenon (for example, an isolated gene fragment) is of interest precisely because it is homologous with a natural phenomenon; it encodes the same information and yet is *not* patentable until the sequence is isolated. The same biological material encoding the same information becomes subject to ownership claims only when it changes state from natural to artificial. Put differently, a length of DNA described as a specific sequence of base pairs is either not patentable or patentable depending, respectively, on whether someone has merely identified it or figured out how to isolate it. Discovery—the very discovery that prompted the material's isolation—then miraculously transmutes into invention; it is the process of material isolation that brings about the conceptual transubstantiation. The Nuffield panel seemed to recognize that this sleight of hand might appear problematic to some. At any rate, a second rationale for patenting, based more on a labor theory of value, was also advanced: namely, that DNA sequences should be patentable only when real work is done (preferably in a lab) to isolate them.[38] The panel was reluctant to concede that computationally identified sequences, using only mechanical computer power, might meet the test of inventiveness necessary for European patenting. The relevant distinction for the panel was no longer the tenuous conceptual boundary between natural and artificial, but the amount of work taken to produce the artificial elements in question. Acknowledging that U.S. law's nonobviousness standard is less demanding on this score, the panel defended the stiffer European requirements of inventiveness and utility as being more appropriate for patenting DNA sequences.

## Conclusion

Disputes over intellectual property law perhaps more than any other area of policy controversy have focused attention on the metaphysical attributes of the new biotechnologies. A tradition of rewarding only socially beneficial inventiveness prevented, in principle, the award of patents for mere discoveries of natural phenomena. Applying the law therefore required lines to be drawn between natural and artificial, discovery and invention, and useful versus merely novel technologies. The inquiry into what can be patented—that is, what constitutes patentable subject matter under the law—took on unprecedented gravity as the life sciences and technologies eroded clear distinctions between animals and machines and converted human DNA sequences into diagnostic and possibly therapeutic instruments. The extension of ownership claims deep into matter that used to be considered human or natural, and hence beyond the reach of private exploitation, created uncertainty and confusion for patent officers, judges, activists, and members of the general public. Taming these anxieties proved to be an unexpected challenge for patent law in Europe as well as the United States, and produced considerably different political dynamics and outcomes.

The U.S. court decisions in *Chakrabarty* and *Moore* succeeded in keeping ethical and political debates out of patent law and thus maintained the boundary between patenting and legislative politics. By contrast, on the European side, such concerns repeatedly overflowed attempts to create a workable patent regime around biotechnological products and processes. There are many reasons for this divergence, but surely an important one is that Europe had to accommodate intellectual property issues through new lawmaking, entailing parliamentary deliberation, whereas the U.S. Supreme Court in *Chakrabarty* took the existing law as given and interpreted it so as to exclude what the majority saw as irrelevant political concerns. *Moore* similarly used the technology of judicial interpretation to allocate property rights between patients and researchers in ways that favored commercialization. Patients' concerns were not ignored but were handled through the all-purpose administrative routine of informed consent.

The discrepancy between 1980 *Chakrabarty* decision and the 2002 Canadian oncomouse decision shows that metaphysical questions could have been raised and addressed even within the framework of statutory interpretation. That U.S. judges did not find the nature-manufacture boundary worthy of deeper analysis points to a thoroughgoing desacralization of the sphere of economic production in the United States, as well as a thorough indoctrination of the judiciary into regarding invention and the market as value-free. Human intervention in any form converts nature into commodity, places it in commercial circulation, and thus takes it out of the realm of the value-laden. The U.S. Patent Act, as construed in *Chakrabarty*, left no room for inquiry into the morality of choosing what to produce; if at all, such concerns would have to be addressed under the heading of usefulness, not in the evaluation of patentable subject matter. In Europe, the Patent Convention's "ordre public" provision and similar language in the EU biopatent directive opened a wider door for ethical concerns. Accordingly, the European discourse of intellectual property rights refused to settle into a technical discourse of legality as readily as in the United States. In its 2002 report, for example, Britain's Nuffield Council on Bioethics recommended that the EPO should seek guidance from the European Group on Ethics for its interpretation of "ordre public."[39]

The political controversies surrounding the patenting of DNA sequences in the United States raises an interesting puzzle. Why did concern surface with regard to these products when equally significant enlargements in the scope of patentability—from nonliving to living matter and from lower to higher organisms—had garnered nothing but praise from U.S. scientists? We must conclude that this step more than any previous one created competitive divisions within the very heart of biotechnological research and development. The patenting of DNA sequences pitted the interests of basic against applied research, public against private sponsors, and large pharmaceutical companies against smaller genomics firms, producing the new transatlantic rifts and alliances addressed by the Clinton-Blair declaration of March 2000. In other words, as in the case of organics labeling, a set of ethical concerns rose to public view in the United States only when they became closely aligned with interest group politics driven by economic concerns. We turn next to the implications of such cleavages within the biotechnology enterprise for relations among science, society, and the state.

# 9

---

## The New Social Contract

Science and technology emerged from the twentieth century's two world wars with their ties to the state immensely strengthened. The new relationship, which some called a "social contract,"[1] was perhaps most explicit in the United States, where no one doubted the utility of publicly sponsored scientific research after World War II's spectacular military and medical successes. Against a backdrop of universally acknowledged needs and unambiguously rendered services, Vannevar Bush, the distinguished MIT engineer and adviser to President Franklin D. Roosevelt, unveiled a plan for the federal government to continue its support for basic scientific research even in peacetime. His 1945 report, *Science—The Endless Frontier*,[2] served as the blueprint for what would become the National Science Foundation (NSF), formed in 1950 to fund basic research in all the natural and social sciences other than biomedicine; biomedical research had been funded since 1930 by the National Institutes of Health, descended from a federal medical research program begun in 1879. It was a simple bargain that Bush and the other architects of NSF envisaged: the government would supply money and agenda-setting prerogatives to the research community; in return, scientists would produce a steady flow of technically trained personnel and discoveries to advance the nation's health, prosperity, and welfare.

In Britain and Germany, too, strong wartime records of scientific and technological achievement influenced peacetime sponsorship of science by the state, with similar expectations of returns on public investment. Both nations created new organizational structures for the support of science in the postwar period. In Britain, the network of research councils was revamped in the mid-1960s, producing a set of specialized funding agencies for social sciences, natural sciences, and medicine. Successive reorganizations kept the basic commitment to science intact, although institutional responsibilities were frequently refined and regrouped; by 2002, seven research councils were active in the field, including the venerable Medical Research Council (MRC), founded in 1913, and the Biotechnology and Biological Sciences Research Council (BBSRC) formed in 1994 to fund the nonmedical biological sciences.

In Germany, the Federal Ministry for Research and Technology was merged with the Ministry for Education and Science in 1994, and the resulting structure was renamed in 1998 the Ministry for Education and Research, with a budget for "knowledge-oriented and cross-program basic research." The other major federal body with responsibilities for science funding is the DFG, whose role in fostering basic research in all fields is similar to that of the U.S. National Science Foundation and the British research councils. In addition, Germany supports several networks of publicly funded research institutes, including the elite Max-Planck Institutes for cutting-edge basic research, the Fraunhofer Gesellschaft for applied engineering research, the Helmholtz research centers for work involving large equipment and infrastructure (or classic "big science"), and a further assorted group of centers comprising the Leibniz network.

All three countries, along with the European Union, were early enthusiasts for biotechnology, and all saw the need to move the results of bench science quickly into commerce in order to reap competitive advantages from their investments in the life sciences. This imperative led in each country to a reexamination of the tacit social contract between science and the state. More specifically, each government had to reconsider a paradox that the social contract of the Vannevar Bush era had managed to mask. How could the ideal of an autonomous, value-free, and disinterested science be reconciled with a science that was at one and the same time economically productive and, especially in the case of the life sciences, a fertile source of medical, agricultural, and environmental innovation for a world straining to overcome the limits to growth?

The resulting tensions and conflicts conspicuously played themselves out at a site that was traditionally the most removed from political influence: universities, the historical locus of "pure" or "basic" science. In each country, state efforts to forge new links between universities and the biotechnology industry redrew the line between science and politics in unexpected ways. The enrollment into the market of what had once been seen as disinterested research produced political after-effects, redefining the status of both science and scientists in the public domain. As science became more useful and commercial, older ideas about science's necessary and well-merited autonomy lost their persuasive force, leading to new charges of cooptation and associated demands for accountability. Elsewhere, policymakers struggled to justify public support for the life sciences in terms of new understandings of the process of innovation and the relations between science and society. Attempts to accommodate these pressures took different national forms, reflecting different understandings about conflicts of interest and of knowledge as a public good.

In this chapter I explore in comparative perspective the transformation of basic, university-based biological research to bio*technological* research and the ramifications of this development for politics around the life sciences. Before turning to the individual country stories, I begin with a broader-brush picture of the changing social relations of science and technology at the turn of the twenty-first century and the implications of those changes for the governance of knowledge production. That picture has contradictory implications for democratic accountability. In each country, governmental efforts to build closer ties between academia and industry opened rifts between the practices of science and the demands of democracy. As a result, nation-specific debates emerged around the observation that science-state-industry collaborations designed for the advancement of public welfare may have had the contrary effect of introducing new zones of nonaccountability into the democratic management of knowledge production and use. Underlying these debates was a growing concern about the capacity of elected officials to discern the public interest and adequately represent it in dealings with science and industry. In building their alliances with science to foster wealth creation, had states lost sight of other values and needs of citizens in contemporary knowledge societies?

## The Changing Context for Science

The success of science as an instrument of national welfare is measured in many forms. The commonest is dollars and cents, either as the absolute amount a nation spends on research or as a fraction of its gross domestic product (GDP).[3] Other measures include the number of patents, published articles, or new scientific journals, whose exponential growth over 250 years led the historian of science Derek de Solla Price to observe, tongue in cheek, in 1963: "It is clear that we cannot go up another two orders of magnitude as we have climbed the last five. If we did, we should have two scientists for every man, woman, child, and dog in the population, and we should spend twice as much money as we had."[4] A different kind of indicator was suggested in 1984 by the political scientist Yaron Ezrahi, who observed the spread of science as a political resource in the United States. All social groups and movements, he noted, regardless of their standpoint on particular issues, seem to embrace science in order to realize their microutopian visions of perfection and progress.[5] Science, in consequence, is ever more widely distributed throughout society, but it has also become a private commodity produced to

227

further *uncommon* ends. By all these measures, the success of science over several hundred years has been phenomenal, but the enterprise of science has changed fundamentally in the process, and so have public expectations of it.

Succinctly put, a discussion of science policy that centered up to the 1960s mainly on preserving science's autonomy and integrity yielded by the 1990s to a debate about the most appropriate ways to regulate the purposes and processes of scientific production. The resulting problems of governance, however, were differently framed on the two sides of the Atlantic, as well as within Europe. Sketching that separation of political and discursive pathways is a necessary prelude to understanding the specific debates that arose around university-industry relations concerning biotechnology in each of the three countries.

### United States: Science under Siege

Pure science is another American Eden. Its loss is perennially mourned; attempts to regain the forsaken state of grace are a recurrent national project, even an obsession.[6] In the politics of science, the rhetoric of the fall harmonizes with that of restoring science's purity, especially when scientific knowledge is imported into complex social and political contexts. Notions of science's platonic specialness abound in the discourse of public policy, expressed in binary oppositions between good or sound science and science that is substandard because it is biased, not objective, or tainted by interests. University-industry relations in the field of biotechnology have provided a ready anchor for these dialectics.

Science, in Robert K. Merton's influential 1942 account, was represented as being detached from interests, universal, and owned in common by the human race.[7] Vannevar Bush's postwar report on federal science policy was consistent with that vision of science's ethical purity and public-spiritedness. *Science—The Endless Frontier* called for governmental funding of science but said nothing explicit about the promotion of technology. That silence itself was telling. The authors apparently assumed that the diffusion of basic knowledge into applications would be linear and unproblematic. With technological innovation commanding potentially huge economic rewards, it was reasonable to think that market considerations alone would sufficiently drive the transfer of scientific knowledge from bench to commerce. No further incentives had to be designed. State efforts to promote innovation could effectively be restricted to the sponsorship of "curiosity-driven" research by scientists who met their peers' criteria of merit. Fundamental knowledge was a public good that might be in short supply unless it was fueled by active support from the state.

Bush's game plan took several years to materialize, and even then it was only imperfectly implemented.[8] But in however modified and scaled-back a fashion the institutional platform for an autonomous, publicly funded science did eventually take shape: the National Science Foundation was established in 1950 to serve as the primary sponsor of basic research. The basic-applied distinction remained the touchstone for distinguishing work done in universities from that in industries, agricultural experiment stations, or other locations concerned mainly with the conversion of scientific knowledge into useful commercial products. For many years no one questioned that the best way to secure the quality and productivity of scientific research was to safeguard the autonomy of scientists, so that they could freely set their own research agendas and maintain their own systems of accountability. The instrument that guaranteed the productivity and integrity of research findings was peer review. Proposal review prior to funding held investigators accountable to disciplinary priorities, theories, and methods; editorial peer review ensured to varying degrees the credibility of published results, as well as their originality and interest.

Historians of U.S. science and technology policy have observed that the postwar consensus on science never was so monolithic as its idealized representation,[9] and signs of wear and tear in the social contract appeared in the 1980s. Recurring incidents of fraud challenged the reliability of peer review in basic science and, with it, the underlying justifications for science's autonomy. What began as isolated, flash-in-the-pan cases reported in the scientific press culminated in a blaze of national notoriety with the long-drawn congressional inquiry into an episode of alleged misconduct in an MIT laboratory headed by the Nobel laureate biologist David Baltimore. He and his colleagues were exonerated after years of investigation by local and national authorities, including at one stage the Federal Bureau of Investigation.[10] It was, while it lasted, a powerful assault on the Mertonian image of pure and disinterested science. In the end Baltimore was personally rehabilitated as president of Caltech, having earlier been forced to resign the presidency of New York's Rockefeller University.

Misconduct charges, however, left their traces in the form of greatly increased federal supervision of research and heightened suspicion among policymakers and the public that all was not in order in the house of basic science. Some saw the Baltimore affair as a sign that legislators were no longer satisfied with the old social contract's simple quid pro quo of money and autonomy in exchange for technological benefits.[11] Science was forced to account more explicitly for the money spent on it. One token of the altered relationship between the parties to the social contract was a reform of NSF's peer review criteria. Investigators were asked to justify proposals, and referees to assess them, not only on the basis of technical merit but also with respect to their broader implications for society; this two-step approach places a greater emphasis on science's social utility.

The idea of "science" as a unitary field of activity also began to break down as it became clear that scientific practices vary from one context to another, not only across disciplines, but—even more important from a political standpoint—across the varied institutional settings in which science is done. In particular, regulatory science, produced to support increasingly pervasive state efforts to protect publics against risk, was seen in many ways as a radically different enterprise from research driven by scientists' collective curiosity. As the product of different institutional and social constraints, regulatory science called forth different approaches to validation. In my 1990 study of U.S. expert advisory committees, for example, I noted that regulatory science is often subjected to what policymakers term "peer review."[12] Yet this exercise turns out, on inspection, to differ in crucial ways from the review of science in research settings. Regulatory review does not operate within disciplinary frameworks with well-settled standards or a core set of experts.[13] It is conducted most often by multidisciplinary committees rather than by specialists in a single field. The role of such bodies is as much to validate the novel methods by which agencies seek to define and investigate risk as to confirm the reliability of the resulting analyses. Like all custodians of scientific standards, expert advisory committees continually engage in boundary work, drawing lines between what will count as good science for policy purposes, and what will not. But the function of such bodies is both political and ontological; put differently, expert committees act as agents of coproduction. Many of the objects defined by regulatory science (GMOs or pre-embryos, for instance) do not actually function as concepts to govern with unless they are grounded in law *and* validated through the processes of scientific review and expert advice.

In practice, the conduct of regulatory science is often delegated to the private sector. Studies designed to test the safety and effectiveness of new technologies in the United States are largely carried out by producers, often under confidentiality provisions, and so are shielded from full public review. There is no rule that all studies done by industry must be made public, and potentially embarrassing information may therefore be kept hidden unless it is formally required to be disclosed. The extent of industry's knowledge-producing burden is often privately negotiated with regulatory agencies and may be affected by economic and political considerations that are not apparent to the public (setting maximum tolerated doses for study animals in cancer research is one common example). Resource constraints may curb audits and inspections of industry labs, leading to uneven or inadequate quality control. Rules exempting confidential business information from disclosure reduce the transparency of product- or process-specific information generated by industry. Finally, the limits of legal and regulatory imagination at any given moment place unseen bounds on industry's overall duty to generate information. Many environmental disasters have disclosed gaps in information and failures to synthesize available knowledge only after the accidental exposure of populations and ecosystems to industrial hazards.

Yet all these empirical observations about the constructedness, context-dependency, and incompleteness of regulatory science, and its consequent function as a political force, have left curiously untouched the discourse of good science that still permeates Western practices of governance. For instance, U.S. policymakers keep returning to peer review as the best means of validating scientific knowledge even when science is used far outside traditional research contexts—as in courtrooms or in supporting regulatory policy. Thus, a 1993 decision of the U.S. Supreme Court, *Daubert v. Merrell Dow Pharmaceuticals, Inc.*,[14] held that peer review was one of the criteria judges should use in deciding whether scientific evidence is admissible in court. In 2003 a unit of the White House Office of Management and Budget (OMB) proposed that scientific information used in formulating regulations with a significant impact on the economy should be subjected to peer review. OMB defended the proposal as a means for ensuring the integrity of policy-relevant science. The proposal was substantially revised after an outcry from scientists, academics, and agency officials, all charging OMB with politicizing peer review, but it illustrated the powerful hold on the U.S. political imagination of the image of science as being detached, and detachable, from the contexts in which it is practiced.[15]

The OMB proposal and similar moves fetishize peer review as a process innocent of any politics of its own; it can then be invoked to construct, or showcase, regulatory science as a domain largely set apart from values and politics. Peer review, as represented in these procedural moves, is imagined not as an arena of micropolitics on its own, a "fifth branch" of government in my terms, but as a kind of inanimate filter or membrane whose incorporation into decisionmaking can remove contaminating values and biases, thereby rendering science apolitical, or less political. In other words, peer review is thought to separate Mertonian "good science" from weak, unreliable, or politically tainted knowledge claims, regardless of context. To call for peer review, then, is to seek to restore a space for objective knowledge within the messy heart of political deliberation. The science produced in this discursively purified space can then be drawn upon as authoritative enough to legitimate the most controversial decisions of state. But the very creation of such a "politics-free" zone is of course an intensely political move, and comparative study reveals it, moreover, to be a move that is distinctively American.

### European Alternatives: New Modes of Production

In science policy as in many arenas of contemporary governance, European discourse has traveled on other roads, beginning and ending at different points from those that seem most natural to Americans. Like their U.S. counterparts, European observers of science and society have questioned whether the categories of basic and applied research hold much meaning in a world where the production and uses of science are tied to each other, as well as to surrounding social and political institutions, through dense networks of relationships. The result, however, has not been to try to return to an imagined prelapsarian state, as in the United States, by rebuilding the dilapidated walls between science and politics. Rather, analysts and policymakers have sought to reconceptualize the design of science-society relations with understandings about the social and political embeddedness of science built in.

European notions of peer review and accountability illustrate these transatlantic differences. Silvio Funtowicz and Jerome Ravetz noted years ago, for example, that the world of policy-relevant science contains knowledge of different degrees of perceived certainty. They characterized three types, each demanding its own approach to quality control: (1) normal science (derived from Thomas Kuhn[16]) for ordinary scientific research; (2) consultancy science for the application of available knowledge to well-characterized problems; and (3) "post-normal science," comprising the highly uncertain, contested knowledge generated in support of health, safety, and environmental decisions.[17] While disciplinary review may be effective within normal and even consultancy science, Funtowicz and Ravetz argued that the quality of post-normal science cannot be assured by standard methods of peer review. Instead, they proposed that work of this kind should be subjected to *extended peer review*, involving not only scientists but the multiple stakeholders affected by science's use in policy.

Looking at the structural characteristics of contemporary scientific activity, the authors of another influential European study concluded that the traditional disciplinary science of Vannevar Bush's "endless frontier" era has been effectively supplanted by a new "Mode 2" of knowledge production.[18] The salient attributes of this new mode, in the view of Michael Gibbons and his colleagues, include the following:

- Knowledge is increasingly produced in contexts of application (that is, *all* science is to some extent "applied" science).
- Science is increasingly transdisciplinary; that is, it draws on and integrates empirical and theoretical elements from a variety of fields.
- Knowledge is generated in a wider variety of sites than ever before, not just universities and industry, but also in research centers, consultancies, and think-tanks.
- Participants in science have grown more aware of the social implications and assumptions of their work (that is, they have become more "reflexive"), just as publics have grown more conscious of the ways in which science and technology affect their interests and values.

The emergence of Mode 2 science, Gibbons and his coauthors argue, has obvious implications for quality control. Besides old questions about the intellectual merits of work, scientists are now also being asked to address questions about the purposes of their research, the marketability of their results, and the conduciveness of their enterprise to social equality and welfare.

Accordingly, these authors say: "Quality is determined by a wider set of criteria which reflects the broadening social composition of the review system. This implies that 'good science' is more difficult to determine. Since it is no longer limited to the judgements of disciplinary peers, the fear is that control will be weaker and result in lower quality work. Although the quality control process in Mode 2 is more broadly based, it does not follow . . . that it will necessarily be of lower quality."[19] In Mode 2 science, then, integrity has for all practical purposes become synonymous with social accountability. The authors of this framework, all seasoned European science advisers and administrators, view the fabric of science as increasingly interwoven with, and hence answerable to, society at large. To insist on a separate space for basic research appears, in their conception, to be a holdover from another era. One implication is that the goal of public policy should be to foster the beneficial interaction between scientific research, technological innovation, and political agenda-setting—all of which in this model are inseparably linked together in a dynamic system of socially embedded production. Correspondingly, the Mode 2 framework leaves little room for reinserting ideas of a "pure" or apolitical science (let alone regulatory science) whose quality can be maintained through depoliticized forms of peer review.

## Biotech and the Academy: Three Styles of Collaboration

The discourses of governance, we have seen, are shifting and fluid, subject to place-specific histories, memories, and intellectual traditions; but the problems of governance are to some extent constant across cultural domains, at least among technologically advanced Western nations. American president, British prime minister, and German chancellor: all are responsible at election time for the felt economic well-being of their electorates, for the public's perceptions of collective safety and security, and for people's satisfaction with the technoscientific projects that confer, or reinforce, a sense of shared national identity and progressive forward movement. Even in domains so seemingly nonpolitical as science and technology policy, we see these countervailing pushes and pulls at work—the forces of convergence that propel governments to speed innovation and create jobs, and the forces of divergence that translate state policy back into the familiar terms of nationhood, invoking the concerns of specific imagined communities.[20] University-industry relations in the life sciences, and public responses to them, have brought these patterns into relief.

### United States: The Loss of Eden

Perhaps nowhere more than in the United States has biotechnology come to be identified as so much a national project. In the agricultural sector, the connections between innovation and nationhood remain veiled to some degree, reflected mainly in numerous surface indicators of being "first": first to release genetically modified organisms such as the Ice Minus bacterium; first to invent commercial products such as Bt corn and cotton, or recombinant bovine growth hormone; first to approve and export GM foods such as the Flavr Savr tomato; and first to reap the economic returns from agricultural biotechnology. There is, too, the rhetoric of "good science," which seeks to distinguish the rational U.S. approach to risk assessment and management of GMOs from the scientifically wrong-headed precautionary stance of America's European trading partners (see chapter 5).

For human biotechnology, which was slower to leave the starting post but did so with more fanfare, the identification with national goals has been more explicit. The Human Genome Project is big science American-style, a product of national talent, ingenuity, and resources, but equally a repository of national dreams that include the spread of American visions of progress to the rest of the world. It is, to begin with, a project for all humankind, or at least that highly diverse subset of it that makes up the U.S. population. The issue of *Science* that announced the complete sequencing of the genome in 2001 carried a telling illustration on its cover.[21] Highlighted against a faded background, in vertically arranged movie-poster fashion, were six faces—all movie-star beautiful—representing three ethnic groups, as well as both genders and the "three ages" of humans. This truly was a project to benefit all people, and not only in America either, as Francis Collins, director of the National Center for Human Genetics Research and a country music singer, made clear in his "This Is a Song":

Chorus: This is a song for all the good people
All the good people whose genome we celebrate
This is a song for all the good people
We're joined together by this common thread.

This is a song for all of the families
Who're looking for answers to come our way
We're grounded in science, in the Genetic Alliance
In faith and in hope for a brighter day.

This is a song for those leaders of science
Who worked in six countries, including the USA,
Germany, China, Japan, France, and England
They worked without resting and gave it away.

This is a song for all ELSI people
Who see the bright promise but also the fears
They press for awareness, for justice, for fairness,
They're part of our conscience, our eyes, and our ears.

It's a book of instructions, a record of history
A medical textbook, it's all these entwined
It's of the people, it's by the people,
It's for the people, it's yours and it's mine.

Just as Collins embodies many American virtues in a single person—elite scientist, talented bureaucrat, and musically minded, regular guy—so his song brings together disparate symbols of Americanness in a single text: family, science, justice, generosity, therapy, internationalism, and, capping it all, Abraham Lincoln's all-encompassing ideal of democracy. If the human genome is the constitution of our race, then, as imagined by the head of NCHGR, it truly is a constitution for all the people. There is nothing exclusionary, secret, or removed from public control in Collins' vision of genome research or its outputs: "It's yours and it's mine."

Enter reality. In 1980 the U.S. Congress passed a piece of bipartisan legislation, the Bayh-Dole Act,[22] which had extraordinary repercussions on the nature of the university, and particularly on research in the life sciences. The law's premises and aims were uncomplicated. While the Bush-era social contract had produced significant discoveries in university laboratories, the sponsors of the act thought that these were in danger of remaining locked away in academic ivory towers, where they might languish indefinitely without producing benefits for society. Incentives were needed to get discoveries hooked up with potential developers, and what better incentive than to let the discoverers themselves share in the fruits of commercialization? The act therefore authorized federal grantees to patent discoveries resulting from research they had done with federal funds; it changed the longstanding presumption that publicly funded work could not be privately owned and exploited.

This simple step transformed the intellectual landscape of the American academy, converting high-powered university labs into de facto incubators for industry, and reconfiguring in many cases even the physical appearance of university towns. Bayh-Dole was read as entailing a duty to commercialize on the part of researchers who contract with the government. Universities, for their part, enthusiastically complied. Between 1979 and 1997 university-held patents rose nearly tenfold, from 264 to 2,436, as compared with a twofold increase in the overall number of patents during that period.[23] Physically, the built environments around major research universities started changing. One dramatic example was the conversion of Kendall Square around MIT in Cambridge, Massachusetts, from a depressed and semiderelict industrial zone into gleaming high-rise homes for biotech and computing start-up companies. In 2002 the British Consulate-General set up its local headquarters near MIT at One Memorial Drive in Cambridge, signaling that diplomatic relations between Britain and the United States were thenceforth to be reconceived as trade and industry relations. The consular website made this message explicit: "The Consulate works with the leaders and innovators on the cutting edge of

235

biotech, science and technology across New England on a wide range of areas of common interest, including the ramifications of new technology on public policy."

A concurrent development was the rising influence of industrial funding on university research. By 2003, industry's share of funding for academic research and development was only 7.7 percent, as compared with 58 percent from the federal government, but this support was growing more rapidly than all other sources.[24] One agreement that garnered intense attention was a 1998 deal struck by the giant biotechnology company Novartis with the Department of Plant and Microbial Genetics at the University of California, Berkeley. Under the agreement, the company provided Berkeley twenty-five million dollars in return for rights to review dissertations prior to publication and to negotiate licenses on roughly a third of the department's discoveries, as well as unprecedented representation on the research committee responsible for overseeing how the money would be spent.[25] Controversial from the start, the deal became even more so when Ignacio Chapela, a member of the contracting department, published his contested study on gene transfer from bioengineered to native varieties of Mexican maize in *Nature* (see chapter 4). Chapela and his student coauthor David Quist had vocally opposed the Novartis agreement, and many saw the furor over their paper more as an act of political revenge than of scientific gatekeeping.[26] Berkeley's eventual denial of tenure to Chapela in 2003 quelled the localized hostilities but did little to alleviate the deeper concerns of critics inside and outside the university. An external review commissioned by the university found that the whole matter had raised serious issues of governance, though Novartis, it seemed, had neither captured the department's scientific agenda nor greatly gained from its investment.[27]

Two decades after the passage of the Bayh-Dole Act, the idea of disinterestedness in U.S. academic biomedical research lay in tatters. Several vigorous lines of criticism emerged. First, observers decried what they saw as unseemly profit-seeking behavior on the part of universities. In their influential article in *Atlantic Monthly* on the "kept university," Eyal Press and Jennifer Washburn described the emergence of an aggressive corporate culture in university technology-transfer offices, leading at times to practices bordering on the unethical. Second, many called attention to the secrecy of academic-industry agreements, which dealt a body blow to the core academic virtue of openness. The atmosphere of secrecy extended beyond the reasonable periods of nondisclosure written into contracts to enable corporate sponsors to file for patents. Self-censorship by scientists through fear of jeopardizing corporate relations was another looming concern. Indeed, it seemed to some that the Mertonian norms had not so much been eroded as stood on their head, as norms to be avoided or ignored by successful scientists. Sheldon Krimsky, a longtime critic of campus-commerce entanglements, noted that in these changed times it was the commercially invested scientist who commanded the highest social prestige.[28] Even ethical training programs for working scientists seemed at times to adopt a presumption of nondisclosure, and to require special justification for the free exchange of information among colleagues and students.[29]

Most troubling of all was the suggestion that closer university-industry relations had led in nontrivial numbers of cases to the distortion of results or to their outright suppression. Press and Washburn were not alone in reporting episodes in which companies had doctored manuscripts beyond limits considered appropriate by the researchers.[30] Leading biomedical journals such as the *New England Journal of Medicine* adopted increasingly stringent disclosure guidelines for conflicts of interest that might have biased the reported results. With the nation's premier biological scientists tied up in multiple commercial relationships, it even became difficult for journal editors to identify referees who could render impartial judgment on the quality of papers submitted to them.

Not everyone agreed that these problems were grave enough to justify public concern. The old social contract, many believed, was idealized in the extreme and naïve to a fault in modeling the links among science, technology, and society. The emerging new social contract for biology and biotechnology, as reflected in the Bayh-Dole Act, some suggested, is more solidly grounded in the realities of production and the shift from a Mode 1 to a Mode 2 system of knowledge generation. The new arrangement, after all, recognizes that knowledge does not evolve into applications without incentives; that new organizational forms like start-up firms are needed to convert intellectual capital into products; and that universities never were so detached and pristine as imagined by some mid-twentieth-century scientists and policy scholars. What concerns us here, however, are the implications of this brand of pragmatism for deliberative decision making about the biosciences and technologies, and on this front U.S. policy garners mixed reviews.

In promoting rapid transfers of knowledge from academy to industry, the Bayh-Dole regime privatized discussion of the larger social purposes of knowledge creation—questions, for example, about the objectives and possible distributive consequences of biological research and development. Choices of where to invest are made in the new dispensation largely out of public view, through the entrepreneurship of individual grantees, investigators, and investors. As a partial consequence, the university-industry complex has lost precisely those virtues of transparency and organized self-criticism that had justified science's autonomy under the postwar social contract. Criticism of the new arrangements expresses itself as a yearning for lost purity, as commentators advocate a reversion to a less commercial culture of research in the life sciences. We will return to the implications and likely effectiveness of this critique at the end of the chapter.

### Britain: U.K. Science, Plc.[31]

British science policy, too, is partly driven by narratives of failure and loss, but those are different from the American myth of science's fall from grace. Possessing one of the highest-ranked research enterprises in the world, Britain nevertheless is widely perceived as doing well in pure science but falling short in translating the results into industrial productivity.[32] There are many tales of science's naïve or ill-considered detachment from commerce. One story of particular relevance to biotechnology concerns the discovery of monoclonal antibodies in the mid-1970s by César Milstein and Georges Köhler. The two Cambridge-based scientists were interested in the functioning of the immune system, and in the course of their research they discovered a way to produce a new kind of biological construct known as a hybridoma. These composite structures fuse together an antibody-producing cell that can recognize and attack specific disease-causing organisms, such as germs and cancer cells, and a tumor cell that keeps replicating indefinitely. Hybridomas proved to have endless applications in modern biomedicine, particularly as diagnostics but also as agents for drug delivery in cancer treatment and other therapies. Monoclonal antibodies became a multibillion-dollar industry, but neither Milstein and Köhler nor their sponsor, the U.K. Medical Research Council (MRC), patented the invention. Their results were published in 1975 as a letter in *Nature*.[33]

Milstein himself saw no utility to the technique and believed that it would be wrong to deny access to other scientists for years while someone figured out what uses hybridomas might possibly have. More interestingly, the National Research Development Corporation (NRDC), the government body with the right of first refusal on inventions arising from publicly funded research, also saw no reason to seek a patent on monoclonal antibodies.[34] A 1976 letter from an NRDC official to the MRC famously observed: "It is certainly difficult for us to identify any immediate practical applications which could be pursued as a commercial venture, even assuming that publication had not already occurred. I would add that the general field of genetic engineering is a particularly difficult area from the patent point of view and it is not immediately obvious what patentable features are at present disclosed in the *Nature* paper."[35] The test of "immediate obviousness"—so natural to an empiricist knowledge culture like Britain's—seems quaint in today's world of aggressive commercial exploitation, where just the isolation of a genetic component is alone almost enough to justify a patent. The hybridoma story reminds us that the 1970s were indeed a different era from the early twenty-first century with respect to university-industry relations. Back then, the utility of a scientific discovery had to be shown to be believed; today the ethos is to stake territorial claims around the smallest biological breakthrough and to market it aggressively in the hope that the claim will eventually prove profitable.

Another kind of loss befell British science during the Thatcher years, resulting from severe cutbacks in university funding that saw support for science dwindle along with resources for all other academic fields. By 1993 Britain was trailing not only the United States but also Japan, Germany, and France in the percentage of GDP spent on science. University researchers formed a new lobbying organization, Save British Science, to stop the attrition in the national science budget and stanch the brain drain occasioned by some of the nation's most talented researchers being lured to greener pastures abroad. By the time New Labour came to power in the 1997 election, the time was ripe to rewrite the script for science policy, along with much else in British politics.

A consultant's report produced for the Labour Party by the Science, Engineering and Technology (SET) Forum provides some indication of the ways in which that challenge was framed shortly before Labour's electoral landslide. The SET analysts rejected two earlier models of science policy as unacceptably linear: science as "engine of change," or "science push" (conforming roughly to the postwar U.S. paradigm); and science as "problem-solver," or "demand pull." In place of either, they proposed a new Phase 3 thinking that would acknowledge a greater degree of institutional complexity in science and the need for proactive state management to secure technological innovation. This analysis came very close to translating into practical terms the Mode 2 framework described above. The Conservatives, according to Labour's SET analysts, had begun to realize the nonlinear character of the problem since at least the early 1990s, following Thatcher's relinquishing of power, but an overly purist free-market ideology still held the opposing party back from effective intervention.

A notable feature of the SET analysts' way of thinking was the inclusion of civil society as a partner in the formulation of science and technology policies. As the authors observed:

> The future is undeniably complex; many more actors are involved. Public understanding of science as a dialogue between science and the public has to be strengthened as part of this process. We need not only SET policy-makers and advisors who understand the Phase 3 agenda but a government with a vision of a fair and socially inclusive society able to work flexibly and responsively with many different participants. Empowering citizens to shape and make the innovative society has to be at the heart of Labour's SET policy.[36]

Successful science policy, in other words, has to be rooted in a reinvigorated and just democracy, in which one of the topics of public deliberation would be science.

How effectively did Tony Blair's government follow through on these prescriptions in relation to biotechnology policy? There is no question that science rose on the political agenda in Labour's first two terms back in power. A symbolic high point was Blair's speech to the Royal Society in May 2002, hailed as the first full speech on record by a British prime minister to that august body.[37] Contradicting a group of scientists in Bangalore, India, who had told him that Europe had "gone soft on science," Blair sang a paean to the achievements of British science and urged that the negative views of a small handful of ideologues not be allowed to impede progress in biological as well as other areas of research. He praised his government's increased support for science, mentioning budgets rising at 7 percent a year. Looking at developments around major academic centers of excellence in the life sciences, however, one is tempted to conclude that more was achieved in making accomplished scientists behave like aggressive entrepreneurs—in much the same way as their successful American counterparts—than in realizing the dream of a democratically robust, inclusive, and participatory process for directing science policy.

To the extent that one man's career can be taken as emblematic of a national trajectory, it is that of Professor Sir Tom Blundell, holder since 1996 of the Sir William Dunn Chair of Biochemistry at Cambridge University and, in 2003, head of the Department of Biochemistry. Trained as an x-ray crystallographer in the famed Oxford laboratory of the Nobel laureate Dorothy Hodgkin, Blundell led a successful researcher's life until 1990, when he became head of the Agriculture and Food Research Council (AFRC). Winds of change were blowing across British science policy in the immediately following years, as exemplified in a 1993 white paper of the Office of Science and Technology that recommended a major overhaul of the state's mechanisms for steering science. The plan included a restructuring of the scientific research councils, and Blundell became the head of the new Biotechnology and Biological Sciences Research Council, formed in 1994 through a merger of the AFRC and parts of the Science and Engineering Council, with an annual budget upward of £215 million.[38] Throughout this time, Blundell continued on the faculty of Birkbeck College. He also became the cofounder of Astex Technology, a drug discovery company based on a technique that Blundell had developed, which was listed in September 2003 as twelfth of the twenty fastest growing technology companies in Britain.[39]

Interviewed at Cambridge, Blundell said that companies like his were able to offer some of the most attractive options to talented researchers. Big pharmaceutical companies seemed to be in trouble at the turn of the century, as evidenced by constant mergers, and many scientists craved both more security and more hands-on control of company policy than are offered by giant firms like GlaxoSmithKline or Pfizer. Blundell stressed that scientists could do very well financially through employment at smaller firms. A Cambridge professor's salary, he noted, is limited to no more than forty thousand pounds annually and offers few privileges; a scientist in a company like his could earn as much again in ten days of work. No wonder, then, that scientists formerly content with the recognition of a Royal Society membership now want to form companies *and* do good science. This is not a matter of private benefit alone in Blundell's assessment. Britain, he said, has little chance of surviving as a nation "unless we live by our wits"; there are no "natural resources" other than scientists' brains in the country, unlike in the United States.[40]

Modern nations, however, cannot live on their wits alone. It takes infrastructure to exploit—and perhaps even produce—the "natural resource" of scientific talent. There is, to begin with, the huge public investment in science education, which the contrarian *London Times* columnist Simon Jenkins once dismissed as a failure because it creates a labor force for big science without engaging or interesting the majority of students.[41] Closer to home, Blundell's company Astex is located in the Innovation Park owned by St. John's College in Cambridge. Besides the park, St. John's operates an Innovation Centre with a usable space of approximately 100,000 square feet, housing some fifty-five companies employing over nine hundred people. One expert on innovation in Cambridge remarked that, with 99 percent occupancy, this was the best investment that St. John's College ever made.[42] Trinity College similarly owns the Cambridge Science Park, founded in 1970 on a derelict wasteland used by the U.S. Army to prepare vehicles and tanks for the D-Day landings in Europe. Following a period of slow growth in the 1970s, the park took off as a home for high-tech companies in the 1980s, especially after the British Technology Group, successor to NRDC, lost its monopoly on patenting publicly funded research in 1985 and venture capital began flowing more freely. These aspects of research policy, by which arguably the rich get richer, have never been widely debated from the standpoint of the public weal.

The area around Cambridge is on most accounts a success story for technology transfer. It was by the beginning of the twenty-first century home to some twelve hundred companies employing about thirty-five thousand people. Cambridge was also host to one of six "genetics knowledge parks" the government announced in January 2002; these publicly funded sites were to bring together clinicians, scientists, academics, and industrial researchers to speed the development of new genetic tests and treatments. To be sure, this regional prosperity drove up property values and made Cambridge considerably less livable for faculty and staff not tied to the high-tech and biomedical boom. The consequences of this kind of lopsided intellectual development in and around a university are difficult to measure and unlikely to yield any easy consensus.

More important for us is the question whether localized job creation in fast-moving technoscientific domains such as biotechnology has actualized the vision of an innovative society, shaped by citizens and oriented toward social justice, as contemplated by Labour's own SET Forum consultants. Clearly, individual creative scientists have lost their inhibitions about linking up with industry to pursue paths of private and, in their view, public gain, but this model of technology transfer is not a substitute for wider societal engagement in policymaking for science. Moreover, whatever one concludes about Britain's success in generating financial support for science, it remains unclear whether the growth of the St. John's and Trinity College science parks is driven by market forces, led by a loosening of controls on private initiative and capital, or (less probably) by a deliberate government policy based on Mode 2 (or, in the SET Forum's terms, Phase 3) thinking.

### Germany: Back to the Future

There are two kinds of stories told in the last twenty years or so about German science and innovation in relation to biotechnology, and both are stories of crisis and barely controlled decline. One is from an outsider perspective. This dwells on the organization of German science and its possible implications for such performance measures as papers or citations per person—a count that has shown Germany lagging behind a number of smaller scientific nations, as well as the United States, United Kingdom, and Canada. Robert May, the former science adviser to the U.K. government, concluded in a 1997 study that the conduct of basic research in a university environment, in the presence of undergraduate and postdoctoral students, might lead to greater productivity over time than in dedicated institutes such as those of Germany's vaunted Max-Planck system: "The peace and quiet to focus on a mission in a research institute, undistracted by teaching or other responsibilities, may be a questionable blessing."[43] The massive overhaul of German higher education in the early years of the twenty-first century will undoubtedly affect any future assessments of this kind, but the consequences of these moves cannot yet be foreseen.

The internal story told by German scientists and science observers well into the 1990s had more to do with the impossible conditions for biotechnological research created through a mix of public hostility and responsive overregulation by a panicky government. An early warning signal was the decision of major German chemical companies to sponsor basic research away from home, especially in the more welcoming climate of the United States. An agreement between Hoechst and Boston's Massachusetts General Hospital to create a new molecular biology department in the early 1980s was emblematic of the troubled relations between industry and basic science in Germany.[44] Other German firms that set up biotechnology facilities in the United States in the same period included Henkel and Bayer, but their focus was not so much on affiliations with U.S. research centers as on building their own facilities or collaborating with some of the leading U.S. biotechnology companies, such as Genentech, Chiron, and Biogen.[45]

Public opposition to biotechnology remained intense throughout the 1980s, and Hoechst again was a focal point for protest. Litigation blocked for years the opening of a factory that the company had built outside Frankfurt to produce genetically engineered insulin, prompting fears of a brain drain from Germany in the life sciences.[46] That controversy led in due course to the passage of the 1990 Genetic Engineering Law, which established a broad-based legal framework for regulating biotechnology and may have helped over time to allay public concern. At the time of its passage, however, the law was seen by scientists as unnecessarily restrictive and more likely to hinder than to promote research in an area where they saw Germany as rapidly losing ground to the United States and Japan.

By the mid-1990s public concern had evolved in new directions. The Green Party dropped its former absolutist position against genetic engineering and gradually accepted the idea of red biotechnology, that is, the techniques' medical applications. Still, stark divisions remained between university science and industrial development. Peter-Hans Hofschneider, a biochemist and Max-Planck Institute researcher, attributed Germany's consistent failure to achieve productive technology transfer in the biotechnology sector to cultural factors that had created, he said, a "two-camps mentality," separating basic biological research in the ivory tower from profit-oriented development in industry.[47] Hofschneider regretted the lack of interinstitutional cooperation that had marked the glory days of German chemistry, from the late nineteenth century through the First World War. In this, he perhaps underestimated the unease that thoughts of a too-close involvement between science and industry still engender among many German intellectuals and politicians. It was, after all, this very coalition that had provided the technological basis for modern chemical warfare and, later, acquiesced in the purging of Jewish scientists from German universities, as well as in the shocking abuses of science and medicine in the Holocaust.[48]

The German government, however, concluded by the latter half of the 1990s that new initiatives were needed to bridge the gap between universities and industry and to reposition German biotechnology as a competitive force in national and international markets. One instrument devised for this purpose was "bioregions"—areas (similar to Britain's genetics knowledge parks) in which a dense concentration of R&D operations, supported by a favorable legal framework, would facilitate innovation and produce cadres of researchers who would not be afraid to cross institutional divides. To stoke the spirit of enterprise, the Federal Ministry for Education and Research conducted a BioRegio competition that produced three winners from among seventeen participants, located respectively in Munich, Rhineland, and the Rhineland-Neckar triangle. A further special region was established in Jena, presumably to boost the economic performance of the former East German states. The winning regions received additional funding from the federal government in expected exchange for their ability to develop private sources of capital. These steps, as the ministry took pains to point out, ended the period of

243

CHAPTER 9

uncertainty for German biotechnology, even reversing to some extent the brain drain of earlier years. By 2002 some 465 new biotech firms were operating in Germany,[49] and Munich in particular had emerged as a mecca for biotechnology comparable to Cambridge in Britain and the Boston-Cambridge area in the United States.

The theme of nation (re)building is hard to miss in official German documentations of these developments. Whether in the graphics on the ministry's website—where a diagram like a protein molecule in the shape of Germany indicates the nationwide coverage of the bioregions—or in the ministry's written reports, the resurgence of biotechnology is told as a phoenix tale of national resurrection and renewal. It is crucial to this story that the developments in the biotechnology sector span the entire nation. The "bio" prefix (attached to words like region, future, and profile) serves in the process as a metaphor that signals Germany's reawakening to "life" after the somnolence of the postwar half-century and the decade-long aftershocks of reunification.[50] *Bio* has become a word with which to reimagine the German community while reinvigorating a stagnant economy and what many critics see as a spent, exhausted welfare state.[51]

And what of the twin themes of politics and democratization in this process of scientific and industrial reconstruction? It is easy to discern in the planned and orderly development of the bioregions a continuation of the consensual and corporatist approach to policymaking characteristic of Germany throughout the postwar period.[52] Even the competition for the bioregions was conducted within a framework of state planning and administratively rationalized regional development. As related in previous chapters, however, debates around the uses of biotechnology continued to surface in both the agricultural and biomedical sectors, and—as signaled by concurrent, politically divisive debates on the importation of stem cells and the immigration of foreign-born computer scientists and engineers in 2002[53]—understandings of Germany's national identity and role in the world remained far from resolved. Approaching technology transfer in the life sciences as a technical problem, the government succeeded in producing new templates for university-industry collaboration and interregional competition. But whether and how these technocratic moves would translate into a deeper politics of the life sciences, engaging citizens as well as scientists and entrepreneurs in discussing the aims of research and development, remained an unresolved question in the early years of the twenty-first century.

## Conclusion

Science and the state are joined together in networks of mutual support in every Western nation, but we have seen that the social contract for science has

244

not been interpreted uniformly across disparate political cultures. Relations between science and the state are conditioned in part by similar presumptions concerning invention and progress. At the same time, the politics of science policy reflects different national understandings of the place of science in society, different experiences of success and failure in technological innovation, and different institutional traditions of enrolling science to serve to social ends. There is, then, no single social contract for science across Western societies, but rather particular historically and culturally situated relationships speaking to varying expectations of how science and politics should be joined together for public purposes. These discrepancies are so fundamental as to constitute, in effect, distinct constitutional positions for science in different polities: that is, the rights and duties of science and scientists vis-à-vis the state are not identical in Britain, Germany, and the United States. In turn, these fundamental dispensations account for cross-national differences in the state sponsorship of biotechnology and in the trajectories followed by research and development in the life sciences.

From the standpoint of public policy, all three countries have used some of the same levers to promote technology transfer—fiscal incentives, decreased regulation, liberalized intellectual property rules, and a gradual backing away from the ideology of pure science. Yet looked at in detail, the three national strategies differ substantially in their tacit modeling of what drives innovation, and correspondingly which policy instruments are most likely to foster the desired partnerships and exchanges of knowledge. U.S. policy centered on the reframing of federal grantees from providers of a scarce public good—pure science—to agents with a positive mission to produce social benefits from the state's investment in their research; both universities and the researchers they employ became transmission agents in the post-Bayh-Dole regime of biotechnology development. British policy, by contrast, bypassed universities in the first instance, adding to the woes of an already strained cultural resource; instead, government policy favored the reengineering of scientists themselves into eager corporate actors, with strengthened ties to the market. This strategy did not so much bridge the ancient British divide between science and industry as produce a new class of prosperous scientist-entrepreneurs whose workplaces appropriately continued to proliferate around university campuses. In Germany, it was the bioregions, representing a new hybridization of science and society, that were the prime beneficiaries of federal policy. These corporatist structures knitted the nation together in a shared enterprise of economic and technological modernization; their overt mission was to restore reunited Germany's national pride and international standing in basic research and its applications.

Unquestionably beneficial from the standpoint of local job creation and growth, these developments are more difficult to evaluate in relation to the democratic governance of science. We have seen that, as the contexts for doing science diversified, questions about the integrity of research increasingly

245

merged with calls for greater public involvement in assessing the aims, costs, and benefits of new science-based technologies. In no scientific sector was this shift more apparent than in the life sciences in the later twentieth century. Yet as we survey three national strategies for moving scientific results more rapidly into the domains of application and commerce, we do not find in any country a complementary attention to new modes of engaging publics in guiding, or even responding to, those state-sponsored transfers of knowledge. This deficit is particularly surprising in the two European countries, where high-level academic and political commentary had drawn attention to a democratic public's indispensable role in securing socially just paths of innovation. The U.S. debate about university-industry relations, for all its liveliness, remained tied to the discourse of scientific purity, with critics pleading for an arguably impossible return to a state of original grace. It is no small irony that all three nations found it easier in the end to industrialize pure science—dismantling the institution that Britain's Michael Polanyi had once celebrated as the very model of a platonic republic[54]—than to incorporate any democratic values other than wealth creation into the public management of research and development.

# 10

Civic Epistemology

The politics of biotechnology since the mid-1970s illustrates the increasingly intimate collaboration between the life sciences and the state, prominently including Britain, Germany, the United States, and the European Union. But how have citizens participated in these relations; and how, in turn, has the rise of science as a space of visible politics affected the role and meaning of citizenship itself? American writing on science policy has largely sidestepped these questions. The social contract discussed in the preceding chapter is generally conceived as a compact between two parties only—science (comprising both academia and industry) and the state. This dyadic representation leaves out of account the publics on whose behalf, or at whose behest, states enter into their agreements with science. Democratic governments are presumed to be capable of discerning their citizens' needs and wants, and of deploying science and technology effectively to meet these. Citizens, once they have elected representative governments, are not seen as retaining independent stakes in steering science; nor are they thought to need an autonomous position from which to oversee the partnership of science with the state. Science, after all, is subject to the transcendental norm of truth-telling, while the institutions of constitutional government guard against abuses of power by the state.

European social theorists have seen citizens as deeply embroiled in the politics of science and technology, but they too have been reluctant to accord imagine citizens as having an active role in the production and use of scientific knowledge. Attention has focused instead on the state's appropriation of the human and biological sciences as instruments of governance, put to use in sorting and classifying people according to standardized physical, mental, and social characteristics. Through these techniques, states and state-like institutions are able both to articulate broad social agendas—for example, to reduce disease, poverty, ignorance, or violence (all defined or redefined, according to standardized biological markers)—and also to discipline people into accepting science-based classifications of themselves and their behaviors, as part of the natural order of things. Naming this phenomenon "biopower," Michel Foucault explored the multiple pathways by which human life has become the stuff that modern institutions of governance seek to control and manage with the aid of science.[1] Others have written compellingly about the effects of classification on the governed, whether by "looping back" to alter people's subjective consciousness of their own identities, or by rendering entire classes of individuals marginal and dispensable when it suits the purposes of the state.[2]

Yet as we survey the democratic politics of biotechnology over a quarter-century or so, these accounts of the relations among science, state, and society seem strangely incomplete. In liberal democracies the public is seldom so completely devoid of agency with respect to the production and application of scientific knowledge and its applications as in the American social contract or under Foucauldian regimes of biopower. Citizens after all are the primary audience for whom the state enacts its scientific and technological demonstrations. As a play could not exist without spectators, so the grand narrative of progress through science and technology demands assenting publics to maintain its hold on the collective imagination, not to mention the collective purse-strings.[3] Not only the credibility of science but the utility of the state's knowledge-producing endeavors must repeatedly be brought home to citizens. Big-ticket, big-science enterprises, such as the space station or the Human Genome Project, could not exist without popular support, and steady public funding for basic research demands the kind of political buy-in that the biomedical sciences have enjoyed for the better part of a century in the United States. Public reactions, as we have seen throughout this book, have played a crucial role in shaping the interactions of government and the life sciences, and comparative analysis suggests that it is the three-cornered relationship among science, state, and society that gives the politics of biotechnology its cultural specificity. These observations point, in turn, to a need for new theoretical resources to bring the missing public back into studies of science and democracy.

There are added reasons for inquiring more deeply into the role of citizens in directing the politics of the life sciences. Unlike atomic physics, biotechnology has not yet won wars, either real or metaphoric, as in the so-called war on cancer; indeed, in the wake of the terrorist attacks of September 11, 2001, biotechnology is more strongly associated with the insidious threat of bioterror than with national defense. Nor has biotechnology scored symbolic victories on the scale of the U.S. space program's moon landing or achieved the near-mythic success of the Green Revolution. By the early twenty-first century, after some three decades of active state promotion, public support for agricultural biotechnology remained ambivalent, while most of the predicted benefits of human biotechnology still lurked tantalizingly around the corner. Biotechnology companies fill short of delivering the economic miracles expected of them. Under these circumstances, it is easy to see how popular assessments of biotechnology's costs and benefits might diverge from those of scientists and the state.

In the absence of universally recognized public proofs, it is the broad acceptance of new technologies that requires explanation. The credibility of state policy must rest in part on foundations that have yet to be explored—on citizens' general proclivity to accept the knowledge claims and demonstrations of efficacy advanced by the state. These patterns of civic engagement with governmental reasoning may also help explain cross-national variation in politics and policymaking around biotechnology.

I argue in this chapter that how publics assess claims by, on behalf of, or grounded in science forms an integral element of political culture in contemporary knowledge societies. There are in any functioning society shared understandings about what credible claims should look like and how they ought to be articulated, represented, and defended—and these understandings vary across well-defined cultural domains such as nation-states. Public reasoning, then, achieves its standing by meeting entrenched cultural expectations about how knowledge should be made authoritative. Science, no less than politics, must conform to these established ways of public knowing in order to gain broad-based support—especially when science helps underwrite significant collective choices. I use the term *civic epistemology* to refer to these culturally specific, historically and politically grounded, public knowledge-ways.

To show how civic epistemologies function within diverse national political contexts, I begin by distinguishing this concept from another one that has been far more frequently used to characterize public knowledge and explain public perceptions of risk, namely, the public understanding of science (PUS or, with the addition of technology, PUST). This notion implicitly takes the "u" of "understanding" as the phenomenon of scholarly interest and, potentially, as the source of cross-cultural variance. Science itself, by contrast, is taken as unproblematic, universal, and invariant, equally understandable in principle in all places and at all times. Failure to understand science then becomes a meaningful dimension of difference among individuals and communities. On this assumption, the PUS model has generated much research to measure the extent of public understanding, usually by asking survey respondents a series of true-false questions concerning specific scientific facts. A particularly interesting example for us was the European Union's attempt to measure public knowledge of biotechnology through the Eurobarometer's ten factual questions, including the much-cited one about whether ordinary tomatoes have genes or only genetically modified ones (see chapter 3).

249

The designers of PUS surveys have tended to assume that, as long as people are properly informed about scientific facts, there should be no cross-cultural variation in their perceptions of science or their receptivity toward technology. By the same token, if differences in social uptake do arise, as they have in relation to biotechnology, then the PUS model attributes them to public ignorance or misunderstanding; lack of understanding, in turn, becomes a deficit that states can correct through better dissemination of knowledge. In this way, the PUS model at once identifies the alleged cause of a problem (public antipathy to science-based developments) and a democratically sanctioned means of rectifying it. Unfortunately for the model, there is little evidence that public ignorance of specific scientific facts correlates in any meaningful ways with collective responses to science and technology; biotechnology is simply one more domain for which this observation holds true.

By turning to the concept of civic epistemology, we move away from *a priori* assumptions about what publics should know or understand of science. Instead, we pose the question, salient in any democracy, how knowledge comes to be perceived as reliable in political settings, and how scientific claims, more specifically, pattern as authoritative. Put differently, civic epistemology conceptualizes the credibility of science in contemporary political life as a phenomenon to be explained, not to be taken for granted. In shifting attention away from individual knowledge or ignorance of facts to how political communities know things in common, the concept of civic epistemology also offers a means of getting at cross-cultural diversity in public responses to science and technology.

In the second part of the chapter, I try to make civic epistemology more analytically tractable by defining it in terms of several criteria that can be qualitatively compared across political cultures. I use these criteria to help make sense of the variations in the politics of biotechnology identified in the preceding empirical chapters. A conclusion that emerges from the three-country comparison is that the civic epistemologies of Britain, Germany, and the United States can be stylized, respectively, as *communitarian, consensus-seeking,* and *contentious.* The final section of the chapter sums up the basis for these characterizations, while recognizing that, as ideal-types, these labels must be involved and applied with the utmost caution.

## Public Understanding of Science

Writing in 1966, the British physicist and science observer John Ziman equated science with "public knowledge." Scientific reality, he said, was constructed through public experiment, creating a common foundation of knowledge and experience. Ziman asserted: "We are all entirely conditioned to accept as absolute and real the public view of things that we can share with other humans."[4] For that "public view of things," Ziman and many of his contemporaries assumed, we naturally turn to science, the one repository of absolute knowledge in modern societies.

Today, there is greater awareness that science itself is a social activity, and that a lot more goes into the creation of shared understandings of the world than the explorations carried on by science. The creation of social order around science and technology involves more than the production of scientific knowledge as an end in itself. Facts and artifacts, as we now recognize, do not emerge fully formed out of impersonal worlds, with cultural values entering into the picture only when a technology's impacts are first felt; nor, by the same token, does sociality enter into the making of science and technology as a secondary player, by side doors only. Science and technology are present in all of the narratives that modern societies weave about the world, as essential threads in the tapestry of social reality. Public questions revolve around how science and technology *ought* to constitute lives, and the answers loop back to shape the means and ends of scientific inquiry.

The notion of "public understanding of science" as it is usually conceived does not do justice to these complexities. To see why not, we need to look more carefully at the ways in which PUS has been defined, measured, and deployed as an instrument of governance. We must look as well at the growing criticism of that concept by social analysts and, in turn, at the limitations of those critiques.

For decades, it has been difficult to discuss the problem of science and technology policy in the Anglophone world without engaging in debates about the "public understanding of science." Major scientific societies both promote it and deplore its absence. The American Association for the Advancement of Science, for example, offers an annual award for journalistic excellence to reporters enhancing the public understanding of science and technology. Britain's Science Museum and Institute of Physics established a journal named *Public Understanding of Science* in the 1990s, subsequently edited at Cornell University in the United States. In 1985, three leading British science organizations established a Committee on Public Understanding of Science (COPUS) to promote better communication of science to the public.[5] Back in the United States, the National Academy of Sciences maintains an Office on Public Understanding of Science (OPUS) whose stated mission "is to foster the mutual responsibility of scientists and the media to communicate to the public, with accuracy and balance, the nature of science and its processes as well as its results." And each year the National Science Foundation publishes in its annual survey of science and engineering indicators an influential chapter—chapter 7—on public understanding of, and attitudes toward, science and technology. Germany was slowest to adopt the PUS model into official policy, but in 2002 a coalition of scientific societies, backed by federal funding, launched a project on Public Understanding of

Sciences and Humanities (PUSH) to promote better dialogue between science and the public. The PUSH agenda was consistent with the broader German definition of *Wissenschaft* ("science") as embracing all knowledge, including the natural, the social, and the humanistic.

These far-flung activities say less about how publics know things in contemporary societies than they do about the presumptions underlying scientists' (and secondarily the state's) expectations of what publics should know. First, though science is favorably regarded by large majorities in Western societies, PUS advocates dwell on the fact that the public does not know as much science as the scientific community deems desirable, even necessary. Second, the gaps in knowledge and understanding are seen as threatening to science; they are said to promote beliefs in pseudoscience and alternative medicine, undermine support for basic research, and (in the United States) offer aid and comfort to proponents of creationism and other scientific heterodoxies.[6] Third, leaders of the scientific community invariably assume that improved communication will raise the level of scientific awareness among the public. Fourth, these arguments and beliefs are used to support programmatic efforts to monitor the uptake of scientific knowledge by the public; NSF's annual survey of public attitudes is the most systematic of these efforts, but comparable surveys have also been undertaken in the Europe Union and a number of its member states. Together, these elements of PUS can be seen in effect as a kind of tacit democratic theory—a theory that presumes ignorant publics are in need of rescue by the state and grants science, a privileged place in forming, and informing, an educated citizenry.

The scientific community's conclusions about what a democratic public needs to know, and how it should be educated, deserve a closer look precisely because they have gained so firm a foothold in official policy circles. Two examples of the PUS agenda are especially illuminating: first, a test of general scientific knowledge regularly used in the NSF surveys, and second, a test of knowledge of biotechnology that has been administered in both Europe and the United States. In the NSF surveys, the U.S. public is asked ten questions designed to survey knowledge from various scientific and technological domains, from elementary physics to genetics and evolutionary biology to the nature of lasers. Although response fractions have changed a little over the years, some broad patterns have remained steady: people have most trouble explaining what a molecule is, and there is highest awareness of the theory of continental drift; about half the respondents do not know that the Earth completes its circuit of the sun in a year, nor indeed that it does so at all; similar numbers think the earliest humans coexisted with dinosaurs; and women consistently do worse than men in answering these questions correctly. The overall rate of right answers varies, not surprisingly, with the respondents' educational level and training in science and technology.

The biotechnology questionnaire, which the NSF indicators report describes as a "pop quiz," used the same ten questions as the *Eurobarometer*. The one that received the most attention from scientists and the media concerned the genetic composition of tomatoes. Respondents were asked to rate as true or false the following statement: "Ordinary tomatoes do not contain genes, while genetically modified tomatoes do." In Europe, when this question was first administered, about a third of the participants said "true," about a third said "false," and about a third took refuge in "don't know." The number giving the correct answer (that is, "false") later rose to 40 percent; in the United States and Canada the corresponding number was measured at 44 percent, but still as less than half of the respondents.[7]

Findings like this have led many to conclude that support for GM products like the Flavr Savr tomato would have been higher if only people had known that all tomatoes contain genes. The engineered tomato would then have been demystified, or so GM proponents think, and acceptance of it would have risen. Differences between the European and American responses, though barely significant, have been construed as evidence of a connection between lack of knowledge and rejection of technology—although, problematically for this view, scientifically educated Americans seem more skeptical of the claimed benefits of biotechnology than their less educated fellow citizens. The persistent inequality of scientific "understanding" between women and men generates similar despondence about women's capacity to understand science and the efficacy of science education for women. Yet the one or two genuinely provocative survey results have received little comment or analysis. Thus, it is interesting to speculate why continental drift should be much more widely appreciated than the nature of molecules, or why women exceptionally do much better than men—by 72 to 58 percent in one survey[8]—on the question whether the father's gene decides a baby's sex.

These weaknesses have not gone unnoticed. Sociologists of knowledge have made valuable inroads on conventional readings of PUS by pointing out that surveys do not just test respondents' understanding of science: they simultaneously construct the respondent as a particular kind of knower, or in this case a nonknower.[9] Brian Wynne, the noted British sociologist and science policy analyst, powerfully argues this point in a series of writings about the public understanding of science.[10] In his well-known study of the disparate cognitive worlds of radiation experts and sheep farmers in the wake of Britain's Chernobyl fallout crisis, Wynne showed that there were competing knowledges about the migration of radiation from soil into grass, from grass into lambs, and from lambs into meat on the table. Radiation experts were not *better* informed than farmers about these matters; they were *differently* informed, and in some ways less so, than the lay people whose lives they were trying to rehabilitate. Failure to take the possibility of such differences in lay and expert knowledge systems into account, Wynne argued, leads to a "deficit model" of the human subject as chronically falling short of some ideal state of knowing as defined by scientists and PUS researchers. That model provides

neither an accurate portrayal of human beings in relation to their understanding of science and technology nor a helpful representation to think with
as we confront the problems of citizenship and democracy in the twenty-first
century.

There is much to say for an approach that characterizes human beings
in terms of integrated belief systems rather than solely in relation to their
"understanding" of science. Instead of a faceless member of the public, whose
engagement with science falls on a linear axis from knowing (for the few) to
nonknowing (for the many), Wynne and others present a more complex and
competent human subject: one who struggles with ambivalence in the face of
competing cognitive and social pressures, robustly copes with ignorance and
uncertainty as well as "the facts," and reserves the right to make moral choices
about the purposes and governance of technology. Wynne's "lay citizens"
refuse to fall easily into the anonymous trio of publics who figure in the U.S.
survey literature on PUS: first, the small and select "attentive public," which
both follows and understands science and technology and constitutes the
public that scientists wish us to be; second, the larger circle of the "interested
public," which follows events but feels inadequately informed; and third, the
still larger penumbra of "residuals" who, possessing neither interest nor competence, stand on the outer margins of civic competence in technological
democracies. Not simply an ignoramus in matters of science, the lay citizen of
PUS's outermost circle is shown by Wynne and others to be a culturally
knowledgeable figure, able to master a more complex phenomenology in
some respects than that of science; in particular, lay citizens may be better
than experts at making room for the unknown along with the known.

Even this richer characterization of the knowledgeable human subject,
however, leaves some puzzles unexplained, especially from the standpoint of
comparative analysis. While legitimately taking issue with the PUS proponents' narrow construction of human intelligence, critics have to be careful
not to fall into their own reductions and simplifications of public epistemologies. Taken out of context, the "lay" subject can become in its way as much an
ideal-type as the "technically illiterate" individual who sits at the heart of the
deficit model. It is unquestionably important to point out that public perceptions of risk are conditioned by different rationalities and knowledges from
those of experts, but the story does not end there. Cumbrian sheep farmers
cannot, after all, provide the template for a global lay citizen any more than
the "attentive" citizen in NSF's surveys of public understanding can represent
a universal personality type. Context matters. How, in particular, does culture
shape the ways in which people see and evaluate their worlds? How does the
pairing of knowledge and power look to citizens conditioned by different relations to ruling institutions and by different experiences of governance? How
can history be accounted for in explaining the diversity of public engagements
with science and technology? And, above all, how can we accommodate a

proactive, dynamic, epistemically active conception of the "public": a collective that neither passively takes up nor fearfully rejects all scientific advances, but instead (as real publics are doing all over the world) shapes, crafts, reflects on, writes about, experiments and plays with, tests, and resists science and technology—so as to produce multiple forms of life around the same techno-scientific developments?

Once the questions are formulated in this way, it becomes clear that the second element of the PUS model requires a grammatical shift, from singular to plural: that is, from a single "public understanding" to alternate "public understandings" of science and technology. Without this shift, there can be no accounting for the disparate ways in which human societies take up the most mundane and widely distributed technological inventions, and, in the process adapt them to fit their locally situated knowledges, values, and preferences. But we need a more fundamental shift as well, from asking what the public knows about particular matters to looking at how knowledge is culturally constituted as a basis for collective action.

## Civic Epistemologies

In the cross-cultural politics of biotechnology, we repeatedly observe different public responses to the possibilities offered by the life sciences. Faced with the same technological alternatives, societies at similar levels of economic and social development often choose to go in different directions, based on divergent framings of what is at stake, and correspondingly different assessments of the risks, costs, and benefits of various possible trajectories. Science and technology take hold of the public imagination in different ways across political cultures—refracted by the culturally specific knowledge-ways that I term civic epistemologies.

Let us begin with a brief working definition of the term. Civic epistemology refers to the institutionalized practices by which members of a given society test and deploy knowledge claims used as a basis for making collective choices. Just as any culture has established folkways that give meaning to its social interactions, so I suggest that modern technoscientific cultures have developed tacit knowledge-ways through which they assess the rationality and robustness of claims that seek to order their lives; demonstrations or arguments that fail to meet these tests may be dismissed as illegitimate or irrational. These collective knowledge-ways constitute a culture's civic epistemology; they are distinctive, systematic, often institutionalized, and articulated through practice rather than in formal rules.

But where should we look for something so dispersed and nebulous as civic epistemology, and how can analysts recognize cultural practices of public knowledge-making? Below, I address this problem in two ways: first through a brief, interpretive case study of the "mad cow" crisis that not only affected the British response to GM foods but also illustrated important aspects of British civic epistemology; and second through a more systematic, comparative account that looks at the components of civic epistemology across the political cultures of Britain, Germany, and the United States, especially in relation to the acceptance or rejection of biotechnology.

## Of Mad Cows and Public Knowledge

On May 6, 1990, the British news media reported one of the worst miscalculations by a government minister in nearly four decades of Queen Elizabeth II's reign. The unfortunate principal was John Gummer, the Conservative minister of agriculture, one of the officials with front-line responsibility for the safety of food. Just a week before, there had been reports that a cat had died of a BSE-like condition, implying that the disease had jumped the species barrier from cattle to cats, probably through pet food made from infected beef. Gummer wished to forestall panic with the confident message, purveyed by the media, that "British beef is safe to eat." The instrument he chose for the purpose was his own four-year-old daughter Cordelia, to whom the minister fed a hamburger before the eager eyes of television cameras and news reporters. His act melded together two age-old repertoires of trust: a father feeding his child, and the state, in loco parentis, reassuring its citizens by public demonstrations of confidence. But the performance backfired, and the manner of its backfiring offers illumination.

Gummer relied, to begin with, on an embodied and personalized notion of trust, as a relationship between a trusting and a trustworthy individual. The witnessing media, and through them the public, were asked to put their faith in John Gummer himself, for who he was and what he stood for: a responsible public servant and minister of the crown. Gummer expected trust because that is what people in his position felt entitled to expect from those they govern. Second, the minister tapped into a commonsense, experiential repertoire for establishing the legitimacy of government policy: anybody can understand a message about food safety delivered by a parent putting food into his child's mouth. How could a parent do this unless he believed the food was perfectly safe? Indeed, a question the news media peppered government officials with throughout the "mad cow" crisis was whether they and their families still ate British beef.[11] By feeding Cordelia and himself in public, Gummer also drew on Britain's cultural commitment to empirical proofs: seeing is believing, his action implied, and what you see is what you get. His gesture was predicated on a conviction that—on issues of great public moment—British citizens are capable of accessing the facts of the matter in direct and commonplace ways. There was no perceived need, as there almost surely would have been in America, for mediating layers of technical expertise to give meaning to the risk of BSE.

The most interesting feature of Gummer's publicity stunt, however, was how very badly it went wrong. His carefully staged enactment of trustworthiness landed in a rich interpretive environment where it produced waves of contradictory meanings, in words and in images. At the level of words, newspapers and television accounts seized on the desperation in the minister's attempt to stage a persuasive public drama. They presented the scene as inauthentic, unsuccessful even as theater, and, in a polity committed to the ideal of public truth-telling, a subversion or avoidance of the facts. An extract from the Phillips Inquiry into the BSE affair is typical: "BBC2 featured television footage of Mr Gummer feeding his four-year-old daughter Cordelia a beef-burger to demonstrate his faith in the safety of beef. Unfortunately the effect was spoilt, as Cordelia would not eat the beef burger as it was too hot. The commentary noted that Mr Gummer was 'happy to chomp his way through a beef burger' but Cordelia was 'less enthusiastic.'"[12] The *Guardian* was more explicit on the theme of dishonesty, calling attention to the gap between staged action and backstage behavior: "In the same month that Daddy Gummer was cramming that burger into her tender mouth, he was secretly discussing whether the entire British herd should be culled."[13]

The critical images were equally arresting. The noted political cartoonist Gerald Scarfe, for instance, saw Gummer as anything but a benevolent public servant, dishing out reassurance to an appropriately deferential public. Scarfe's caustic pen drew an altogether more sinister figure. Above the caption, "What have we been fed?" a black-clad, male figure is shown force-feeding a helpless child in a pink dress. Depicted midway between china doll and living human, the girl stiffly succumbs to the man's superior physical strength. Her head is held back (as it was not in real life), the hamburger pressed into "her tender mouth." The image is violent, disturbing, with barely concealed overtones of child abuse. In the background stands the drooping figure of a chef with a frying pan containing a smoking steak and the legend, "How do you like your beef done? Burnt please." A pyre of cattle carcasses burns in the still greater distance. No one who saw this cartoon could have imagined a long and happy future ahead for an official who had allowed himself to be portrayed in this way, nor for the government he represented.

As an artist, Scarfe continues the tradition of Britain's great eighteenth-century political critic, James Gillray, whose fluid lines and distorted figures mercilessly lampooned state power in a time of growing imperial ambition. For Gillray, as for many Britons, physical gluttony was a metaphor for political excess, the gluttony of power. In numerous cartoons, Gillray captured political actors and events through the frame of unchecked appetite. In one famous image, entitled "A Voluptuary under the Horrors of Digestion," graceful lines and pretty, pastel colors belie the viciously accurate portrayal of the future George IV in a state of gross relaxation, picking his teeth after a massive meal. In another well-known cartoon, "Monstrous Craws, at a New Coalition Feast," the theme again is the Prince of Wales's boundless appetite and dissipation, with his parents, the king and queen, shown pathetically participating in a notorious parliamentary buy-out of their son's enormous debts.

The situations in which Britain's rulers are called to account for their performance have changed beyond recognition since Gillray's time. Expertise—its construction, public manifestation, and defense—were not matters of state in the eighteenth century; today, these issues are ubiquitous in public life. The elements that make up a nation's political imagination, however, arguably remain more stable over time, woven into institutional forms and practices with which publics assesses their rulers' credibility, whether the issues are scientific, technical, moral, or, as so often, a mix of all these. In a culture built on roast beef, a hamburger could never be just a sandwich. As Edmund Burke, observed in 1770, speaking of the restive colonies in America, "When men imagine that their food is only a cover for poison, and when they neither love nor trust the hand that serves it, it is not the name of the roast beef of Old England that will persuade them to sit down to the table that is spread for them."[14]

Poor John Gummer! Unreflectively reared in an insulated tradition of public service, he did not realize soon enough that embodying the state can be risky business. Governing bodies—political or royal, real or symbolic—can as readily attract satire as deference. Display alone is no guarantee of authenticity. The reservoir of common sense that Gummer appealed to allows people to judge public behavior by more complex criteria than the minister realized, including, as Wynne argued in the case of skeptical Cumbrian sheep farmers after Chernobyl, memories of past state failure and abuse of power. Trying to occupy a polity's imaginative field with new visual symbols requires, in any case, extreme sensitivity to the objects that already inhabit it, not only the reality of photographs in this case, but also the gloss on reality provided through articulate political cartoons. The hapless U.K. agriculture minister did not recognize the heterogeneous resources for truth-testing encompassed in his nation's civic epistemology. In turn, his ploy to win public trust through a personal demonstration of rectitude misfired massively.

## How Publics Know: Comparative Perspectives

John Gummer's crisis of credibility underlines a number of more general points about the concept of civic epistemology. It is a composite notion, comprising the mix of ways in which knowledge is presented, tested, verified, and put to use in public arenas. The concept has meaning only if we conceive of public life, in part, as a proving ground for competing knowledge claims and as a theater for establishing the credibility of state actions. In technology-intensive societies, the construction of governmental credibility necessarily encompasses the public production of scientific knowledge. Citizens are implicated not only as the passively disciplined subjects of biopower, as Foucault perceptively observed, but also as the watchful audience on whose behalf public knowledge is produced and deployed, as Yaron Ezrahi convincingly proposed.[15] In wealthy democracies, moreover, the state has no monopoly on knowledge. Corporate actors and even private citizens have the competence and resources to test or contest the state's expert claims, and to produce alternative forms of knowledge when their interests call for active cognitive representation.

Looking across the case studies of biotechnology in previous chapters, we can identify six constitutive and interrelated dimensions of civic epistemology, on each of which cross-national differences proved salient. These are in each country: (1) the dominant participatory styles of public knowledge-making; (2) the methods of ensuring accountability; (3) the practices of public demonstration; (4) the preferred registers of objectivity; (5) the accepted bases of expertise; and (6) the visibility of expert bodies. Differences among the United States, Britain, and Germany on these dimensions are summarized in table 10.1. The following text elaborates on the table, offering examples from the preceding case studies.

But first some words of caution. A tabulation such as this offers conceptual clarity but at enormous risk of reductionism. The boxes in the table aim to capture some recurrent tendencies in each political culture without suggesting that these stylistic features are rigidly fixed, uncontested, changeless over time, or evenly distributed across all sectors of society. They represent at most

TABLE 10.1
Civic Epistemologies—A Comparative View

|  | United States<br>Contentious | Britain<br>Communitarian | Germany<br>Consensus-seeking |
|---|---|---|---|
| Styles of public knowledge-making. | Pluralist, interest-based | Embodied, service-based | Corporatist, institution-based |
| Public accountability (basis for trust) | Assumptions of distrust;<br>Legal | Assumptions of trust;<br>Relational | Assumption of trust;<br>Role-based |
| Demonstration (practices) | Sociotechnical experiments | Empirical science | Expert rationality |
| Objectivity (registers) | Formal, numerical, reasoned | Consultative, negotiated | Negotiated, reasoned |
| Expertise (foundations) | Professional skills | Experience | Training, skills, experience |
| Visibility of expert bodies | Transparent | Variable | Nontransparent |

several deep-seated patterns, like rest positions in a length of elastic or a piece of bent wood, to which the practices of ruling institutions and actors continually return, in part because they are held in place through time-honored legal, political, and bureaucratic practices. Like any aspects of culture, the attributes of civic epistemology have to be performed and reperformed to maintain their hold as living, breathing instruments. It follows that radical breaks and disjunctures can always occur in theory, but shocks of exceptional severity may be needed to precipitate them.

## Participatory Styles of Knowledge-making

The politics of biotechnology over the last several decades can be represented as the politics of public knowledge. Industry and government both recognized that problems of knowledge must be solved as a precursor to gaining social acceptance for GM technologies, and many institutional experiments were conducted to answer questions like the following: what knowledge is needed to achieve policy closure; whose responsibility is it to produce it; in what forms should it be expressed and codified; how should its validity be determined; and how can it be challenged? Knowledge of the risks, costs, and benefits of biotechnology remains contested despite all efforts to render it unproblematic, and decision-making processes have altered in various ways in order to achieve sustainable levels of consensus. From these efforts, we can discern some patterned approaches to knowledge-making that appear to be stably characteristic of national political cultures, indeed sufficiently so to be termed "styles" of public knowledge-making.

The contexts in which public knowledge relevant to biotechnology is produced and validated are constrained in each nation by established institutional routines. This has meant, in the United States, a primary reliance on interested parties—industry, academic researchers, environmentalists—to generate relevant facts and claims. The U.S. regulatory system, like that of other Western countries, depends on manufacturers to supply information related to risks, but in the case of biotechnology these data-production burdens proved quite low to start with for both GM agriculture and food. On two occasions, the Cornell monarch butterfly study and the Berkeley study of gene transfer in Mexican corn, academic researchers showed that issues many had accepted as relevant to safety evaluation had not been adequately investigated in the framework of formal regulatory assessment. In another case— the detection of StarLink corn in food products—it was an environmental group's private initiative that led to the unraveling of EPA's policy of granting split approvals for GM products with multiple uses.

In both Britain and Germany, knowledge production was more broadly conceived than in the United States and conducted with more active involvement by the state. To some degree, a wider conception of what information is needed flowed from the initial framing of biotechnology as being about more than mere products and their safety. In skeptical, post-BSE Britain, the state found it necessary to rebuild expert credibility through new institutional forms, including the Agriculture and Environment Biotechnology Commission, whose influence led to the four-year farm-scale trials (an exercise with no other national counterparts), and also to the three-pronged public evaluation of the commercialization of GM crops. These institutions and processes set in train a more sweeping exposé of uncertainties, including those on the social and behavioral aspects of technological development, than in either of the other two countries. Yet, though pluralist in composition and open to many public inputs, the AEBC and the groups steering the components of the public debate reflected traditional British conceptions of the public servant: as persons of proven standing whose right to participate in knowledge-making for the state could not be seriously questioned. In opening up the GM debate to more diverse points of view, British institution-building practices did not altogether relinquish their reliance on trustworthy bodies to discern those viewpoints, and publicly represent them.

In Germany, fact-finding took place in discrete stages, under very different presumptions of transparency and public accountability. Extensive social inquiry took place along with the investigation of policy alternatives in the lead-up to the promulgation of new laws, most notably the Genetic Engineering Law and the Embryo Protection Law of 1990. The parliamentary Enquiry Commission on genetic engineering, a uniquely German institution, enabled the joint participation of legislators and technical experts in a wide-ranging discussion of the foundations and purposes of lawmaking. Once the initial rules for biotechnology were formally in place, however, implementation, with its attendant technical analysis, retreated to the relatively invisible sphere of expert decision making. Abandonment of the public hearing for deliberate release of GMOs was one indicator of the firm boundary that Germany draws between expert knowledge and public values. It was as if the law's most important function was to separate discursively hybrid public spaces into realms of cleanly technical deliberation and more impassioned political (hence civic) engagement. Perhaps not surprisingly, German intellectual debates over biotechnology *after* the enactment of laws remained resolutely focused on normative rather than factual issues; in no area of biotechnology development did German public controversies focus centrally on the need for new knowledge, as in the continuing U.K. (and to a lesser extent U.S.) debates on GM crops and their commercialization.

## Public Accountability

Generating knowledge claims is only the first stage in the enterprise of securing a regime of reliable public knowledge. In democratic societies, the holders of policy-relevant knowledge must find ways of persuading onlooking publics of their credibility, whether or not, in given cases, publics actively use the opportunities for questioning what they are told. Devices for holding policymakers and experts accountable exist in every modern knowledge society, but the nature and significance of these processes differ from one political system to another.

The U.S. accountability system owes its special flavor to the extraordinary prevalence of litigation. In scientific as in other areas of policy disputation, the adversary process remains the dominant route for establishing credibility. Truth, in this template, emerges only from aggressive testing in a competitive forum. Indeed, lawsuits accompanied each step in the development of biotechnology, from the deliberate release of GMOs to patenting living organisms and developing risk assessment guidelines for specific classes of products. Yet, in contrast to the case of chemical regulation in the 1980s,[16] biotechnology litigation did not deconstruct the uncertainties surrounding the commercialization of GM products in agriculture or medicine to anything like the same extent. Episodes like the StarLink and Prodigene disasters led to massive liability claims, to be sure, but the relatively low threshold for establishing the safety of GMOs in the product framework withstood legal challenge and protected regulatory science against stricter scrutiny.

The more insulated regulatory processes of both Britain and Germany historically depended on greater trust in expertise. The manner in which trust was produced and institutionalized, however, differed between the two countries and proved unequally resistant to the political controversies surrounding GMOs. In Britain, even after the Thatcher years, when many believed that core civil service values had been eroded, policy officials remained relatively immune to political pressures and thus were seen to be at greater liberty to act in the public interest. Experts for their part earned respect and trust through years of serving the common good, gradually acquiring credibility as recognized public servants. The BSE scandal rocked that credibility to its foundations, leading as we have seen to a complete overhaul of food and agriculture policy institutions. Nonetheless, even the new and reconstituted bodies continued to be staffed by persons of demonstrated personal commitment to public issues, although now a wider spectrum of such people was convened to deliberate more openly on a larger set of issues.

Trustworthiness in Germany is more a product of institutional affiliation than of proven personal service to citizens or the state. Enquiry commissions, expert committees, and lists of witnesses involved in administrative hearings are all broadly balanced according to a tacit understanding of the map of interests and positions considered essential to fact finding and deliberation on any given issue. Each participant in such processes represents an institution or a recognized collective interest: a political party, a church, a professional or occupational group, a patients' association, a public interest organization, or a relevant academic discipline such as law or genetics. Institutional representation rather than a record of public service is key. As a result, whereas in Britain the personal integrity of the members gives credibility to the bodies they constitute, in Germany it is the preexistence of trustworthy institutions that underwrites the credibility of the individuals who represent them.

## Demonstration Practices

Not only the representatives of specialized skills or knowledge but also the facts and things they speak for have to acquire credibility in the public eye. The tendency in the United States has been to conduct substantial socio-technical experiments whose apparent success not only validates the immediate venture they stand for but bolsters the overall narrative of technological optimism. The American state's instrumental uses of technology to prove itself in war and peace are well documented. Agricultural biotechnology has been a beneficiary of this style of self-legitimation. For instance, on corporate websites and in advertising leaflets, images of fertile fields planted with straight, undisturbed rows of plants offer compelling visual evidence that biotech has conquered nature's disorder. The fact that Americans have (unknowingly[17]) been consuming GM crops for years with no known adverse health effects has been advanced by biotechnology enthusiasts as further proof of the technology's success. To counter European skepticism toward GM crops, both U.S. multinationals and the U.S. government have stressed biotech's potential to conquer malnutrition and world hunger.

Ezrahi suggests that such demonstration projects are essential to the functioning of modern liberal democracies.[18] While this is a persuasive observation about America's particular democratic settlement, in which public claims are continually tested by skeptical citizens and journalists, it is worth noting that the very idea of public demonstrations as a space of experiment is a culturally particular, not universal, way of engaging citizens. It assumes that disclosure and transparency are *possible*, and that people have the will, the means, and the competence to evaluate the claims and proofs presented to them.[19] Technology serves the democratic will far more readily than pure science in this kind of demonstrative environment, for polio vaccines and moon shots offer clearer indications of a state's capacity to set and meet concrete goals than any particular addition to the storehouse of scientific knowledge. We may recall that after World War II it was the success of weapons and drugs that provided the strongest justification for creating the U.S. National Science Foundation, and the promise of cures for a panoply of genetic diseases built support for the Human Genome project.

In Britain, despite the vagaries of public funding, elite science has held a more secure place in public esteem, not needing constant affirmation through technological demonstrations to merit recognition and support.[20] The dominant policy culture shares with British science a pragmatic, empirical orientation, producing skepticism about claims that appear to go beyond the observable facts of nature or society.[21] In the early years of environmental regulation, British policymakers thus were more resistant than their U.S. counterparts to claims of chemically induced cancer, a condition for which adequate empirical evidence was felt to be lacking. As far as possible, regulatory policy was based only on what could be shown with near absolute certainty.[22] This bias led to what seemed like underregulation by comparison with U.S. and German approaches. Yet, once the science became reasonably secure, as for instance with respect to tobacco smoking, ozone depletion, or climate change, British policymaking did not generate the persistent controversies that plagued regulatory science in the United States.

In connection with biotechnology, the preference for empirical proofs was tied to greater hesitation about environmental release and eventually to a wider recognition of the unknowns surrounding the commercial use of GMOs. The science review conducted during 2002–2003 showed U.K. experts voicing many of the same concerns as British citizens about the unproven status of claims of environmental safety. In the absence of persuasive information, lay and expert sensibilities converged toward the conclusion that science could not rule out the adverse scenarios that might unfold if more GM crops were released into the environment. The science panel's July 2003 report stressed the nonhomogeneity of GM technology and, in line with longstanding British administrative practice, emphasized the need for case-by-case evaluation of the risks and benefits of particular genetic modifications.

In Germany, high entry barriers and cautious regulatory procedures kept some of the more urgent regulatory questions at bay longer than in either of the other countries. For instance, by the early twenty-first century, commercialization of GM crops was not yet an issue in Germany; accordingly, governmental approval, when it came in 2004, occurred without anything like the nationwide debate provoked in Britain. Early decisions to label foods as GM-free preempted potentially troublesome consumer and policy controversies; tellingly, as of 2001 no company had selected Germany as the place to file a novel foods petition for a GM food. Eschewing most boundary-testing conflicts, German expert committees like the Central Commission for Biological Safety (ZKBS) were able to conduct their business in routinized fashion, largely undisturbed by skeptical public oversight; experts needed only to achieve internal consensus and to report on their decisions as *faits accomplis*. Taken as axiomatic in this scheme was the capacity of appropriately constituted expert bodies to achieve reasoned solutions to the technical problems before them; that presumption of deliberative rationality in turn formed the cornerstone for public legitimacy.

### Objectivity

Objective knowledge is by definition reliable public knowledge, for such knowledge looks the same from every standpoint in society; it is untainted by bias and independent of the claimant's subjective preferences. Objectivity is a powerful resource for states. It allows governing bodies to claim the cognitive high ground, a place from which they can be seen to be acting for the benefit of all without bowing to any particular interests or knowledge claims of the governed. As with accountability, however, the practices for securing objectivity—or the appearance of it—differ from one political culture to another.

In the United States, a preferred method for displaying objectivity in public decisions has been to clothe the reasons for allocative choices as far as possible in the language of numbers. A preference for quantitative analysis goes back to at least the early years of the twentieth century, when civic groups such as mothers campaigning for children's issues recognized the force of numbers as a strategic tool for making their troubles visible, and so political.[23] At the same time, federal agencies needing to insulate potentially explosive distributive decisions against congressional skepticism found refuge in the neutrality of enumeration. Already in the interwar period, cost-benefit analysis was a powerful legitimating instrument for the U.S. Army Corps of Engineers.[24] The energetic expansion of social regulation during the 1970s produced a knock-on burst of methodological development, as institutions like the Environmental Protection Agency struggled to make costly decisions both comprehensible to the public and robust against challenges from regulated industries. Once again, quantification proved to be an invaluable resource, as governmental decision makers sought to weigh and compare incommensurables such as property values, loss of species, chronic health risks, lost jobs, and the costs of retrofitting plants and equipment with the latest in pollution control technologies.[25]

One method above all gained political currency throughout the 1980s, especially in the context of environmental regulation: the array of largely quantitative techniques known as "risk assessment." Originating in the finance and insurance industries, risk assessment was adapted for use in hazardous industries during the expansion of nuclear power in the 1960s and 1970s. Further adaptations led in the 1980s to the widespread adoption of risk assessment as a method for analyzing biological hazards, from environmental cancer to ecosystem disruptions and threats to global sustainability. U.S. regulators were at first careful not to claim too much objectivity for risk assessment, referring to it as a heuristic, a method of systematic analysis, even as an art rather than a science. Repeated court challenges hardened the agencies' practices, however, and a 1980 decision of the U.S. Supreme Court in a case involving the occupational standard for benzene forced federal regulators to carry out some form of quantification in support of their health, safety, and environmental regulations.[26]

As risk assessment became the preferred method for making regulatory judgments appear objective, so too it gradually took on the mantle of science. In 1983 the influential U.S. National Research Council promoted this shift by defining risk assessment as a largely scientific component of regulatory decision making that should precede, and be separated from, value judgments that were considered appropriate only at the later stage of "risk management." No sooner was the boundary between facts and values drawn in this fashion than critics emerged to challenge it. Study after study and commentary after commentary called attention to the profoundly normative character of risk assessment, showing that it is a far from objective method: indeed, that it is a highly particular means of framing perceptions, narrowing analysis, erasing uncertainty, and defusing politics.[27] Nonetheless, the discourse of science that gathered around risk assessment proved irresistible to regulators and modernist managers of every stripe, all of whom found risk assessment to be an invaluable tool for hiding judgment and making the complexity of biopower politically defensible as well as administratively tractable.[28] By the mid-1990s the terms "risk assessment" and "sound science" were inseparably linked in the U.S. regulatory discourse on biotechnology. Confidence in the objectivity of risk assessment underlay the U.S. decision in 2003 to bring a case in the World Trade Organization against Europe on its alleged moratorium against GM crops and foods. The action charged the EU, in effect, with backing away from the objectivity of risk assessment, and so committing offenses against science and reason, as well as violating the WTO treaty.

Regulators in Britain and Germany, too, have accepted risk assessment as a principled approach to ordering knowledge and weighing alternatives, and risk analysis occupies a central place in both countries' practices for coping with the possible impacts of biotechnology. Yet in neither setting is the method alone seen as sufficient to establish the objectivity of regulatory judgments. More is needed to produce the detachment from positional interests— the "view from nowhere"—that is the ultimate goal of objective science, and to some extent also of authoritative regulation. In both European countries, appropriate political representation remains part and parcel of the process of risk analysis, consciously built into the design of expert committees and consultative processes. In Germany, for example, it would be almost unthinkable to pack committees with instrumentally selected experts as the Bush administration was charged with doing in 2003;[29] and in post-BSE Britain greater diversity on biotechnology advisory committees has become an article of faith, along with more transparency in committee proceedings.

But even where the scientific and social hybridity of risk judgments is generally conceded, as in Britain and Germany, practices for ensuring lack of bias remain important and continue to vary, reflecting different cultural constructions of civic epistemology. On the whole, objective knowledge is sought in Britain through consultation among persons whose capacity to discern the truth is regarded as privileged. Though British expert advisers can and do represent social interests to some extent, ultimately it is the excellence of each individual's personal discernment—the capacity to see the distinctions that matter—that ensures something recognized as objectivity. Needless to say, this faith in expert discernment could hardly exist in a cultural context where common norms of seeing and believing were felt to be lacking, as in the United States. In Germany, by contrast, expert committees are often constituted as microcosms of the potentially interested segments of society; knowledge produced in such settings is objective not only by virtue of the participants' individual qualifications, but even more so by the incorporation of all relevant viewpoints into the output that the collective produces. The appearance of a view from nowhere is achieved by resolutely embracing, within the German decision-making body, the views from everywhere (or everywhere that matters for the issue at hand).

## Expertise

Experts are indispensable to the politics of knowledge societies. They tame the ignorance and uncertainty that are endemic to modernity and pose threats to modernity's democratic and managerial pretensions. Faced with an ever-changing array of issues and questions—based on shifting facts, untested technologies, incomplete understandings of social behavior, and unforeseen environmental externalities—how can governments ever know enough to act wisely or publics be persuaded that their governments are behaving responsibly? The unknown threatens continually to engulf the known, and action becomes impossible unless ground rules are laid for how much and what kinds of evidence are needed to justify collective action. It falls to experts, to satisfy society's twinned needs for knowledge and reassurance under conditions of uncertainty. Testing, for instance, is a device designed by experts to establish the safety of technologies whose actual performance can be known only through subsequent use, in other words, *after* they have been shown to be safe enough to use.[30] Experts are the people to whom publics turn for answers to the crucial regulating question, "How safe is safe enough?" In the politics of biotechnology, as on any issue of public moment, the credibility of experts is therefore as crucial to democratic governance as is the legitimacy of officials; only the rules of credibility, unlike the rules of constitutional delegation, are in all modern states almost entirely unwritten. They are cultural properties and as such are a source of cross-cultural variation.

Assumptions about what makes expertise legitimate differ, if subtly so, across the three countries compared in this book. Specialized knowledge is indispensable everywhere, of course, but knowledge alone is not synonymous with expertise. The expert is a social kind, in Ian Hacking's terms, a kind of person who not only provides information but satisfies the desire for order in the management of uncertainty. Experts therefore have to be accountable as well as knowledgeable. How do they meet this double demand?

Looking across the three countries, the primary source of variation is in the balance each strikes between the expert's formal qualifications and personal or institutional experience. Professional skills and standing count for more in the United States than the intangible qualities of individual character or institutional credibility. In a meritocracy that prides itself on impersonal tests of intelligence,[31] anyone can become an expert by climbing the ladder of professional success. What such a person stands for outside the professional sphere is of lesser consequence. Of course, it is understood that too much emphasis on technical qualifications alone could lead to imbalance in expert bodies. Instructively, the U.S. Federal Advisory Committee Act corrects for this eventuality through its requirement that committees should be balanced in terms of the views they represent. Nonetheless, disinterestedness is deemed to be a prerequisite for expertise. Allegations that experts have been captured by political interests continually erupt in the United States. Characteristically, the controversy over Ignacio Chapela's data on gene transfer in Bt-corn was interpreted, first and foremost, as a battle among political interests, penetrating in this case to the heart of the academy.[32] The equally divisive debate over Arpad Pusztai's data in Britain was seen as a failure in the management of science, bearing on the competence of particular individuals and institutions; political interest and capture by industry were not the salient themes in the British controversy. Similarly, the charges about the control of President Bush's bioethics committee by religious conservatives not only lacked British and German parallels, but *could* not, on this account, have been easily replicated in either country.

Experience is more heavily weighted as a defining element of expertise in Britain and Germany, but it is not in each case the same kind of experience. To a remarkable extent British expertise remains tied to the person of the individual expert, who achieves standing not only through knowledge and competence, but through a demonstrated record of service to society. It is as if the expert's function is to discern the public's needs and to define the public good as much as it is to provide appropriate technical knowledge and skills for resolving the matter at hand. In this cultural setting individuals ranging from Prince Charles and Baroness Warnock to Julie Hill of the Green Alliance and various academic social scientists can all emerge as authoritative policy actors. They all possess the transcendental capacity for discernment. It is hardly surprising, then, that a private body such as the Nuffield Council on Bioethics can achieve standing in policy deliberations simply through the credibility of its members and panels, without any official imprimatur.

Such reliance on personal credentials is rare in Germany unless it is also backed by powerful institutional supports. Respect for institutions may relate to, and is certainly consistent with, a horror of charisma and skepticism toward a politics of the person that are lasting legacies of Nazi rule. To be an expert in Germany, in any event, one ideally has to stand for a field of social engagement larger than one's own particular position in society. But the nature of expertise reflects something deeper and of historically longer standing about what counts as right reason in the German public sphere. Expertise encodes a belief that it is possible to map the terrain of reason completely, and that it is the function of legal and political processes to make sure the map is correctly configured and accurately reflected in public decision making. An expert then becomes almost an ambassador for a recognized nation or city, fiefdom or place, from among the allowable enclaves of reason. Rationality, the ultimate foundation of political legitimacy in Germany, flows from the collective reasoning and shared representations produced by authoritatively constituted expert bodies, acting as microcosms of society.

One sees numerous reflections of this approach in the German politics of biotechnology: the expansion of the ZKBS's membership in the early years of regulating deliberate release, the rejection of the disorderly public hearings on deliberate release that blurred the line between reason and passion, the hybrid constitution of parliamentary Enquiry Commissions as a prelude to lawmaking, and the almost agonized debate that followed Schröder's unorthodox creation of a second, cabinet-appointed ethics commission to look at biomedical issues. A paradoxical consequence of this map-making approach to delivering public reason is that bodies, once representatively constituted, leave no further room for *ad hoc* citizen interventions. They become perfectly enclosed systems, places for a rational micropolitics of pure reason, with no further need for accountability to a wider, potentially excluded, and potentially irrational, public.

## Visibility

Finally, not only the composition of expert bodies but the procedural environment in which they operate differs interestingly across the three countries, with important consequences for what citizens can see or know of the basis for public decisions. Consistent with ideas of pluralist politics and a notion of objectivity detached from individuals, U.S. expert committees are required to conduct a large part of their deliberations under the public gaze—on the assumption that close scrutiny from all interested quarters is the best way to wash out personal bias and subjectivity. Much of the work product of expert groups, along with the underlying reasons, is available for inspection during and after meetings. Openness, moreover, is guaranteed by federal laws that apply across the board to all governmental committees. British expert bodies are substantially more variable in their visibility practices, and it is generally left to the particular body to craft its rules of access consistent with its own, contingently developed notions of how to communicate with its publics. German expert committees, like their British counterparts, are not subject to overarching legal requirements of transparency. On the whole expert dissent, including on matters of science, tends to be least visible in Germany. It is as if, having constituted expert groups as perfect microcosms of relevant rationality, German politics sees no particular reason for a further layer of transparency, allowing irrelevant outsiders to look in.

# Conclusion

The concept of civic epistemology allows us to explore in richer detail a paradox that has resurfaced in the comparative politics of science and technology for nearly a quarter-century; biotechnology is simply the latest example. Why, in progressive, rational, Enlightenment societies, do the same scientific facts and technological artifacts so often elicit such different political responses? One conventional answer—that gaps arise because the public does not understand science—begs the question or, worse, trivializes it. The framework of "public understanding of science" diminishes civic agency, erases history, neglects culture, and privileges people's knowledge of isolated facts (or their ignorance of such facts) over their mastery of more complex frames of meaning. It reduces human cognition to a one-dimensional scale. It makes no allowance for the multivalency of interpretation. As a first move, therefore, we should stop thinking in terms of a singular public understanding of science and acknowledge the possibility of multiple understandings.

The notion of civic epistemology factors in this multiplicity and goes further. It redresses some of the obvious weaknesses of PUS's "deficit model" of the modern citizen, but without setting up in its place an equally problematic model of lay wisdom. The idea of civic epistemology takes as its starting point that human beings in contemporary polities are knowledgeable agents, living their lives in relation to governments, and that any democratic theory worth its salt must take note of the human capacity for knowing things in common. Collective knowing is a feature of political life that needs to be studied in its own right; it is reductionist in the extreme to assume that societal knowledge is simply the sum total of a population's understanding of a few isolated scientific facts. The public knowledge-ways that I term civic epistemology cannot be reduced to binary differences in knowledge and perception between laypeople and experts. We must speak instead of more grounded and systematic, shared approaches to sense-making without which no polity would be able to create public knowledge, let alone to maintain confidence in it, or use it as a basis for acting.

Are there, then, as many civic epistemologies as there are discrete national or political communities to host them? If so, the term would be so locally contingent as to offer little explanatory potential. By operationalizing civic epistemology in terms of five criteria, however, I have suggested that cross-cultural variations are not in fact infinite, but that they can be grouped in accordance with some widely recurring patterns of collective judgment. Britain, Germany, and the United States, in particular, represent three modalities of producing public knowledge that can be roughly characterized as *communitarian* (resting on shared perceptions), *consensus-seeking* (built through negotiation), and *contentious* (resolved through conflict), respectively. These are dominant tendencies only, and elements of all three modalities can be found in each country, though not perhaps at similar depth or breadth of institutionalization. More important for our purposes, each epistemological tradition carries consequences for democratic governance: entailing, for instance, different understandings of what transparency means, or what constitutes adequate citizen participation in technical decisionmaking. In a world tending toward denser global linkages, it is increasingly important to ask how such differences arise, and how they are sustained within political cultures.

Civic epistemology finally is a conceptual tool for planting the politics of science and technology firmly in the social world, where it rightfully belongs. For the greatest weakness of the "public understanding of science" model is that it forces us to analyze knowledgeable publics in relation to their uptake of science and technology rather than science and technology in relation to their embeddedness in culture and society. That move makes a mockery of political analysis. The ignorant, innumerate, and illiterate publics that PUS research confronts us with would not be capable of carrying out the housework of democracy. In their place, comparative analysis shows us full-blooded cognitive agents who test and appraise public knowledge claims, including those of experts, according to culturally sanctioned criteria of competence, virtue, and reasoning. Citizens seeking collective knowledge may hesitate and doubt, need help from experts, or hold questionable views on some technical issues; and they may have genuine trouble grasping the meaning of mathematical probabilities. But as members of sophisticated civic cultures, they cannot be characterized in the one-dimensional terms that the PUS model is predicated on. Civic epistemology provides a richer analytic frame for accounting for the forms of public trust and reason displayed by modern polities—and the political or policy consequences that flow from them. These powerful knowledge-ways cannot be excluded from contemporary social and political theory without reducing its explanatory power.

# 11

## Republics of Science

The reception of biotechnology into the political and policy systems of Germany, Britain, and the United States displays the everyday workings of three democratic nations in a time of large historical transitions. Posing linked and simultaneous challenges for science, power, and legitimacy, the politics of biotechnology offers a textbook example of the perplexities of governance in twenty-first-century knowledge societies. In previous chapters we observed one after another national struggle to work out in detail how to assess risks, clarify values, engage citizens, create property rights, foster technological innovation, and put out the wildfires of public distrust or antipathy around issues raised by rapid advances in the life sciences and technologies. Often gripping in their own right, these micropolitical stories show how profoundly the human ability to manipulate life has insinuated itself into age-old human concerns about preserving social order and securing moral certainty. Through the lens of civic epistemology, we have also seen how publics in different political cultures evaluate the claims made on behalf of novel science and technology, and how—under conditions of uncertainty—they decide whether the claims are robust enough to warrant support for particular forms of innovation. In this concluding chapter, we turn to the implications of the preceding comparisons for democratic theory more generally.

The attempt to build working routines of governance around biotechnology has called into question some of the founding assumptions of liberal democracy: that representative governments are capable of discerning citizens' preferences and of acting to promote them; that democratic institutions are knowledgeable enough to regulate the directions of scientific and technological change wisely; and that citizens have the opportunity to participate meaningfully in decisions that seem increasingly to call for specialized knowledge and expertise. Comparative study helps us evaluate national responses to these predicaments and, on that basis, to assess the texture and quality of democracy in contemporary industrial societies. The three country studies elaborated in this book not only show different institutional and procedural traditions at work in accommodating biotechnology; they also allow us to reflect on what is gained or lost in the settlements that each nation has achieved.

The substantive observations gleaned from the preceding chapters are regrouped here under three thematic headings that connect the specifics of science policy and politics to wider issues of democracy. The first has to do with the liveliness of political culture as a factor in the reception of biotechnology. How, we may ask, did culturally conditioned framings of the problem of biotechnology arise, persist through time, and affect new issues as they appeared on political and policy agendas? The second theme concerns the relationship between democratic processes and scientific and technological change. Under this rubric we inquire into how the classic political categories of participation, deliberation, and representation were energized or transformed through national attempts to make policy for biotechnology. In what ways, more specifically, did publics in each nation shape the social meanings of genetic modification, and how did these meanings become aligned with official discourses of risk, ownership, ethics, or innovation? Were there any notable failures of democracy, and, if so, how were these distributed across the three countries? The third major theme revisits the sources of political accountability in technologically advanced societies. Where do authoritative ideas for governing science and technology come from, how are national governments held responsible for the management of science, and how do processes of authority-creation vary across the political cultures of Europe and the United States?

A reorganization of the book's substantive findings along these lines highlights substantial cross-national divergences and inevitably leads us back to questions about the purposes of comparison that were laid out in chapter 1. Are differences among nations inevitable and, if so, is that necessarily a cause for worry? If national responses to science and technology are conditioned by the autonomous, self-replicating, deeply institutionalized dynamics of political culture, then what can states or citizens hope to learn from each other's experiences? What, finally, can cross-national comparison teach us with respect to the prospects for harmonizing political and ethical responses to biotechnology across competing cultures of knowledge and power? Answers to these questions must necessarily be partial and open-ended, but they may nevertheless serve as springboards for further inquiry.

I argue in this chapter that cross-cultural learning from the first quarter-century of the genetics revolution is both possible and necessary, but not in the form of universal prescriptions for improving upon current administrative practice. This book does not aim to be a how-to manual for constructing better risk assessment guidelines for GM foods or standard operating procedures for public consultation. Nor does it provide a blueprint for how national governments should best foster innovation in the life sciences. At the same time, the comparative findings offered in preceding chapters and summarized here are intensely relevant to the design of frameworks for governing biotechnology. Some implications of my analysis are spelled out below; others will occur to the sympathetic reader who is prepared to indulge the ideas presented here and to speculate on their possible ramifications.

## The Stickiness of Frames

On May 13, 2003, the U.S. Trade Representative's office announced that the United States and several cooperating countries had filed a case at the World Trade Organization against the European Union's "illegal, non-science based moratorium" on biotech food and crops, which was "harmful to agriculture and the developing world."[1] In July of the same year the EU drew up new regulations on the labeling and traceability of foods containing genetically modified ingredients, claiming that European consumers now had a transparent, reliable means of choosing between alternative sources of food. Nevertheless, in August the United States called for a WTO dispute settlement panel to address the GMO issue. David Byrne, the EU commissioner for health and consumer protection, deplored the U.S. action, stating, "Only a month ago we updated our regulatory system on GMOs in line with the latest scientific and international developments. Clear labelling and traceability rules are essential to help restore consumer confidence in GMOs in Europe."[2] Few starker reminders could be found that the global march of biotechnology had not automatically brought policy convergence in its wake. Indeed, in this case regulatory polarization seemed, to some, a more fitting description of what had occurred.[3] Here were two of the world's economic superpowers disagreeing not only about whether and how to promote biotechnology in agriculture, but, even more astoundingly, about what counts as science for regulatory purposes and how that science should be deployed in controlling the fruits of biotechnology.

The roots of disagreement run deep, as we have seen. They were planted, in the 1980s, with divergent framings of the risks of biotechnology between the United States and the EU, as well as among EU member states. Scientific accounts of the risks of biotechnology became enmeshed at that time with political institutions and practices through a process of "coproduction."[4] This approach to thinking about the unfolding of science and technology in society stresses the myriad ways in which knowledge about the world both conditions and is conditioned by choices about how people wish to live in it. The sociopolitical history of biotechnology amply demonstrates that natural order sustains and is sustained by social order. Human encounters with the life sciences and technologies repeatedly became occasions for the manufacture or redesign of politically significant institutions, identities, representations, and discourses. Political practices, social norms, and ideas of nationhood were produced together with new biological concepts and constructs, each bolstering the other's firmness. The coproduction framework, understood in this way, helps account for the stickiness of frames.[5]

Three separate framings of biotechnology emerged in the three countries—as product, process, and program—each resting on its own scientific, administrative, legal, and political arrangements.[6] The product-based approach particularly took hold in the United States, where it went hand-in-hand with a scientific account of genetic engineering as a highly specific intervention, grounded in molecular biology, promising untold benefits, and entailing largely predictable (and, as it happens, negligible) adverse consequences for human health and the environment. Britain and the EU, by contrast, adopted a process-based approach, which brought together more ecologically oriented expert perspectives with a normative posture that admitted more uncertainties and called for a precautionary approach to regulation. Germany took precaution yet one step further by highlighting political and ethical as well as scientific unknowns—in particular, the possibility of a programmatic alliance between science and the state that might lead to abuses of power unless the development of biotechnology was tightly controlled. These fears were reflected in numerous German legislative and regulatory enactments that sought to create categorical bright lines and guard against ambiguous or conceptually messy situations. For much of the period covered by this study, the German politics of biotechnology was most notable for its resistance to experimentation with new forms of life, in nature, society, or the state.

The disparate initial settings of the technical and political dials not only were coproduced in each nation but were contingently readjusted as new issues appeared on the horizon. Comparative investigation reveals enduring connections between public articulations of risk and institutional features of governance in each country. We find neither blind technological determinism nor rigid path dependency, but a more complex and subtle interplay of novel technoscientific possibilities with rooted expectations concerning relations among citizens, science and industry, and the state.

It is not surprising that the framing of biotechnology as a stream of commercial products was most readily accepted in the United States, where, in areas other than national security, the market often out-competes the state as the more powerful model of legitimate social organization. A preference for market solutions as an alternative to state control grew during the 1980s, as the deregulatory fervor of the Reagan era was incorporated into routine administrative practices. With the downfall of communism and the "end of history," in Francis Fukuyama's well-known phrase,[7] the ideology of the market gained additional political force. Promarket and antiregulatory tendencies manifested themselves across the entire range of governmental action, or more accurately inaction, on biotechnology, from the failure to enact comprehensive federal legislation in the 1970s to the relative laxity of the Coordinated Framework in the 1980s, and from the permissive patenting decision in *Diamond v. Chakrabarty*[8] to the facilitation of university-industry technology transfer through the 1980 Bayh-Dole Act. At the same time, a chronic aversion toward incurring opportunity costs, expressed through a *laissez-faire* policy toward private initiative and risk-taking,[9] significantly lowered the threshold barriers to biotechnological innovation—even though, as in the Star-Link and Prodigene episodes, the consequences to both producers and the state sometimes carried enormous price tags and eventually led to stricter controls.

The product framing of biotechnology both reflected and reinforced America's historical record of seeing technology as an instrument of progress and nature as ripe for appropriation through human ingenuity.[10] This framing casts citizens as insatiable consumers of technology, perennially scanning the horizon for new goods and services to meet an ever-diversifying array of needs. Courts and ethics commissions, as well as Congress and regulatory agencies, constantly reaffirmed this construction of the citizen as an entrepreneurial adopter of the new biotechnologies. Thus, *Chakrabarty's* easy erasure of the distinction between living and nonliving "compositions of matter" facilitated innovation; the decision sharply contrasts with the Canadian Supreme Court's reluctance, more than twenty years later, to take a similar step with regard to the genetically modified oncomouse. Support for liberal individualism, and a concomitant faith in the intrinsic goodness of social and technological experimentation, also shine through judicial decisions such as *Moore v. Regents of the University of California*,[11] granting medical researchers unencumbered property rights in human tissue, and *Johnson v. Calvert*, positioning the gestational surrogate mother as a biocapitalist exercising commodity rights over her own body.[12] A focus on individual rights and a preference for utilitarian logics consistent with market values characterized the pronouncements of official bioethics, most notably so in the Clinton-era presidential commission's repudiation of human cloning on grounds of risks to the clone.[13] Only the ethics council nominated by President George W. Bush charted a course away from utilitarianism; we will return to the democratic implications of that move below.

In Britain, by contrast, a proactive state policy toward biotechnology, endorsed by the Tories under John Major and enthusiastically embraced by the succeeding Labour government, had to contend with a sharply divided public response: on the one hand, a relatively uncontested, science-friendly legal regime developed around research on embryos and embryonic stem cells, and numerous collaborations arose between university scientists and pharmaceutical companies; on the other hand, agricultural biotechnology suffered from a severe backlash against the paternalistic "nanny state," necessitating nationwide political consultation on a scale not encountered in either of the other two countries. Developments in both areas illustrate the long-term persistence of framing biotechnology as a process meriting special public concern, even though those concerns were successfully contained through official policy actions in the case of embryo research and pharmaceutical biotechnology.

Though seemingly divergent in their outcomes, the politics of red and green biotechnology in Britain owe their flavor to several systemic aspects of that nation's political culture. Expert trustworthiness and credibility were invoked in both contexts to shore up state policy, but the dissolution of the Ministry of Agriculture, Fisheries and Food in the wake of the BSE ("mad cow") crisis left policymakers bereft of the expert "safe hands" that normally stamp their approval on official action. The creation of the new Food Standards Agency in 2000, with the administratively seasoned Sir John Krebs at its helm,[14] exemplified the attempted build-up of confidence around tried and true individuals in the aftermath of BSE. Yet Krebs's lackluster performance and his eventual resignation in 2004 suggest that the Blair government misjudged the kind of experience needed to communicate effectively with the British public on food safety issues. In the context of embryo research as well, established processes of authorizing trustworthy bodies—both individual and collective—positioned Mary Warnock and, later, the Nuffield Council on Bioethics to play crucial roles in carving out a protected space for British reproductive science and its biomedical offshoots. The appointment of Suzi Leather, the first FSA deputy chair and a known consumer advocate, to head the potentially controversial Human Fertilisation and Embryology Authority illustrates the continuing cultural production of, and reliance on, embodied expertise.

These practices of political authorization worked in harmony with a related feature of Britain's dominant civic epistemology: the preference for empirical demonstrations. In the contexts of both embryo research and agricultural biotechnology, appeals were repeatedly made to proofs that ordinary people could see and understand. Failure to meet these demands for empirical justification generated uneasiness about the safety of GM crops even before the outbreak of the BSE crisis. British policyadvisers were notably reluctant to embrace the U.S. position that most GM products are substantially equivalent to their unaltered counterparts, and hence safe. Skeptical voices from academia and environmental groups were added to U.K. policy deliberations in an effort to build a stronger consensus in support of agricultural biotechnology, but this effort to broaden politics led to a more extensive unpacking of scientific unknowns than in the United States. Farm-scale trials of GM crops and the national debate on commercializing them were two of the more noteworthy results.

By contrast, a mutually reinforcing alliance of social and scientific authority kept the barely visible, less than fourteen-day-old pre-embryo firmly demarcated from the embryo proper, allowing the former to be treated as unproblematically devoid of human traits, and hence as a suitable object of research. In a coproductionist turn, however, this splitting of the developing human embryo into two distinctive legal and biosocial entities was feasible only with the full mobilization of the state as an agent of biopolitics. The line of demarcation had to be made morally and scientifically tenable, and this in turn meant that the state had to call upon all of the reserves of authority that allow a government to construct reality on behalf of its citizens. Creating and maintaining a space for embryo research required flawless performance by an official ethics commission, carefully orchestrated parliamentary approval, the transfer of statutory authority to HFEA, and the shoring up of HFEA's regulatory powers by the House of Lords against later legal challenge. Arguably, too, this process necessitated the emergence of a new social kind—the ethics expert—originally exemplified in the person of Mary Warnock.

Germany, too, adopted the process frame along with Britain and the EU, acknowledging that genetic modification in and of itself calls for special oversight in all of its domains of application. But taming the risks of biotechnology in Germany proceeded in tandem with taming recalcitrant historical and political memories,[15] both before and after the fall of communism in 1989 and the reunification of the divided state in October 1990. Key to the resolution of early political debates around biotechnology was the reaffirmation of Germany as a *Rechtsstaat*, a place where the rule of law enjoys supreme respect. In turn, this construction of the German state demanded principled behavior on the part of government and strict adherence to basic constitutional norms such as respect for human dignity. Relations among science, technology, and the state as played out in this context could not help taking on a programmatic character. They were, in effect, staging grounds for continuing struggles over the definition of postwar German identity.

Looking at regulatory solutions to biotechnological problems in Germany, we are repeatedly struck by an apparent yearning for moral and legal absolutes. Difficult problems were avoided altogether or cast into well-regulated categories of exception. Thus, the Basic Law was construed to affirm the unrestricted right to life of the human fetus, even though reasons were found to allow women to have abortions on grounds including psychic distress. Surrogacy was forbidden, as was the creation of spare embryos; like hidden marauders, some vanishingly small population of these accidental beings languished in an extralegal borderland, where even their numbers could not be accurately recorded. Embryonic stem cells were not permitted to be made in Germany.

They *could* be imported, but only if they were created before a date clearly stipulated by law. Without such firm lines of demarcation, it seems that the state is always in danger of sanctioning a *Dammbruch*—a breaking down of high dams with who knows what consequences for public morality, law, and order. This fear of lawlessness stands in sharp contrast to British and American confidence that the law can, at need, build robust stopping prints on the slippery slopes of moral judgment. Ambiguity exists, of course, in Germany, as it must in any functioning rule system, but the *Rechtsstaat* cannot officially tolerate its presence.

Inhibited from making controversial choices without clear legal supports, the German state exercises its discretionary power through a network of anonymous expert agencies, such as the Central Commission for Biological Safety, that see themselves as making purely technical administrative decisions within the parameters of the law. Just as morally ambiguous life forms are denied legal existence, so the regulatory process seems uncomfortable with ambiguous forms of political expression. The elimination of the public hearing requirement for deliberate releases of GM plants originally included in the Genetic Engineering Law makes sense in this perspective. It turned out to be an uncomfortably hybrid process, blurring the lines between procedure and substance, and between expressive or symbolic politics and the reasoned political debate of Jürgen Habermas's idealized public sphere.

It is against this backdrop of uneasy rule-following that one has to read the Schröder government's attempts to reprogram the relations between the life sciences and the state. Compromises were made on many fronts so as to facilitate the growth of biotechnology: in the explicit abandonment of a "just say no" attitude to agricultural biotechnology by Renate Künast, Green minister for agriculture and consumer affairs; in the approval of GM foods and crops following the adoption of EU labeling and traceability rules; in the appointment of a bioethics council by the executive branch to create a more moderate policy on stem cells and other aspects of biomedical research; and in the active sponsorship of bioregions to promote university-industry technology transfer. Yet although these steps indicated a moderation of older policy hard lines, they left intact the state's central role as the sponsor and regulator of biotechnology. Schröder's bioethics council, for example, reaffirmed its adherence to the Habermasian model of the public sphere at its very first open meeting, and members privately described the council's need to be representative, in a transparent way, of all reasonable views on biomedical ethics.[16] Künast's opening the door to GM crops was counterbalanced by the promise of large public subsidies to organic agriculture. Even the "market" for new public-private linkages in the bioregions remained, at bottom, a state-run enterprise geared toward generating competition. Put differently, the programmatic relations among science, technology, and the state in Germany persisted into the new century, with this difference: that a postwar politics of moral anxiety, fearful of risky innovation, was supplanted by an equally powerful German bureaucratic tradition aiming for rational management of the inventive process.

I have suggested throughout that the politics of biotechnology at the EU was subordinated in key respects to that of the member states. Basic questions about the acceptability of biotechnology's products and the allowable forms of debate concerning them remained national in character. Yet to leave the EU political scene out of account would be to miss an important element of the turn-of-the-century politics of globalization. Through its revised regulation on deliberate release and its later enactments on labeling and traceability of GM foods, the EU in effect carved out a space for "coexistence" that applied literally to GM and non-GM products, but also metaphorically to different judgments by member states about the suitability of GM products for their national farms and tables. In bioethics, similarly, the debate on the Sixth Framework Programme for research was in part a debate on the possibility of normative coexistence, between those who abhor research on, and appropriation of, human biological materials, and those who believe in the feasibility of regulated, stepwise innovation. The brewing conflicts with the United States in the WTO can be seen against this backdrop of European identity negotiation not simply as a battle over free trade, but over alternative models of political—and ontological—coexistence.

## Varieties of Democratic Experience

Making peace with biotechnology was not, in any of the three countries, simply a matter of applying old political routines to new agenda items; nor was it a case of creaky legal and political institutions playing catch-up with rapid developments in science and technology. Through their attempts to accommodate biotechnology, each democratic society tested, and to some extent reconstituted, its core understandings of what is political about science and technology and how to organize and conduct the associated processes of technopolitics. In this sense the politics of biotechnology proved to be constitutive of aspects of democratic politics writ large. The results point in each country to lingering difficulties in opening up the bilateral social contract between science and the state to active civic engagement

In all three countries, the life sciences posed some very special political problems. Metaphysical disorder, or confusion about how to classify things, was an inevitable by-product of genetic modification, producing attendant confusion in the practices of governance. Biotechnology disrupted ancient classifications and transgressed boundaries that had for centuries been accepted as given in Western legal and political thought. Distinctions between nature and artifice, animate and inanimate, living and nonliving, body and property suddenly became problematic, and thus in principle political, in many areas of decision making. The consequent attempts to reinterpret the law and to craft new administrative practices provide a lens for looking into a millennial transition: one in which life and its material components have become at once a means of production, a commodity, a therapeutic agent, and an object of regulation.[17] This is as well a constitutive moment for politics, as citizens project onto the state the obligation to articulate what should be held sacred about life, while asserting contradictory visions of how to bring life under firmer human control.

Public involvement in the politics of biotechnology, and the life sciences more broadly, can be separated into three comparative strands to highlight the democratic implications of this era of technopolitics: *representation*, or the means by which some voices are made audible in the political and policy process, and how political inclusion in turn affects the framing of issues; *participation*, or who actually takes part in politics, and who does not; and *deliberation*, or the discourses in which political debate is conducted, together with their limits and achievements. Under each heading, we observe cross-national similarities and divergences, with implications for public policy.

## Representation

Representation is a two-faced word in the politics of knowledge societies. It refers both to the self-presentation *of* the public within and before governing institutions, and the presentation *by* the public to those bodies of matters that are seen to be of collective significance. The first meaning of representation is the stuff of traditional democratic theory, but the second, which directly affects the framing and implementation of policy agendas, is of increasing importance. The two meanings, moreover, are not unrelated, as we frequently observe in debates on the life sciences.

The politics of biotechnology proceeded to start with as a case of business as usual in each country, although in each this pattern was eventually disrupted to varying degrees. In the United States, leading research scientists, banding together at Asilomar, called attention to the threats of recombinant DNA research and laid conceptual foundations for regulation that were widely imitated throughout the world. Propelled by genuine concerns for public welfare, the Asilomar conferees nonetheless exemplified the classic American pattern of interest group politics. Molecular biologists stumbled into action through their fear of displaying the lack of foresight and suffering the loss of esteem that nuclear physicists had experienced in developing the atomic bomb. But trying to forestall charges of heedlessness, the Asilomar participants crafted powerful narratives that influenced U.S. politics and catered to molecular biologists' interests for years to come: narratives of scientific self-regulation and social responsibility; of molecular biology as the science with the most to say about the safety of GM techniques; of health risk as the issue of largest concern; and of physical and biological containment as the primary instruments for risk control. The channeling effects of these early stories become starkly apparent only through comparison with political systems in which other social actors helped create the initial framings around biotechnology.

In Britain, the first state responses to biotechnology took the form of top-down measures, informed and assisted by a network of largely invisible expert committees.[18] Unlike pluralist politics, however, Britain's consensual, elitist traditions of government offered the means for accommodating all reasonable voices at the policy table—or at least those voices that the state deemed reasonable. This greater inclusiveness encompassed ecologists and environmentalists, who succeeded during the 1990s in canvassing a wider range of unknowns around the dissemination of agriculture biotechnology than were acknowledged in U.S. assessments. Even before BSE became a household word, British scientists and policymakers favored a more precautionary approach to regulating biotechnology than their American counterparts, and the idea of targeting only the products of genetic technologies for regulation never took hold. The public at large played little or no role in these developments, however, neither volunteering nor being asked for its reactions to genetic modification or any of its applications.

The overt breakdown of trust over BSE marked a turning point of sorts in the politics of biotechnology, but what that crisis revealed was not so much the British public's antipathy to science and industry as the decayed state of rapport between the rulers and the ruled in one of the world's most economically and technologically successful knowledge societies. Coming on top of disenchantment with the long years of Conservative government, and the ever-present anxiety over the pace and course of Europeanization, the BSE affair highlighted the fragility of the ties that bind British citizens to their state. It was, as I have observed elsewhere, a moment of "civic dislocation" and profound self-questioning—about the ends of government and the ends of technology.[19] Restoring trust, as the Labour government was forced to recognize after the GM food panic, required a reinvention of the grammar of government—the ground rules according to which the state would learn to communicate more effectively with its citizens, and vice versa. Reinventing government, in turn, demanded a reinvention of citizenship. The three-pronged GM debate of 2003 not only drew science, economics, and politics into a months-long, nationwide conversation, but it became an occasion for producing the very public to whom the state could hold itself accountable.

Yet all this innovative fever also showed that old habits of governance die hard. At the highest levels of government, Britain continued to subscribe to elements of the "nanny state" that engaged citizens and NGOs had powerfully challenged through the debates on biotechnology. In Tony Blair's triumphalist

2002 speech on science to the Royal Society, in Sir John Krebs's resistance to public concerns about GM foods, and in the government's 2004 decision to approve the commercial growing of Bt-maize, one sees little acknowledgment of the humility and precaution that British activists and academics had begun to demand of the state and its corporatist (or, in my terms, programmatic) alliance with science and technology.

Representation in Germany proceeded as in the United States through traditional democratic channels, but there the task of speaking for the public fell to political parties rather than to interest groups. Early debates on biotechnology were dominated by the intensely vocal Greens, who (though always in the minority) effectively mobilized every available political channel—legislation, litigation, direct action, the media—to assert their vehement opposition to all forms of genetic manipulation. As elsewhere in Europe, the early strategy of the German Greens was to frame biotechnology as a novel, and unnatural, industrial process, and to keep alive the specter of overreaching by state and industry. The resulting morally charged political environment was inhospitable to experimentation, scientific, social, or political. It discouraged risk-taking, whether in labeling food, forming novel forms of kinship, or creating borderline biological entities such as pre-embryos and stem cells. Only when the Greens were drawn inside the networks of power, as in the Red-Green coalition of the Schröder period, did shifts develop away from the party's absolutist positions. Ironically, those years saw, if anything, an intensification of reunited Germany's reliance on biotechnology as a tool for addressing the nation's economic woes. After years of speaking out against just such an alliance,[20] the Greens therefore participated in redrawing the map of Germany in terms of bioregions—signaling the state's programmatic commitment to biopolitics as an instrument of national economic regeneration.

## Participation

Different traditions of representation in the three countries helped to determine who participated in deliberations on biotechnology, and with what consequences for the production of new knowledge and new forms of social action. The dual function of representation—speaking for publics and framing issues for policy—established the preconditions for different patterns of political inclusion and exclusion, of people as well as topics for deliberation. Most dramatically, perhaps, the rapidly achieved focus on products in the United States foreclosed wide debate on the social and ethical implications of GM technologies, particularly in agriculture, and blocked public notification, through labeling, of the production history of GM foods. The resulting absence of politics, which policymakers and industry interpreted as public acceptance of GM technologies, seemed more plausibly to reflect public ignorance. A September 2003 poll conducted by the nonprofit PEW Initiative on

Food and Biotechnology found, for example: "Although it has been estimated that 70–75% of processed foods in grocery stores contain GM foods, just 24% of Americans believe they have eaten GM foods while 58% say they have not, suggesting that Americans continue not to recognize the extent to which GM foods are present in foods they eat every day."[21] The containment of intellectual property conflicts within the Patent and Trademark Office and the courts had a similarly restrictive effect on participation while opening up the economic frontier for biotechnology.[22] The organics movement partly overcame the discursive and institutional barriers against questioning the process of genetic modification, but only by establishing a profitable market niche for non-GM products. Their victory left intact the basic framing of genetic modification as an unproblematic means of production, relegating *non*-GM production to marked status—literally and figuratively—under the organic label.

In the area of biomedical applications, a history of ethical breakdowns in medical experimentation, as well as the establishment of the Ethical, Legal, and Social Implications program at the National Center for Human Genome Research, opened up an institutional space for expressions of moral concern. Yet struggles over the direction of ELSI attested to the power of the very research community that had shaped the debate on ethics from the 1970s onward. The development of an intellectual property regime around biotechnology proceeded under the aegis of the law with few political impediments. Substantial changes in the U.S. social contract for science ushered in by the Bayh-Dole Act further reduced the public's capacity to supervise the applications of basic research in numerous areas of the life sciences, while rendering university-based science less self-critical about its own aims and ends, and less considerate of any public interests that stood in the way of science's "endless frontier."

In Britain, the dynamics around green and red biotechnology to some degree reversed the U.S. state of affairs, as public sensitivity to food safety forced the state to reconsider its institutional apparatus for regulating the processes of agricultural biotechnology. Academic and NGO representatives gained entry into newly constituted advisory committees charged with expanding the range of issues deemed relevant to governmental policy. In this production sector, as in no other, British politics produced, however fleetingly, a nationwide debate on the purposes of technology rather than simply accepting genetic modification as a progressive step in human development. By contrast, research policy on embryos and stem cells continued to be shaped by a more traditional coalition of state officials and their elite advisers—all intent on preserving freedom for research in a domain where British science enjoyed an international competitive advantage. Similarly, a loosening of social and legal constraints on commercialization forged closer relations between university science and corporate interests in the pharmaceutical sector without generating significant public concern. In view of the experience with GM crops, one may still wonder whether and how long such carefully managed settlements are likely to endure; British investments in social research on genetics and genomics point to official awareness of such instability.

German participation illustrates yet a third pattern, characterized by relatively high degrees of mobilization and informed public debate in the lead-up to major legislative actions, but then an effective closing down of politics under the auspices of functionally specialized expert committees. A striking consequence of this approach has been official silence (or intentional nonrecognition?) with respect to uncertainty. It is as if—having painfully crafted the legal framework for biotechnological research and development—neither German politicians nor German citizens wish to maintain the state of high alertness demanded by traditions of interest group activism and litigation in the United States. Participation and extended deliberation stop when the law is laid down—whether at the door of parliament or, as in the case of abortion, with a decree from the constitutional court in Karlsruhe. Neither in the agricultural nor in the biomedical sector did German actors, including scientists, take the lead in testing or contesting the limits of legislative settlements once they were firmly in place. Scientific uncertainty, which so often surfaced as a public issue in both Britain and the United States, was accorded little or no political role in Germany.

## Deliberation

Germany, Britain, and the United States all aspire to be deliberative as well as representative democracies. The legitimacy of political action in all three countries depends in theory not only on the amount of participation, but also on its quality. Mere voting, empty of deliberative content, is regarded as insufficient for democratic governance, and public referenda of the sort so readily conducted in California are widely deplored. Yet each of these Western nations is plagued by bouts of pessimism about the possibility of genuine civic engagement. Concerns about the public understanding of science expressed on both sides of the Atlantic bear witness to this self-doubt, as do worries about a democratic deficit in the European Union, moral panics (over food safety, for example) in Britain, and the persistent despair in the United States over the role of the "phantom public"—articulated as far back as the 1920s by the influential American journalist Walter Lippmann.[23] What light does comparative analysis shed on the capacity of democratic polities to deliberate on the profoundly unsettling implications of the biological revolution?

Across all three countries and in almost all the issue areas surveyed in this book, we are struck by the dearth of meaningful debate on the metaphysical aspects of biotechnology: that is, debate about the kinds of entities, and associated forms of life, that the technology has sought to, or should seek to, create. Genetic modification is at its core, a means of bringing novel entities into the world, and an engaged deliberative politics might have been expected to focus in the first instance on the desirability of these new productions. How *should* we use the power to intervene in nature's processes? Yet time after time the official discourses of policymaking channeled debate toward looking only at the impacts of technologically generated things whose existence, purpose, and value were barely questioned in public or private forums. Can recombinant bovine growth hormone (rBST) harm human health or Bt-corn damage nontarget insect species? Does GM food pose a risk to health, and are efforts to restrict their flow in trade contrary to international agreements? Are cloned animals weaker or more vulnerable to disease and age than their sexually propagated counterparts, and are risks like these a sufficient basis for prohibiting human cloning? Can patents be lawfully awarded on gene fragments, novel bacteria, and genetically altered animals? Corporate discourse, for its part, stoutly denied that the creative processes of biotechnology are novel enough to raise new issues for political deliberation. The problems listed above were all considered amenable to expert resolution. What need, then, for the turbulence of politics?

Excluded from official regulatory forums, questions about biotechnology's breaks with nature found their way into a variety of oppositional discourses, ranging from destroying research plots to filing *amicus* briefs, conducting media campaigns, dumping grain before politicians' doors, and finding creative ways around the rules that prevented GM products from being labeled as such. Public intellectuals such as Leon Kass and Francis Fukuyama in the United States, or Jürgen Habermas and Peter Sloterdijk in Germany, offered personal visions, their voices sometimes amplified through that odd late-twentieth-century governance mechanism, the national ethics commission. The resulting patchwork of strategies and articulations led to changes in normal politics in some instances, but, interestingly, the most notable participatory victories were achieved in places where existing deliberative institutions and discourses were weakest and least grounded in prior practice.

Thus, environmentalists scored some of their greatest successes with relatively new institutions like the European Parliament (on gene patenting and research ethics, for example) and with industrial sectors, such as supermarkets, that were newcomers to GM politics and directly answerable to consumer preferences. Britain's historical lack of experience with broad public engagement opened the door to the remarkably innovative, three-pronged deliberative process on the commercialization of GM crops; this exercise led to perhaps the widest-ranging exploration of scientific unknowns around biotechnology since the much less democratic Asilomar conference on rDNA research. In the United States, GM opponents made themselves most clearly heard in the nascent organic food market, for which federal regulatory principles had yet to be devised in the 1990s. By contrast, the securely established U.S. Food and Drug Administration resisted demands for a broad-based assessment of the value of GM foods until European opposition forced at least a partial retreat from the naturalizing doctrine of "substantial equivalence" between GM and non-GM products.

Only the specter of new kinds of humans produced through cloning—and the corresponding destruction of embryonic stem cells as potential humans—led in all three countries to something approaching full-blown public deliberation on the ontological consequences of biotechnology, but even on these explosive issues the results were curiously constrained. In Britain, a successful dehumanizing of the pre-embryo prevented a slide into the polarized abortion debate that hemmed in U.S. research, but it also inhibited the exploration of eugenic and feminist concerns that surfaced in Germany around embryo manipulation. Yet Germany's intellectually more robust debate paved the way to an exceptionally stringent law whose black-and-white prohibitions blocked subsequent innovation and forced uneasy moral compromises on stem cell research. In the United States, bioethics initially developed as a facilitator of medical research, with utilitarian overtones. The conservative-tending bioethics council appointed by President George W. Bush adopted an allegedly transparent vocabulary to illuminate "descriptive reality" and facilitate moral reasoning. In charting this novel discursive path, however, the council denied the framing power of its own language and reinscribed a highly questionable fact-value distinction into the domain of ethical analysis. And—in a far from neutral move—the council's proposed terminology broke the emerging links between stem cell research and the politically compelling discourses of therapeutic and reproductive rights.

## In the Hinterlands of Power

Democratic engagement with biotechnology, we have seen, was shaped and constrained by national approaches to representation, participation, and deliberation. These approaches in turn selectively delimited who spoke for people and issues, how those issues were framed, and how they were reflected upon in official policymaking processes. Almost as a corollary, certain matters remained in each country stubbornly outside the reach of popular control, marked off from ordinary politics; various forms of institutional boundary work either rendered those issues invisible or else designated them as appropriate only for expert analysis. Democratic control, in other words, was sometimes set aside in favor of other culturally sanctioned norms of political legitimacy. What were these ineffable sources of authority, and how did they vary across national lines?

Our comparisons indicate that there were in each country preferred models of virtuous decision making that supplemented, to some degree, formal democratic processes for the governance of biotechnology. Most clearly visible in the controversies about deliberate release (chapter 4), these hidden reserves of power included the authority of science in the United States, of expertise in Britain, and of institutional rationality in Germany. These legitimating devices carved out deliberative spaces that were not subject to ordinary rules of political accountability. Identifying them advances the project of political theory, for it shows how in any well-ordered democratic system certain relatively unquestioned social practices serve as a backstop to normal politics—much in the way that, in Thomas Kuhn's account of scientific revolutions, the invisible assumptions of dominant scientific paradigms provide a safe haven within which normal science enjoys its day-to-day recognition, meaning, and legitimacy.[24] Neither absolute nor universally at play, the paradigmatic elements of political culture demand recognition, and, once exposed to view, they are not above critique.

In the United States, it is the transcendental cognitive authority of science that most powerfully complements political authority. So fundamental, even constitutional, is the status of science within politics that it is tempting to refer to the United States as a *Wissenschaftsstaat* (a state of science), parallel to Germany's self-designation as a *Rechtsstaat* (a state of law). Science was repeatedly invoked to close, or foreclose, political debates on biotechnology, especially in the agricultural sector, but factual claims alone were never sufficient. The capacity of science to trump political dissent is channeled through a variety of expert institutions, from the National Academy of Sciences to advisory bodies attached to regulatory agencies. Through such bodies, science underwrote the initial product framing that drove U.S. regulation, provided the grounds for assuming safety in agricultural and food biotechnology, offered persuasive arguments against human cloning and for stem cell research, and supported the rewriting of the social contract so as to bring university research into closer alliance with commercial production.

Of course, to perform these legitimating functions, science itself has to stand apart from the contaminating touch of politics, and there is no dearth of actors prepared to do the boundary work that makes this separation seem real. The Mertonian vision of disinterested science is kept alive even by corporate science's most ardent American critics; not for them the Mode 2 analyst's dream of a thoroughly socialized, politically embedded science. Moments of slippage do occur, as happened in biotechnology patent disputes and in recurrent controversies about the ethical status of university-industry relations. But in all these debates, it is a particular faulty development, corporation, or person that draws negative attention; the ideal of value-free science retains its firm hold on the national imagination, of cheerleaders as well as skeptics.

Science in Britain enjoys a no less preeminent intellectual status than in the United States, but as a guarantor of political legitimacy science lacks power unless it speaks through socially accredited expert bodies, both individual and collective. Much energy is devoted in Britain to producing experts whose right to speak on behalf of the public will be virtually unquestioned. Their authority is not a matter of skills and knowledge alone, but of those attributes coupled to significant demonstrations of social responsibility. British experts generally earn the right to represent the public through successive episodes of personal testing that elevate them to the ranks of the "great and the good" and endow them, in effect, with common vision: the power to see for the people, with an encompassing gaze that goes beyond the ordinary person's capacity for foresight. When this system of authorization malfunctions, as it massively did in the case of BSE, reforms are targeted toward incorporating a wider range of expertise into ruling institutions, thereby shoring up the credibility of the state. Underlying notions of what constitutes expertise are not necessarily questioned. It hardly needs restating that embodied political authority in Britain does not reside in scientists alone, but in all those who have demonstrated the ability speak for the people—and so to do the state some service. Even Prince Charles, the much lampooned heir to the throne, exerted wide-ranging moral and political influence when he framed the uncertainties around GM crops in ways that resonated with the majority of his subjects. Contingently, on this single issue, Charles overcame his image of know-nothing traditionalist and functioned for a time as a people's expert.

In place of transcendental knowledge or embodied authority, the German decision-making system has reposed its trust in institutional reasoning as an important supplement to democratic politics. More than in either of the other countries, visible and unruly political interventions, such as the Greens' partial capture of the parliamentary Enquiry Commission on genetic engineering, tend to go underground in Germany once an issue is felt to be under comprehensive legal control. Matters then move from the sphere of politicized debate to that of administrative implementation, guided by appropriately constituted expert committees. The status of these committees in relation to policy legitimation, however, is subtly different from that of similar bodies in the United States and Britain. Whereas U.S. and U.K. bodies offer advice, German committees function as miniworlds of reason: they aim to encompass within them the capacity to articulate all relevant arguments and to produce a consensus that is binding on society because it is, in effect, the consensus that society would have reached if it had been able to deliberate in common. Issues of membership and representation become crucial for bodies such as these, and it is perhaps no accident that German legislation frequently demands not only the designation of members representing relevant social groups, but also the identification of alternates for each designated member. The absence of any viewpoint would create a vacuum detrimental to the proper functioning of the whole.

# The Ends of Comparison: Theory and Practice

The early reception of biotechnology into the political life of three leading industrial democracies was not, I have argued, simply a matter of advances in science and technology, but a product of complex entanglements among knowledge, technical capability, politics, and culture. To make sense of the resulting national settlements, one must look not only at the discoveries and commodities that materialize out of research laboratories and industrial facilities; one must equally take into account the scripts for human development and collective choice that emanate from a nation's political and social institutions, and from its citizens. For politically engaged scholars of science and technology in society, following scientists around (as some sociologists of knowledge have urged[25]) provides only a tiny peephole on the power of science. It is just as necessary to chart the trajectories of the myriad other social actors whose values and expectations interpenetrate with those of scientists and inventors, creating the conditions in which scientific ideas are translated into material and social realities. Scientific cultures are at one and the same time political cultures.[26] Describing one without the other leads to partial vision, thin description, and inadequately informed critique. Any responsible ethnographer of modernity has to find an interpretive vantage point on both science and politics, and on their increasingly thick networking.

Comparison, particularly across national political systems, provides one such angle of vision. Setting the experiences of one country against another's offers salutary reminders of the degree to which even the homogeneous West is not univocal in its responses to science and technology. Democracy, too, is not a singular form of life but a common human urge to self-rule that finds expression in many different institutional and cultural arrangements. In tracing biotechnology's complex passages from invention to regulation to social uptake (or, in some cases, rejection), we gain deeper insights not only into what modern cultures want to make out of life—their designs on nature—but also the resources with which they wish to reflect and deliberate on life's very meaning. We observe not a predetermined, linear march of technoscientific progress, but a series of contingent, culturally specific accommodations, in which the continual intertwining of knowledge with politics produces outcomes that are as rich as they are strange.

There are, of course many similarities among the three countries compared in this book. I have chosen to highlight the cross-national differences in part because they may not be easily apparent to actors operating exclusively within the perimeter of a given political culture. Comparison in this respect promotes cultural self-awareness. It is by looking outward at others' ordering practices that one most readily appreciates the markers of one's own. Only by examining the politics of biotechnology across nations does the traveling explorer of modernity come to recognize, and perhaps regret, such "missing" elements of political culture as proactive citizen engagement in Britain, robust scientific debate in Germany, and respect for embodied virtue in the United States. That these absences exist, however, is not a necessary or sufficient argument for trying to import them from abroad. The distinctive features of political culture are not so easily transferable. They are embedded in a rich matrix of experience and practice, and efforts to graft them onto other settings may fail or entail a higher price than enthusiasts for such transfer would find acceptable.

Any expectation that other policy systems can be used as models to imitate rests on a notion of cultural deficit that this book should render seriously problematic. Political cultures are not, like the characters in the beloved American children's classic, L. Frank Baum's *The Wonderful Wizard of Oz*,[27] deficient along particular dimensions that can be objectively identified and easily compensated for: a brain for the Scarecrow, a heart for the Tin Woodman, and courage for the Cowardly Lion. Let us not forget that Baum's title character, when unmasked as a fraud and an impostor, disavows his magical powers like some latter-day Prospero and debunks his own capacity to make complex beings more whole than they already are. Handing out a diploma to the Scarecrow, a testimonial clock to the Tin Woodman, and a medal of honor to the Cowardly Lion, the Wizard of Oz only emphasizes the ironic insufficiency of these externally procured organ transplants. The enduring work of moral and social generation, he implies, requires inner resources of invention and reinvention—a lesson that holds no less true for today's vibrant knowledge societies than it did in 1900 for Baum's populist fictional heroes.

Comparison, then, should aspire less to prescribe policy solutions than to explain outcomes; it should offer interpretive criticism rather than clinical diagnosis, or *Verstehen* (understanding) rather than *Erklärung* (causal explanation). To state the purposes of comparative politics most simply, we can do no better than return to the words of Baroness Warnock before the House of Lords in the 1988 debate on the Embryo Bill. She said on that occasion, "[T]here is a very noble prayer frequently uttered in my college chapel that we may be given the grace to distinguish things that differ."[28] With one small but important modification, Warnock's prayer can serve as the starting point for the comparatist's creed. The observer of other cultures should also be given the grace to distinguish things that differ, but always keeping in mind the meanings that those distinctions have for people in the observed cultural spaces. The aim of comparison is to reveal, with critical detachment but epistemic charity, what gives significance to another culture's distinctions and differences, not forgetting in the process to reflect on the commitments encoded in one's own. It is not the divine prerogative of producing universally valid principles of knowledge or governance that comparison should strive for. It is to make visible the normative implications of different forms of contemporary scientific and political life, and to show what is at stake, for knowing and reasoning human beings, in seeking to inhabit them.

# Appendix: Chronology

***Scientific and Policy Landmarks in the Development of Biotechnology***

| Year | Discovery or Development |
|------|--------------------------|
| 1865 | Gregor Mendel discovers laws of genetics |
| 1900 | Rediscovery of Mendel's work |
| 1944 | Oswald Avery, Colin MacLeod, and Maclyn McCarty demonstrate that DNA is the hereditary material |
| 1953 | James Watson and Francis Crick describe the double-helix structure of DNA |
| 1972 | Stanley Cohen and Herbert Boyer develop recombinant DNA technology |
| 1973 | Gordon Conference discussion held on safety of recombinant DNA (rDNA) research |
| 1974 | Belmont report on the use of human subjects in research issued |
| 1975 | Conference on safety of rDNA research held in Asilomar, California |
| 1976 | U.S. National Institutes of Health *Guidelines for Research Involving Recombinant DNA Molecules* issued |
| 1977 | Frederick Sanger, Allan Maxam, and Walter Gilbert develop DNA sequencing |
| 1978 | Louise Brown is born, the first baby produced through in vitro fertilization technique developed by British researchers Patrick Steptoe and Robert Edwards |
| 1980 | U.S. Supreme Court decides in *Diamond v. Chakrabarty* that living organisms can be patented |
| 1982 | GenBank database for DNA sequences established by NIH |
| 1983 | First human disease gene—for Huntington's disease—is mapped with DNA markers |
| 1984 | Alec Jeffreys invents DNA fingerprinting in Britain; Kary Mullis invents polymerase chain reaction (PCR) in United States; |

report of the Warnock Committee into Human Fertilisation and Embryology published in Britain

1987    Advanced Genetic Sciences conducts first field trial of a genetically modified organism, frost inhibiting *Pseudomonas syringae*

1988    U.S. National Institutes of Health issues report on *Mapping and Sequencing the Human Genome*; Human Genome Organization (HUGO) formed; first transgenic higher organism (the "Harvard mouse") patented in United States

1989    Human Genome Project (HGP) is launched in United States; Calgene conducts first field trials of genetically modified tomatoes

1990    Germany enacts Genetic Engineering Law and Embryo Protection Law; Britain enacts Human Fertilisation and Embryology Act; EU issues directives on contained use and deliberate release of genetically modified organisms, and on worker safety

1996    First human gene map established; Dolly, the first cloned mammal, born in Roslin, Scotland

1997    Birth of Dolly announced in *Nature*, seven months after actual event

1998    Human embryonic stem cell lines established by researchers at Johns Hopkins and University of Wisconsin

2000    Draft version of human genome sequence completed

2001    Draft version of human genome sequence published

2003    Finished version of human genome sequence published; HGP ends with all goals accomplished; Dolly the sheep dies of progressive lung disease; Britain conducts GM Nation, a public debate on the commercialization of genetically modified crops

Source: Adapted in part from *Nature* 422, April 24, 2004 (foldout at pp. 836–837).

# Notes

1. The town was first mentioned in 1289 as "Janshagen."

2. In 1999 Daimler Chrysler was sued for compensation by a group of about a hundred Polish women, half of whom had worked as forced laborers at the Genshagen plant during the war. In 2000 these so-called Genshagerinnen received the AnStiftung's Dresden Peace Prize, awarded since 1996 under the umbrella of the foundation's "projects against forgetting." See also Daimler-Chrysler Special Report, "Moral Responsibility: Confronting the Past," http://www.daimlerchrysler.com/index_e.htm?/specials/zwangs/zwarb3_e.htm (visited March 6, 2002).

3. In attendance at the Genshagen meeting were the Bishop of Angers and representatives of the Catholic, Protestant, Jewish, and Islamic faiths.

4. Stem cells are a particular cell type derived from the blastocyst stage of embryonic development. These cells are therapeutically promising because they have the potential to differentiate into any of the roughly 200 cell types needed for human development. For a detailed review of the stem cell debate, see chapter 7.

5. House of Lords, *Stem Cell Research—Report*, http://www.parliament.the-stationery-office.co.uk/pa/ld200102/ldselect/ldstem/83/8301.htm, February 13, 2002 (visited April 2004).

6. For an argument that such moments are increasingly commonplace in ordering the relations between science and the state, see Sheila Jasanoff, "In a Constitutional Moment: Science and Social Order at the Millennium," in Bernward Joerges and Helga Nowotny, eds., *Social Studies of Science and Technology: Looking Back, Ahead* (Dordrecht: Kluwer, 2003), pp. 155–180.

7. Simon Jenkins, "This Constitutional Cloud-Cuckoo Land," *The Times (London)*, January 24, 2001, p. F 16.

8. Alex Berenson and Nicholas Wade, "A Call for Sharing of Research Causes Gene Stocks to Plunge," *New York Times*, March 15, 2000, p. A1.

9. Ernst and Young, *Eighth Annual European Life Sciences Report 2001* (London: Ernst and Young International, 2001), pp. 66–67.

10. The idea of such a thoroughgoing transformation can be traced back at least as far as Daniel Bell, *The Coming of Post-Industrial Society* (London: Heinemann, 1973).

For more recent elaborations, see Helga Nowotny, Peter Scott, and Michael Gibbons, *Re-Thinking Science: Knowledge and the Public in an Age of Uncertainty* (Cambridge: Polity, 2001); Gernot Bohme and Nico Stehr, *The Knowledge Society: The Growing Impact of Scientific Knowledge on Social Relations* (Dordrecht, NL: Reidel, 1986). The concept's growing importance was reflected in the theme of the joint meeting of the European Association for the Study of Science and Technology and the Society for Social Studies of Science, "Signatures of Knowledge Societies", held in Bielefeld, Germany, in October 1996.

11. See, for example, Edward Yoxen, *The Gene Business: Who Should Control Biotechnology?* (London: Pan Books, 1983); Sheldon Krimsky, *Genetic Alchemy: The Social History of the Recombinant DNA Controversy* (Cambridge: MIT Press, 1982); Susan Wright, *Molecular Politics: Developing American and British Regulatory Policy for Genetic Engineering, 1972–1982* (Chicago: University of Chicago Press, 1994).

12. This idealized view of science's independence from politics is presented in Robert K. Merton, "The Normative Structure of Science," in Merton, *The Sociology of Science: Theoretical and Empirical Investigations* (Chicago: University of Chicago Press, 1973), pp. 267–278; Michael Polanyi, "The Republic of Science," *Minerva* 1 (1962): 54–73. That view, however, has been questioned in recent work in the sociology and politics of science. See Yaron Ezrahi, *The Descent of Icarus* (Cambridge: Harvard University Press, 1990); Sheila Jasanoff, *The Fifth Branch: Science Advisers as Policymakers* (Cambridge: Harvard University Press, 1990); Chandra Mukerji, *A Fragile Power: Scientists and the State* (Princeton: Princeton University Press, 1989).

13. Lee Silver, *Remaking Eden* (New York: Perennial, 2002); Geoffrey C. Bowker and Susan L. Star, *Sorting Things Out: Classification and Its Consequences* (Cambridge: MIT Press, 2000); Michel Foucault, *The History of Sexuality*, vol. 1 (New York: Vintage, 1990 [1976]).

14. I am elaborating here on Benedict Anderson's highly productive definition of nationhood as "imagined community" in *Imagined Communities*, 2nd ed. (London: Verso, 1991). Unlike Anderson, however, I see possibilities for civil society as well as the state to engage actively in projects of reimagining nationhood. I also see episodes of significant scientific and technological change as important moments in reimagining nationhood.

15. On January 27, 2004, Tony Blair narrowly survived a revolt within his own party on the issue of charging fees to university students. The government won the vote by 316 to 311, although its normal majority was 161 votes. Many Labour parliamentarians considered the fees a retreat from the party's historical commitment to egalitarianism in education.

16. See the five-part series of articles by Stephen Mulvey, BBC News Online, including "The EU's Democratic Challenge" (source of the quoted line) and "EU Values—United in Diversity," http://news.bbc.co.uk/2/hi/europe/3224666.stm (visited January 2004). Also see Richard Bernstein, "Europe's Lofty Vision of Unity Meets Headwinds," *New York Times*, December 4, 2003, p. A1.

17. One sign of the renewed questioning of America's identity was the spate of books produced in this period on the extent and limits of U.S. power. See Serge Schmemann, "The Only Superbad Power," *New York Times Book Review*, January 25, 2004, p. 12

18. Robert Kagan, *Of Paradise and Power: America and Europe in the New World Order* (New York: Knopf, 2003).

19. Juan Díez Medrano, *Framing Europe: Attitudes to European Integration in Germany, Spain, and the United Kingdom* (Princeton: Princeton University Press, 2003).

20. Thomas Bernauer, *Genes, Trade, and Regulation: The Seeds of Conflict in Food Biotechnology* (Princeton: Princeton University Press, 2003).

21. This rejection of explanatory asymmetry is central to current work in science and technology studies. See in particular David Bloor, *Knowledge and Social Imagery* (Chicago: University of Chicago Press, 1976). Also see my discussion of symmetry in relation to political explanation in Sheila Jasanoff, "Beyond Epistemology: Relativism and Engagement in the Politics of Science," *Social Studies of Science* 26, 2 (1996): 393–418.

22. Svetlana Alpers, *The Art of Describing: Dutch Art in the Seventeenth Century* (Chicago: University of Chicago Press, 1983), p. xxvii.

23. The concept of *Verstehen* or "understanding" was initially elaborated by the German philosopher Wilhelm Dilthey. As adapted and used by Max Weber, the term is particularly associated with interpretive sociology, that is, the kind of sociology that attempts to make sense of human motives and intentions through systematic introspection. In terms more familiar to contemporary sociologists of science and technology, such introspection would be characterized as "reflexivity," or the capacity of individuals and institutions to inquire into the assumptions underlying their beliefs. For a brief discussion of understanding in Weber's sociology, see Hans H. Gerth and C. Wright Mills, eds., *From Max Weber: Essays in Sociology* (New York: Oxford University Press, 1946), pp. 56–57.

### CHAPTER 1: WHY COMPARE?

1. Whether in producing reliable scientific knowledge or in testing technological systems, the evaluator must continually make judgments about whether a given phenomenon is the same as or different from the one with which it is to be compared. On the inevitable contingency (i.e., nonuniversality) of such judgments, see Barry Barnes, *T. S. Kuhn and Social Science* (London: Macmillan, 1982); H. M. Collins, *Changing Order: Replication and Induction in Scientific Practice* (London: Sage Publications, 1985); Trevor Pinch, "'Testing—One, Two, Three . . . Testing!': Toward a Sociology of Testing," *Science, Technology, and Human Values* 18 (1993): 25–41.

2. Philip Kitcher, *Science, Truth, and Democracy* (Oxford: Oxford University Press, 2001); Ian Hacking, *The Social Construction of What?* (Cambridge: Harvard University Press, 1999).

3. On the constructedness of nature, see William Cronon, ed., *Uncommon Ground: Rethinking the Human Place in Nature* (New York: Norton, 1996). See also Sheila Jasanoff and Marybeth Long Martello, *Earthly Politics: Local and Global in Environmental Governance* (Cambridge: MIT Press, 2004).

4. On boundary drawing in the policy process, see Jasanoff, *The Fifth Branch*; David Guston, "Stabilizing the Boundary between U.S. Politics and Science," *Social Studies of Science* 29 (1999): 87–112. On the power of policy discourse, see Langdon Winner, *The Whale and the Reactor: A Search for Limits in an Age of High Technology* (Chicago: University of Chicago Press, 1986).

5. The point about rational politics yielding to aesthetic impulses is forcefully argued by Ezrahi in *The Descent of Icarus*. On the role of science in creating "subpolitics" and micro utopias, see, respectively, Ulrich Beck, *Risk Society: Towards a New Modernity* (London: Sage, 1992); and Yaron Ezrahi, "Science and Utopia in Late 20th Century Pluralist Democracy," in Everett Mendelsohn and Helga Nowotny, eds., *Nineteen Eighty-Four: Science between Utopia and Dystopia, Sociology of the Sciences Yearbook VIII* (Dordrecht, NL: Reidel, 1984), pp. 273–290. On science's loss of monopoly in representing contemporary reality, see Ezrahi, "Science and Political Imagination in Contemporary Democracies," in Sheila Jasanoff, ed., *States of Knowledge: The Co-Production of Science and Social Order* (London: Routledge, 2004), pp. 254–273. For a recent empirical study of expertise being appropriated by political interests, see U.S. House of Representatives, Committee on Government Reform (Minority Report), *Politics and Science in the Bush Administration*, Washington, DC, August 7, 2003, http://www.house.gov/reform/min/ politicsandscience/index.htm (visited April 2004).

6. The theme of eroding sovereignty has frequently been linked to technological change and ecological interdependence. See Richard Falk, *This Endangered Planet: Prospects and Proposals for Human Survival* (New York: Vintage Books, 1971); Joseph Camilleri and Jim Falk, *The End of Sovereignty? The Politics of a Shrinking and Fragmenting World* (Aldershot, Hants: Edward Elgar, 1992); Jens Bartelson, *A Genealogy of Sovereignty* (Cambridge: Cambridge University Press, 1995); Thomas Bierstecker and Cynthia Weber, eds., *State Sovereignty as Social Construct* (Cambridge: Cambridge University Press, 1996). For a partially contrary view, see Eugene Skolnikoff, *The Elusive Transformation: Science, Technology, and the Evolution of International Politics* (Princeton: Princeton University Press, 1993).

7. See, for example, Klaus Eder and Maria Kousis, eds., *Environmental Politics in Southern Europe* (Dordrecht: Kluwer, 2001), especially Eder, "Sustainability as a Discursive Device for Mobilizing European Publics," pp. 25–52.

8. For case studies illustrating these dynamics, see Jasanoff and Martello, *Earthly Politics*.

9. For a recent canonical example, see the two-volume work of The Social Learning Group, *Learning to Manage Global Environmental Risks: A Comparative History of Social Responses to Climate Change, Ozone Depletion, and Acid Rain* (Cambridge: MIT Press, 2001).

10. Merton, "The Normative Structure of Science."

11. Merritt Roe Smith and Leo Marx, *Does Technology Drive History? The Dilemma of Technological Determinism* (Cambridge: MIT Press, 1994).

12. See, for example, David Vogel, "The Hare and the Tortoise Revisited: The New Politics of Consumer and Environmental Regulation in Europe," *British Journal of Political Science* 33 (2003): 557–580.

13. The competence of agencies responsible for social regulation has long been an issue in American political science and policy studies. Representative works include Stephen G. Breyer, *Breaking the Vicious Circle: Toward Effective Risk Regulation* (Cambridge: Harvard University Press, 1993); and Marc K. Landy, Marc J. Roberts, and Stephen R. Thomas, *The Environmental Protection Agency: Asking the Wrong Questions from Nixon to Clinton*, expanded ed. (New York: Oxford University Press, 1994). For a comparative study addressing agency competence, see Peter M. Haas, Robert O. Keohane, and Marc A. Levy, eds., *Institutions for the Earth: Sources of Effective International Environmental Protection* (Cambridge: MIT Press, 1993).

14. Thus, in Kingdon's widely accepted model of agenda-setting, "policy," "politics," and "problems" are presumed to flow in three separate streams, which only occasionally join together. John W. Kingdon, *Agendas, Alternatives, and Public Policies*, 2nd ed. (New York: Longman's, 1995). Authors who are more sensitive to the social construction of policy agendas have not addressed the problem this framework raises for comparative, including cross-national, policy analysis or evaluation. See, for example, Roger W. Cobb and Charles D. Elder, *Participation in American Politics: The Dynamics of Agenda-Building* (Baltimore: Johns Hopkins Press, 1972); Joseph Gusfield, *The Culture of Public Problems: Drinking-Driving and the Symbolic Order* (Chicago: University of Chicago Press, 1981).

15. Significant contributions include Stephen Kelman, *Regulating America, Regulating Sweden: A Comparative Study of Occupational Safety and Health Policy* (Cambridge: MIT Press, 1981); Graham Wilson, *The Politics of Safety and Health: Occupational Safety and Health in the United States and Britain* (Oxford: Clarendon Press, 1985); Ronald Brickman, Sheila Jasanoff, and Thomas Ilgen, *Controlling Chemicals: The Politics of Regulation in Europe and the United States* (Ithaca: Cornell University Press, 1985); Joseph Badaracco, *Loading the Dice: A Five Country Study of Vinyl Chloride Regulation* (Cambridge: Harvard University Press, 1985); Jasanoff, *Risk Management and Political Culture* (New York: Russell Sage Foundation, 1986); David Vogel, *National Styles of Regulation: Environmental Policy in Great Britain and the United States* (Ithaca: Cornell University Press, 1986).

16. Brendan Gillespie, Dave Eva, and Ron Johnston, "Carcinogenic Risk Assessment in the United States and Great Britain," *Social Studies of Science* 9 (1979): 265–301; Sheila Jasanoff, "Cultural Aspects of Risk Assessment in Britain and the United States," in Branden B. Johnson and Vincent Covello, eds., *The Social and Cultural Construction of Risk* (Dordrecht: Reidel, 1987), pp. 359–397; Jasanoff, *Risk Management and Political Culture*.

17. For particularly cogent examples of such convergences in regulatory outcome, see the case studies of chemical regulation in Brickman et al., *Controlling Chemicals*.

18. On the resurgence of the state as a focus of analysis in this period, see particularly Peter Evans, Dietrich Rueschemeyer, and Theda Skocpol, eds., *Bringing the State Back In* (New York: Cambridge University Press, 1985).

19. One has to be careful in interpreting the role of the courts in U.S. culture in view of statistics that indicate a sharp reduction over the past decade in the percentage of cases proceeding to trial. It remains the case, though, that entry barriers to litigation are lower in the United States than in other industrial nations, and lawsuits are prods to economic and social change, even if the results in the vast majority of cases are achieved through one or another form of settlement.

20. Badaracco, *Loading the Dice*. See also David Vogel, "Consumer Protection and Protectionism in Japan," *The Journal of Japanese Studies* 18 (1992): 119–154.

21. The inadequacy of traditional notions of delegation and expertise as legitimating devices became particularly evident in the decisionmaking of the U.S. Environmental Protection Agency (EPA). See Jasanoff, *The Fifth Branch*; and Jasanoff, "Science, Politics, and the Renegotiation of Expertise at EPA," *Osiris* 7 (1991): 195–217. Of course, even numbers may not provide adequate protection. As recently as 2001, EPA's capacity to make rational decisions protecting public health and safety on the basis of quantitative risk assessment was under sharp attack by industry. In *Whitman v. American Trucking Associations, Inc.*, 531 U.S. 457 (2001), the Supreme Court unanimously

affirmed both the legality of EPA's decisionmaking under the Clean Air Act and the constitutionality of the scope of administrative discretion conferred by this most basic piece of environmental legislation.

22. Like all generalizations, this one needs to be qualified and will be softened in later chapters of this book. It is worth noting here, however, that Europeans have been more concerned in recent years to establish the independence of the expert from political interests than to establish the formal validity of methods. See, for example, Commission of the European Communities, *European Governance: A White Paper*, COM (2001) 428, Brussels, July 27, 2001, http://europa.eu.int/eur-lex/en/com/cnc/ 2001/com2001_0428en01.pdf (visited April 2004).

23. Theodore M. Porter, *Trust in Numbers: The Pursuit of Objectivity in Science and Public Life* (Princeton: Princeton University Press, 1995).

24. For more on the notion of social kinds, see Ian Hacking, "World-Making by Kind-Making: Child Abuse for Example," in Mary Douglas and David Hull, eds., *How Classification Works: Nelson Goodman among the Social Sciences* (Edinburgh: Edinburgh University Press,1992), pp. 180–213.

25. Within the growing body of literature that concerns itself with the nature of so-cial categories, the following works have proved to be especially salient because they display the historical and social contingency and, in some cases, the political force of such categorization: Michel Foucault, *The Archaeology of Knowledge* (New York: Harper and Row, 1976 [1972]); Bloor, *Knowledge and Social Imagery*; Donna Haraway, *Primate Visions: Gender, Race, and Nature in the World of Modern; Science* (New York: Routledge, 1989); Bruno Latour, *We Have Never Been Modern* (Cambridge: Harvard University Press, 1993); Ian Hacking, *Rewriting the Soul: Multiple Personality and the Sciences of Memory* (Princeton: Princeton University Press, 1995); Bowker and Star, *Sorting Things Out.*

26. Jasanoff, ed., *States of Knowledge.*

27. Kingdon, *Agendas, Alternatives, and Public Policies.*

28. See particularly Theda Skocpol, *Protecting Soldiers and Mothers: The Political Ori-gins of Social Policy in the United States* (Cambridge: Harvard University Press, 1992).

29. AIDS activism in the United States provides an especially instructive case of a social movement that at once asserted and redefined itself through engagements with the process of knowledge-making about a devastating disease. See Steven Epstein, *Impure Science: AIDS, Activism, and the Politics of Knowledge* (Berkeley: University of California Press, 1996).

30. Doug McAdam, Sidney Tarrow, and Charles Tilly, *Dynamics of Contention* (Cambridge: Cambridge University Press, 2001), p. 22.

31. It was arguably this kind of theoretical deficiency that kept the majority of Western political analysts from observing or predicting the sudden collapse of the communist regime in the Soviet Union in the 1989. Similar peripheral blindness may also account for the failure of U.S. authorities to foresee (or at any rate to act to pre-vent) the attacks of September 11, 2001, by Islamic terrorists.

32. Earlier work exploring the framing of biotechnology as a policy issue includes Sheila Jasanoff, "Product, Process, or Programme: Three Cultures and the Regulation of Biotech-nology," in Martin Bauer, ed., *Resistance to New Technology* (Cambridge: Cambridge Uni-versity Press, 1995), pp. 311–331; Herbert Gottweis, *Governing Molecules: The Discursive Politics of Genetic Engineering in Europe and the US* (Cambridge: MIT Press, 1998).

33. On the comparative regulation of chemical risks, see Brickman et al., *Controlling Chemicals*; Wilson, *The Politics of Safety and Health*; Badaracco, *Loading the Dice*; Jasanoff, *Risk Management and Political Culture*; Vogel, *National Styles of Regulation*.

34. Some recent studies of comparative and international environmental policy-making have recognized the political significance of discourse. See, for example, Maarten Hajer, *The Politics of Environmental Discourse* (Oxford: Oxford University Press, 1995); Karen Litfin, *Ozone Discourses: Science and Politics in Global Environmental Cooperation* (New York: Columbia University Press, 1994).

35. Spirited exchanges about reading, or misreading, culture have erupted around the work of postcolonial scholars. See particularly Gananath Obeyesekere, *The Apotheosis of Captain Cook: European Mythmaking in the Pacific* (Princeton: Princeton University Press, 1992); and Marshall D. Sahlins, *How "Natives" Think: About Captain Cook, For Example* (Chicago: University of Chicago Press, 1995). Also see Clifford Geertz, *Available Light: Anthropological Reflections on Philosophical Topics* (Princeton: Princeton University Press, 2000).

36. A search of the Harvard University library catalogue would quickly persuade the visitor from Mars that political culture is located largely in the past (much of the work is historical) and predominantly not in the West. Even in Western nations, political culture is often most discernible to scholars at levels below the nation state, such as in individual states of the U.S. federal union. In these respects scholarship unreflectively gives back the progressive history of the Enlightenment, which shows humankind on a long march away from the specificities of culture toward the universality of science and reason.

37. Jasanoff, ed., *States of Knowledge*.

38. On Boyle's scientific achievements and their social implications, see Steven Shapin and Simon Schaffer, *Leviathan and the Air-Pump: Hobbes, Boyle, and the Experimental Life* (Princeton: Princeton University Press, 1985); Shapin, "Pump and Circumstance: Robert Boyle's Literary Technology," *Social Studies of Science* 14 (1984): 481–520; Shapin, *A Social History of Truth* (Chicago: University of Chicago Press, 1994). On the relationship of the scientific revolution to the rise of liberal democracy, see Ezrahi, *The Descent of Icarus*.

39. Michel Foucault, *The Order of Things: An Archaeology of the Human Sciences* (New York: Pantheon, 1970); Helga Nowotny, "Knowledge for Certainty: Poverty, Welfare Institutions and the Institutionalization of Social Science," in Peter Wagner, Björn Wittrock, and Richard Whitley, eds., *Discourses on Society: The Shaping of the Social Science Disciplines* (1990), vol. 15, pp. 23–41; Theodore M. Porter, *The Rise of Statistical Thinking 1820–1990* (Princeton: Princeton University Press, 1986). On statistics as a means of disciplining industrial risks, in particular, see Beck, "From Industrial Society to the Risk Society." On the role of images and inscriptions in cementing centralized state power, see also Anderson, *Imagined Communities*; Bruno Latour, "Drawing Things Together," in Michael Lynch and Steve Woolgar, eds., *Representation in Scientific Practice* (Cambridge: MIT Press, 1990), pp. 19–68.

40. Mary Douglas, *Purity and Danger: An Analysis of Concepts of Pollution and Taboo* (London: Routledge and Kegan Paul, 1966).

41. Brian Wynne, "Public Uptake of Science: A Case for Institutional Reflexivity," *Public Understanding of Science* 2 (1992): 321–337. See also Irwin and Wynne, *Misunderstanding Science? The Public Reconstruction of Science and Technology* (Cambridge: Cambridge University Press, 1996).

42. Jasanoff, "Science, Politics, and the Renegotiation of Expertise."

43. William Butler Yeats, "Among School Children," in *The Collected Poems of W. B. Yeats* (New York: Macmillan, 1956), pp. 212–214.

44. Erving Goffman, *Frame Analysis: An Essay on the Organization of Experience* (Cambridge: Harvard University Press, 1974).

45. For a discussion of frames in structuring the symbolic politics of social movements, see Sidney Tarrow, *Power in Movement: Social Movements and Contentious Politics*, 2d ed. (Cambridge: Cambridge University Press, 1998), pp. 106–122. Also important is work building on Goffman's sense of public action as theatrical performance, with actors intentionally staging their roles. See in this connection Stephen Hilgartner, *Science on Stage: Expert Advice as Public Drama* (Stanford: Stanford University Press, 2000).

46. History is replete with examples of events of similar magnitude and severity but producing huge differences in political uptake and social response. Consider the effects of the Nazi Holocaust in comparison with Stalin's massacres in the Soviet Union, the 1994 genocide of Tutsis by Hutus in Rwanda versus the later Serbian atrocities in Kosovo, or the 1984 chemical disaster in Bhopal, India, compared with the 2001 destruction of the World Trade Center.

47. The first consequence of this framing of events was the actual war fought against Afghanistan by the United States and its allies in late 2001, but the metaphor soon had repercussions beyond that initial theater of war in justifying Israel's operations in the Palestinian territories in 2002 and the U.S. war on Iraq in 2003. Ian Hacking, speaking in a very different context, nicely captures the potential of such language to shape the imagination and direct future action: "Metaphors influence the mind in many unnoticed ways. The willingness to describe fierce disagreement in terms of the metaphors of war makes the very existence of real wars seem more natural, more inevitable, more a part of the human condition. It also betrays us into an insensibility toward the very idea of war, so that we are less prone to be aware of how totally disgusting real wars really are." *The Social Construction of What?*, p. viii.

48. The most influential extension of Goffman's ideas to the policy domain is Donald A. Schön and Martin Rein, *Frame/Reflection: Toward the Resolution of Intractable Policy Controversies* (New York: Basic Books, 1994).

49. Gusfield, *Public Problems*.

50. Expositions of actor-network theory may be found in Michel Callon, "Some Elements of a Sociology of Translation: Domestication of the Scallops and Fishermen of St. Brieuc Bay," in John Law, ed., *Power, Action, and Belief: A New Sociology of Knowledge?* (London: Routledge, 1986), pp. 196–233; Wiebe E. Bijker, Thomas P. Hughes, and Trevor Pinch, eds., *The Social Construction of Technological Systems* (Cambridge: MIT Press, 1987); John Law and John Hassard, *Actor Network Theory and After*, *Sociological Review Monographs* (Oxford: Blackwell, 1999).

51. This is not a purely rhetorical point nor one without relevance to the comparisons made in this book. German voters have repeatedly refused to accept speed limits on the Autobahn in spite of concerns about drunk driving. American courts have imposed liability on innkeepers for alcohol-related traffic accidents, whereas British courts have hesitated to extend liability in this direction.

52. Medrano, *Framing Europe*.

53. Ibid., p. 6.

54. On practices in IVF clinics, see, for example, Frances Price, "Now You See It, Now You Don't: Mediating Science and Managing Uncertainty in Reproductive Medicine," in Irwin and Wynne, *Misunderstanding Science?*, pp. 84–106; Charis Cussins, "Ontological Choreography: Agency through Objectification in Infertility Clinics," *Social Studies of Science* 26 (1996): 575–610. On embryo "adoption," see Anne Zielke, "Im Disneyland der Kindermacher," *Frankfurter Allgemeine Zeitung*, Feuilleton no. 67, March 20, 2002, p. 49.

55. Francis Collins, the head of the U.S. Human Genome Project, included this as one of the ten most surprising facts to emerge from the completed mapping of the human genome in a speech to the American Association for the Advancement of Science, Washington, DC, February 19, 2001.

56. Thomas Gieryn, *Cultural Boundaries of Science: Credibility on the Line* (Chicago: University of Chicago Press, 1999); see also Gieryn, "Boundaries of Science," in Sheila Jasanoff et al., eds., *The Handbook of Science and Technology Studies* (Thousand Oaks, CA: Sage, 1995), pp. 393–456.

57. Latour, *We Have Never Been Modern.* See also Donna Haraway, *Simians, Cyborgs, and Women: The Reinvention of Nature* (New York: Routledge, Chapman, and Hall, 1991).

58. Zygmunt Bauman, *Modernity and Ambivalence* (Ithaca: Cornell University Press, 1991).

59. Susan Leigh Star and James R. Griesemer, "Institutional Ecology, 'Translations' and Boundary Objects: Amateurs and Professionals in Berkeley's Museum of Vertebrate Zoology, 1907–39," *Social Studies of Science* 19 (1989): 387–420.

60. For extensive discussion of the boundary work done by courts in relation to science and technology, see Sheila Jasanoff, *Science at the Bar: Law, Science, and Technology in America* (Cambridge: Harvard University Press, 1995).

61. Boundary work by expert advisory committees is especially important in establishing the dividing line between science (which is institutionally defined to include everything in the committee's purview) and nonscience. This in turn has important consequences for who participates or does not participate in the analysis of policy options. For examples, see Jasanoff, *The Fifth Branch.*

62. The notable exception is Mary Douglas, *How Institutions Think* (Syracuse: Syracuse University Press, 1986).

63. Roger Friedland and Robert A. Alford, "Bringing Society Back In: Symbols, Practices, and Institutional Contradictions," in Walter W. Powell and Paul J. DiMaggio, eds., *The New Institutionalism in Organizational Analysis* (Chicago: University of Chicago Press, 1991), p. 251.

64. Clifford Geertz, *The Interpretation of Cultures: Selected Essays* (New York: Basic Books, 1973), p. 314.

65. Brigitte von Beuzekom, *Biotechnology Statistics in OECD Member Countries: Compendium of Existing National Statistics*, STI Working Papers 2001/6 (OECD, 2001).

66. Robert Bud, *The Uses of Life: A History of Biotechnology* (Cambridge: Cambridge University Press, 1993).

67. The development of molecular biology as a field, under active patronage from the Rockefeller Foundation in the United States, is a central component of this history. See Lily E. Kay, *The Molecular Vision of Life: Caltech, the Rockefeller Foundation, and the Rise of the New Biology* (New York: Oxford University Press, 1993). Molecular biology itself has been attributed to the confluence of three scientific perspectives: the biochemical

and physical, the genetic, and the structural. For a concise and helpful review, see Lawrence Busch et al., *Power, and Profit: Social, Economic, and Ethical Consequences of the New Biotechnologies* (Oxford: Blackwell, 1991), pp. 66–81.

68. Evelyn Fox Keller, *The Century of the Gene* (Cambridge: Harvard University Press, 2000), pp. 1–2.

69. The seminal articles describing the structure of the DNA double helix were James D. Watson and Francis H. C. Crick, "Molecular Structure of Nucleic Acids: A Structure for Deoxyribonucleic Acid," *Nature* 171 (1953): 737–738, and "Genetical Implications of the Structure of Deoxyribonucleic Acid," *Nature* 171 (1953): 964–967. Both Watson and, years later, Crick published engaging and deeply personal accounts of their discovery (in Crick's case, more a reflection on it): Watson, *The Double Helix: A Personal Account of the Discovery of the Structure of DNA* (New York: Atheneum, 1968); Crick, *What Mad Pursuit: A Personal View of Scientific Discovery* (London: Weidenfeld and Nicolson, 1989). For informative historical reviews of the golden age of molecular biology and the rise and development of biotechnology, see Horace F. Judson, *The Eighth Day of Creation* (New York: Simon and Schuster, 1979); James D. Watson and John Tooze, *The DNA Story: A Documentary of Gene Cloning* (San Francisco: W.H. Freeman, 1981); Bud, *The Uses of Life*.

70. The genes of some viruses are composed of the closely related ribonucleic acid (RNA).

71. Crick, *What Mad Pursuit*, p. 76.

72. Ibid., p. 62. This off-hand phrase signifies the extent to which the structure of DNA has become in about fifty years part of the everyday conceptual currency of the industrial world. The structure of DNA has been described and illustrated in countless books, articles, newspaper and magazine reports, and lately websites. A particularly accessible version can be found in National Research Council, *Mapping and Sequencing the Human Genome* (Washington, DC: National Academy Press, 1988).

73. Watson and Crick, "Molecular Structure," p. 738.

74. Stanley N. Cohen et al., "Construction of Biologically Functional Bacterial Plasmids in Vitro," *Proceedings of the National Academy of Sciences* 70 (1973), pp. 3240–3244. Cohen and Boyer subsequently patented the process of cutting and recombining DNA, as further discussed in chapter 8.

75. For an informative overview of twenty years of European governmental support for biotechnology, see Mark Cantley, "The Regulation of Modern Biotechnology: A Historical and European Perspective," in *Biotechnology*, vol. 12 (Legal, Economic and Ethical Dimensions) (New York: VCH Weinheim, 1995), chapter 18. As an enthusiast for the technology, Cantley was highly critical of European incoherence and hesitation and laudatory of what he saw as greater U.S. coherence and resolve. See also Gottweis, *Governing Molecules*.

76. Ian Wilmut et al., "Viable Offspring Derived from Foetal and Adult Mammalian Cells," *Nature* 385 (1997): 810–813.

77. Cloned human beings famously figured in Aldous Huxley's *Brave New World: A Novel* (London: Chatto and Windus, 1932). So compelling was the book's vision of a world in which identical human beings were created to serve particular social purposes that the phrase "brave new world," quoted by Huxley from Shakespeare's *The Tempest*, acquired second life as code for a biologically manipulated and controlled future.

78. Foucault, *The History of Sexuality*, vol. 1, pp. 135–145. See also Giorgio Agamben, *Homo Sacer: Sovereign Power and Bare Life* (Stanford: Stanford University Press, 1998), pp. 1–8; Michael Hardt and Antonio Negri, *Empire* (Cambridge: Harvard University Press, 2000), pp. 22–41.

79. For a vigorous argument that these possibilities should be pursued only under tight regulatory supervision, see Francis Fukuyama, *Our Posthuman Future: Consequences of the Biotechnology Revolution* (New York: Picador, 2002).

80. Kay, *The Molecular Vision of Life*.

81. Garland E. Allen, "Modern Biological Determinism: The Violence Initiative," in Michael Fortun and Everett Mendelsohn, eds., *The Practices of Human Genetics* (Dordrecht: Kluwer, 1999), pp. 1–23.

82. James C. Scott, *Seeing Like a State: How Certain Schemes to Improve the Human Condition Have Failed* (New Haven: Yale University Press, 1998).

83. It has become conventional to date the disenchantment with science's universalistic claims and the growing awareness of science's social foundations to Thomas Kuhn's enormously influential work, *The Structure of Scientific Revolutions* (Chicago: University of Chicago Press, 1962). The reality is, of course, more complex. For a review of relevant academic literature, see Jasanoff et al., eds., *Handbook of Science and Technology Studies*.

84. Spencer Weart, *Nuclear Fear: A History of Images* (Cambridge: Harvard University Press, 1988). See also Dorothy Nelkin and Susan M. Lindee, *The DNA Mystique: The Gene as a Cultural Icon* (New York: Freeman, 1995).

85. Historical studies examining the role of scientists and doctors in propagating and enforcing Nazi racial beliefs include Benno Müller-Hill, *Tödliche Wissenschaft: die Aussonderung von Juden, Zigeunern und Geisteskranken 1933–1945* (Reinbek bei Hamburg: Rowohlt, 1984); Robert Proctor, *Racial Hygiene: Medicine under the Nazis* (Cambridge: Harvard University Press, 1988); Peter Weingart, Jürgen Kroll, and Kurt Bayertz, *Rasse, Blut und Gene: Geschichte der Eugenik und Rassenhygiene in Deutschland* (Frankfurt: Suhrkamp, 1992).

86. An influential fictional work on this theme was Michael Crichton, *Jurassic Park: A Novel* (New York: Knopf, 1990); it was made into a highly successful film in 1993, with several sequels, none of which as successfully engaged with the book's major theme of scientific, technological, and financial hubris.

87. An early pessimistic appraisal of biotechnology was produced by the prominent activist and social critic Jeremy Rifkin; see particularly Rifkin, *Algeny* (New York: Viking Press, 1983). Important critiques of modernity that pay attention to the role of the life sciences include Jacques Ellul, *The Technological Society* (New York: Vintage Books, 1964); Michel Foucault, *Power/Knowledge: Selected Interviews and Other Writings 1972–1977* (New York: Pantheon, 1980); Beck, *Risk Society*; Bauman, *Modernity and Ambivalence*. On the general theme of technological pessimism, see Leo Marx, "The Idea of 'Technology' and Postmodern Pessimism," in Smith and Marx, eds., *Does Technology Drive History?*, pp. 237–257; Yaron Ezrahi, Everett Mendelsohn, and Howard Segal, eds., *Technology, Pessimism, and Postmodernism* (Dordrecht: Kluwer, 1994). Historical work on the rise of genetics that stresses themes of standardization and control includes Lily E. Kay, *Who Wrote the Book of Life: A History of the Genetic Code* (Stanford: Stanford University Press, 2000); and on agricultural biotechnology, Busch et al., *Plants, Power, and Profit*. An idiosyncratic and peculiarly American manifestation of individual

disenchantment with technology (including genetics) at the end of the century was the fifty-thousand-word document circulated by the hermit and convicted bomber Theodore Kaczynski; see *The Unabomber Manifesto* (San Francisco: Jolly Roger Press, 1995). Much more surprising to technology enthusiasts was a bleak assessment of a future out of human control by Bill Joy, the cofounder and chief scientist of Sun Microsystems, who is often described as a visionary. See Joy, "Why the Future Doesn't Need Us," *Wired* 8.04 (April 2000): 1–11, http://www.wired.com/wired/archive/8.04/joy.html (visited April 2004).

88. This point has been most consistently and passionately argued in more than a decade of work by the Indian activist and author Vandana Shiva. See, for example, Shiva, *Monocultures of the Mind: Perspectives on Biodiversity and Biotechnology* (London: Third World Network, 1993); *Biopiracy: The Plunder of Nature and Knowledge* (Toronto: Between the Lines, 1997); *Yoked to Death: Globalisation and Corporate Control of Agriculture* (New Delhi: Research Foundation for Science, Technology and Ecology, 2001).

89. Jasanoff, *States of Knowledge.*

90. George E. Marcus has used this term to describe anthropologists' need to describe new cultural formations that are constituted piecemeal in many intersecting locations. See Marcus, *Ethnography through Thick and Thin* (Princeton: Princeton University Press, 1998), pp. 79–104.

91. One of the most influential works on this topic is Jürgen Habermas, *Legitimation Crisis* (Boston: Beacon Press, 1975).

## CHAPTER 2: CONTROLLING NARRATIVES

1. The seriousness of the Windscale and Three Mile Island events can be judged from the fact that both were classified as level 5 on the seven-point International Nuclear Event Scale.

2. Angela Liberatore, *The Management of Uncertainty: Learning from Chernobyl* (Amsterdam: Gordon and Breach, 1999).

3. Rachel Carson, *Silent Spring* (New York: Houghton Mifflin, 1962).

4. For accounts of some of these events and their impact on policy, see Michael Reich, *Toxic Politics: Responding to Chemical Disaster* (Ithaca: Cornell University Press, 1991); Sheila Jasanoff, ed., *Learning from Disaster: Risk Management after Bhopal* (Philadelphia: University of Pennsylvania Press, 1994).

5. Michael J. Sandel, "The Case against Perfection," *Atlantic* (April 2004): 51–62.

6. Weart, *Nuclear Fear.*

7. Mary Wollstonecraft Shelley, *Frankenstein, or, The Modern Prometheus* (1818). The name "Frankenstein" belonged to the Swiss medical student who created the monster in Shelley's story, not to his creation. The slippage of the name from the medical experimenter to the creature he made testifies to the enduring fear that science will release monsters beyond human control.

8. Edison produced a fifteen-minute-long silent film. Lost for many years, it was eventually rediscovered and restored. Stills from the film may be viewed at http://www.lrsmarketing.com/adventures/Frankenstein/stillsfrank.htm (visited April 6, 2002). Boris Karloff played the monster in the 1931 film version directed by James Whale. Many

consider this the story's classic cinematic rendition. The last major film of Frankenstein in the twentieth century was the disappointing 1994 version directed by Kenneth Branagh.

9. The environmental theme never found its Shelley or Huxley, but a little-known science fiction story of the mid-1950s told how a biological experiment gone awry in China resulted in the global destruction of grain species, accompanied by famine and complete breakdown of the norms of civilization around the world. John Christopher, *The Death of Grass* (London: Michael Joseph, 1956).

10. See, for example, Aldo Leopold, *Game Management* (New York: Scribner's, 1933); Donald Worster, *Nature's Economy: A History of Ecological Ideas* (Cambridge: Cambridge University Press, 1977); Daniel Botkin, *Discordant Harmonies: A New Ecology for the Twenty-First Century* (Oxford: Oxford University Press, 1990).

11. For vivid accounts of this and other effects of British imperialism on the natural world, see James Morris, *Pax Britannica: The Climax of an Empire* (New York: Harcourt, Brace and World, 1968), pp. 77–78.

12. *Buck v. Bell*, 247 U.S. 200 (1927), p. 208. Although one of America's most revered jurists, Holmes was scientifically and politically a man of his times.

13. For more detailed accounts of these events, see Daniel J. Kevles, *In the Name of Eugenics: Genetics and the Uses of Human Heredity* (Berkeley: University of California Press, 1985); Robert Proctor, *Racial Hygiene*; Troy Duster, *Backdoor to Eugenics* (New York: Routledge, 1990); Weingart, Kroll, and Bayertz, *Rasse, Blut und Gene*.

14. The landmark article alerting U.S. medical researchers to the absence of adequate ethical safeguards on their practices was Henry K. Beecher, "Ethics and Clinical Research," *New England Journal of Medicine* 274 (1966): 1354–1368.

15. The impetus for the Pugwash Conferences was the "Russell-Einstein manifesto" jointly issued in 1955 by the noted British philosopher and peace activist Bertrand Russell and Albert Einstein. The first conference, which gave its name to the series, was hosted by the Canadian philanthropist Cyrus Eaton at his birthplace in Nova Scotia. In October 1995 the Pugwash Conferences and the organization's president, the former nuclear physicist Joseph Rotblat, were jointly awarded the Nobel Prize for Peace.

16. For an informative history of activism and debate on the safety of nuclear power, see Brian Balogh, *Chain Reaction: Expert Debate and Public Participation in American Commercial Nuclear Power, 1945–1975* (New York: Cambridge University Press, 1991).

17. The members constituted a contemporary *Who's Who* of molecular genetics research: Paul Berg, chairman, David Baltimore, Herbert W. Boyer, Stanley N. Cohen, Ronald W. Davis, David S. Hogness, Daniel Nathans, Richard Roblin, James D. Watson, Sherman Weissman, Norton D. Zinder.

18. Among many accounts of this history, the following are especially informative: Judith Swazey et al., "Risks and Benefits, Rights and Responsibilities: A History of the Recombinant DNA Research Controversy," *Southern California Law Review* 51 (1978): 1019–1078; Clifford Grobstein, *A Double Image of the Double Helix: The Recombinant DNA Debate* (San Francisco: Freeman, 1979); Krimsky, *Genetic Alchemy*; Donald S. Frederickson, "Asilomar and Recombinant DNA: The End of the Beginning," in Kathi E. Hanna, ed., *Biomedical Politics* (Washington, DC: National Academy Press, 1991), pp. 258–292.

19. Paul Berg et al., "Potential Biohazards of Recombinant DNA Molecules," *Science* 185, 4148 (1974): 303.

20. For a study of the discursive construction of biotechnology by its adherents, see Gottweis, *Governing Molecules*.

21. Interview, Sheldon Krimsky, Cambridge, MA, July 26, 2004. See also Jon Beckwith, *Making Genes, Making Waves: A Social Activist in Science* (Cambridge: Harvard University Press, 2002).

22. For confirmation of this statement, see Donald S. Frederickson, *The Recombinant DNA Controversy: A Memoir: Science, Politics, and the Public Interest 1974–1981* (Washington, DC: ASM Press, 2001). Reviewing the book, the bioethicist Thomas Murray wrote that Frederickson emerges as "first and foremost a scientist, deeply imbued with the values and perspectives of that profession, occasionally dismissive of those who challenge scientific prerogatives." Presumably, this attitude influenced his and NIH's efforts to beat back the legislative bills introduced in Congress, including fourteen bills and resolutions in 1977 alone. Murray, *Journal of the American Medical Association* 286 (2001): 2331–2332.

23. The dispute arose under the U.S. Patent Act, 35 U.S.C. Sec. 100 *et seq.* The intersections of biotechnology and patent law are comprehensively discussed in chapter 8.

24. *Diamond v. Chakrabarty*, 447 U.S. 303 (1980), at 306.

25. Ibid. at 307.

26. Ibid. at 309.

27. Ibid. at 317.

28. *PBC Brief*, quoting Dr. George Wald, Harvard biochemist and Nobel laureate, and Dr. James F. Crow, population geneticist at the University of Wisconsin, respectively.

29. Jasanoff, *Science at the Bar*.

30. *Foundation on Economic Trends v. Heckler*, 756 F.2d 143 (D.C. Cir. 1985).

31. "Report of the Committee to Review Allegations of Violations of the National Institutes of Health Guidelines for Research Involving Recombinant DNA Molecules in the Conduct of Studies Involving Injection of Altered Microbes into Elm Trees at Montana State University," Washington, DC, December 15, 1987.

32. For a brief history of the legal standoffs during this period, see Jasanoff, *Science at the Bar*, pp. 138–159.

33. Among those most distressed was Maxine Singer, one of the chief architects of the Asilomar conference. Singer, "Genetics and the Law: A Scientist's View," *Yale Law and Policy Review* 3 (1985): 315–335. For further discussion of her and other scientists' views, see Jasanoff, *Science at the Bar*, pp. 153–155.

34. James D. Watson, "In Defense of DNA," *New Republic* 170 (1977): 11. see also "Trying to Bury Asilomar," *Clinical Research* 26 (1978): 113.

35. National Research Council, *Field Testing Genetically Modified Organisms—Framework for Decisions* (hereafter referred to as *Field Testing*) (Washington, DC: National Academy Press, 1989).

36. Among the agencies participating in the Coordinated Framework, EPA's positions tended consistently to diverge from those of the FDA and USDA in a more risk-averse, or precautionary, direction.

37. Henry I. Miller et al., "Risk-Based Oversight of Experiments in the Environment," *Science* 250, 4980 (1990): 490–491.

38. See, for example, NRC, *Field Testing*, Executive Summary, pp. 3–4.

39. Henry I. Miller, "The Big Fed Freeze," *National Review On Line*, April 4, 2002, http://www.nationalreview.com/comment/comment-miller040402.asp (visited July 2002).

40. For details of this arrangement, see Brickman et al., *Controlling Chemicals*, pp. 81–82; Wilson, *The Politics of Safety and Health*, pp. 112–119.

41. On the politics of naming, see Les Levidow and Joyce Tait, "The Greening of Biotechnology: GMOs as Environment-Friendly Products," *Science and Public Policy* 18, 5 (1991): 271–280.

42. Existing regulations were deemed defective not only because of their limited scope but because they referred to the no longer existent GMAG. See *The Impact of New and Impending Regulations on UK Biotechnology*, report of a meeting sponsored by the Department of the Environment, the Health and Safety Executive, and the Bio-industry Association (hereafter cited as *Impact*) (Cambridge: Cambridge Biomedical Consultants, 1990), p. 12 (remarks of Richard Clifton, health and safety executive).

43. Bernard Dixon, "Who's Who in European Antibiotech," *Bio/Technology* 11 (1993): 44–48.

44. I am indebted to Les Levidow for calling my attention to this point. MAFF was restructured in the wake of the BSE crisis to become part of the new Department for Environment, Food and Rural Affairs.

45. Letter from Sir Donald Acheson, Chief Medical Officer, to Mr. D. H. Andrews, MAFF, March 23, 1988. See also U.K. Government, *The BSE Inquiry: The Report*, vol. 4, introduction, sec. 1, http://www.bseinquiry.gov.uk/report/volume4/chapterb.htm#886837 (visited April 10, 2002).

46. Royal Commission on Environmental Pollution, *The Release of Genetically Engineered Organisms to the Environment* (hereafter referred to as *Release of GEOs*), Thirteenth Report (London: HMSO, 1989). Although the commission spoke of genetically engineered organisms (GEOs), the term genetically modified organism (GMO) eventually took over as the international standard term for organisms produced by genetic engineering. In this chapter, I follow the later international usage.

47. Ibid., p. 21.

48. Ibid., p. 20.

49. On imperialism, botany, and agriculture, see Kavita Philip, "Imperial Science Rescues a Tree: Global Botanic Networks, Local Knowledge, and the Transcontinental Transplantation of Cinchona," *Environment and History* 1 (1995): 173–200; William Storey, *Science and Power in Colonial Mauritius* (Rochester: University of Rochester Press, 1997); Richard Drayton, *Nature's Government: Science, Imperial Britain, and the "Improvement" of the World* (New Haven: Yale University Press, 2000). See also Morris, *Pax Britannica*.

50. See comments of Richard Clifton and Douglas Bryce in *Impact*, note 42, pp. 15, 24.

51. See chapter 1, note 85.

52. On this theme, see particularly Bauman, *Modernity and Ambivalence*.

53. Müller's remarks in an interview are quoted in Ian Buruma, *Wages of Guilt: Memories of War in Germany and Japan* (London: Vintage, 1994), p. 89.

54. The ministry was subsequently renamed the Federal Ministry for Education and Research (Bundesministerium für Bildung und Forschung, BMBF).

55. Sheila Jasanoff, "Technological Innovation in a Corporatist State: The Case of Biotechnology in the Federal Republic of Germany," *Research Policy* 14 (1985): 23–38.

56. For sociological investigations of the gap between public understanding and (apparent) public acceptance of science, see Irwin and Wynne, eds., *Misunderstanding Science?*

57. See generally Gottweis, *Governing Molecules*, pp. 237–245.

58. *Report of the Parliamentary Commission of Enquiry on Prospects and Risks of Genetic Engineering,* German Bundestag, Bonn, January 1987. All quoted material is taken from the official English-language version of the report, hereafter cited as *Prospects and Risks.*

59. Die Grünen, *Erklärung zur Gentechnologie und zur Fortpflanzungs—und Gentechnik am Menschen,* Hagen, February 15–16, 1986.

60. Gottweis, *Governing Molecules*, pp. 229–262.

61. *Prospects and Risks,* p. 316.

62. Ibid., p. 315.

63. Etel Solingen, "Between Markets and the State: Scientists in Comparative Perspective," *Comparative Politics* 26 (1993): 31–51; at p. 43.

64. Wright, *Molecular Politics.*

65. *Prospects and Risks,* p. 354b. This view of McClintock's work was no doubt shaped by the feminist historian Evelyn Fox Keller's enormously influential, though controversial, biography, *A Feeling for the Organism: The Life and Work of Barbara McClintock* (San Francisco: W. H. Freeman, 1983).

66. Constitution of the United States, article I, sec. 8, cl. 8.

67. I have developed this point at length in *Science at the Bar.*

68. Numerous philosophical texts have treated these issues. See, for example, Jonathan Glover, *What Sort of People Should There Be? Genetic Engineering, Brain Control and Their Impact on Our Future World* (Middlesex, UK: Penguin, 1984). For a more institutional approach, grounded in an analysis of judicial practice, see Sheila Jasanoff, "Ordering Life: Law and the Normalization of Biotechnology," *Politeia* 17, 62 (2001): 34–50.

69. Polanyi, "The Republic of Science," 54–73.

70. The guarantee derives from article 5, paragraph 3, of the German Basic Law. See *Prospects and Risks,* p. 284.

### CHAPTER 3: A QUESTION OF EUROPE

1. There is an enormous and growing literature on globalization from many disciplinary perspectives, including international relations, public policy, sociology, anthropology, and cultural studies. Recent exemplary texts include Robert O. Keohane and Joseph S. Nye, Jr., *Power and Interdependence* (New York: Longman, 2001); Nye and John D. Donahue, eds., *Governance in a Globalizing World* (Washington, DC: Brookings Institution Press, 2000); Will Hutton and Anthony Giddens, eds., *Global Capitalism* (New York: The New York Press, 2000); Ann Cvetkovich and Douglas Kellner, eds., *Articulating the Global and Local: Globalization and Cultural Studies* (Boulder: Westview Press, 1997); Mike Featherstone, *Undoing Culture: Globalization, Postmodernism and Identity* (London: Sage Publications, 1995).

2. The extensive literature on the European Union includes Medrano, *Framing Europe*; Kjell Goldmann, *Transforming the European Nation-State* (London: Sage,

2001); Alberta Sbragia, ed., *Euro-Politics: Institutions and Policymaking in the "New" European Community* (Washington, DC: Brookings Institution, 1992).

3. Resentment was triggered in part by the perception in several countries that the euro's introduction had caused unacceptable price hikes. Even in Germany, where the Deutschmark was abandoned with surprisingly little public nostalgia, the new currency was soon dubbed the *Teuro*, based on the German word *teuer*, meaning expensive.

4. For a highly skeptical instant opinion, see Andrew Moravcsik, "If It Ain't Broke, Don't Fix It," *Newsweek*, March 4, 2002, p. 15. See also Bernstein, "Europe's Lofty Vision of Unity Meets Headwinds."

5. Speech by Romano Prodi, president of the European Commission, Opening Session of the Convention on the Future of Europe, Brussels, February 28, 2002, http://european-convention.eu.int/docs/speeches/181.pdf (visited May 2002).

6. The concept of "state formation" is of course widely used to describe the emergence of the nation state as a form of government. In the case of the EU, however, part of the conceptual and juridical challenge is to find the right designation for the entity that is coming into being. The EU is sometimes referred to as a superstate and sometimes as supranational. Neither term quite captures the fact that the EU stands at once above its member states in some of its regulatory powers and below them in its need for legitimation through the processes and representatives of national politics. For one view stressing the continued independence of the member states and the importance of subsidiarity, see Jack Straw, "By Invitation," *Economist*, July 10, 2004, p. 40.

7. Jasanoff, ed. *States of Knowledge*.

8. Thomas Fuller, "A Blunt Appraisal of EU's Laggards," *International Herald Tribune*, January 13, 2004, p. 1.

9. Personal communication, Angela Liberatore, Brussels, May 14, 2002.

10. Anderson, *Imagined Communities*.

11. Neil Walker, "The White Paper in Constitutional Context," in Christian Joerges, Yves Mény, and J.H.H. Weiler, eds., *Mountain or Molehill? A Critical Appraisal of the Commission White Paper on Governance* (Florence: European University Institute, 2001), pp. 33–53; see particularly p. 37.

12. For a study of these conflicts as played out in a single European agency, see Claire Waterton and Brian Wynne, "Knowledge and Political Order in the European Environment Agency," in Jasanoff, ed., *States of Knowledge*, pp. 87–108.

13. The commission is not the only actor to have struggled with the question of a day for Europe. In May 1998, for example, the Berlin-Brandenburg Institute at Genshagen, a body devoted to bettering Franco-German relations, held a conference on "Sites of Memory in European Perspective." Although the meeting's main purpose was to review recent French and German historiography on the creation of public memorials, the participants perhaps inevitably fell to discussing whether there could be a day of European commemoration. Papers from the meeting were published in Alexandre Escudier, Brigitte Sauzay, and Rudolf von Thadden, eds., *Gedenken im Zwiespalt: Konfliktlinien europäischen Erinnerns* (Göttingen: Wallstein, 2001).

14. May 9, 1950, was the date of the "Schuman declaration." On this day, French Foreign Minster Robert Schuman proposed to create the European Coal and Steel Community, an organization to pool these two major national resources. Schuman and his advisers saw this union as a means of securing peaceful relations among the

continent's historically warring nations. For the web citations, see "Europe Day, 9 May," http://europa.eu.int/abc/symbols/9-may/index_en.htm (visited August 2003).

15. These were the six members of the ECSC: Belgium, France, Germany, Italy, Luxembourg, and the Netherlands.

16. The threat of internationalization differs according to the particular form of democratic theory one espouses (liberal, deliberative, republican), but it never altogether disappears. See Goldmann, *Transforming the European Nation-State*, pp. 142–145.

17. Christian Joerges, "'Economic Order'—'Technical Realisation'—'the Hour of the Executive': Some Legal Historical Observations on the Commission White Paper on European Governance," in Joerges et al., *Mountain or Molehill*, pp. 128–129.

18. David Earnshaw and David Judge, *The European Parliament* (Houndmills, Hampshire: Palgrave Macmillan, 2003).

19. Medrano, *Framing Europe*.

20. The steam engine, for example, preceded the theoretical elaboration of thermodynamics. Crosbie Smith and M. Norton Wise, *Energy and Empire: A Biographical Study of Lord Kelvin* (Cambridge: Cambridge University Press, 1989). Similarly, telegraphy predates the development of field theory. Bruce J. Hunt, "Michael Faraday, Cable Telegraphy, and the Rise of Field Theory," *History of Technology* 13 (1991): 1–19; Simon Schaffer, "Late Victorian Metrology and Its Instrumentation: A Manufactory of Ohms," in Robert Bud and Susan E. Cozzens, eds., *Invisible Connections: Instruments, Institutions and Science* (Bellingham, WA: SPIE Optical Engineering Press, 1992), pp. 23–56. I am indebted to Michael A. Dennis for these examples.

21. Commission of the European Communities, *European Governance: A White Paper* (hereafter referred to as *White Paper*).

22. In particular, Beate Kohler-Koch argued that the *White Paper's* commission-centric views came through in the choice of instruments it favored, as well as its criteria for judging success. Kohler-Koch, "The Commission White Paper and the Improvement of European Governance," in Joerges et al., *Mountain or Molehill*, pp. 177–184.

23. Philippe C. Schmitter, "What Is There To Be Legitimized in the European Union . . . and How Might This Be Accomplished?", Political Science Series, Institute for Advanced Studies, Vienna, May 2001.

24. The allusion to the otherworldliness of an unidentified flying object, or UFO, is presumably not accidental.

25. Walker, "The White Paper," *Mountain or Molehill*, p. 35.

26. *White Paper*, p. 35.

27. Joerges et al., *Mountain or Molehill*.

28. *White Paper*, p. 17.

29. Paul Magnette, "European Governance and Civic Participation: Can the European Union be Politicised?" *Mountain or Molehill*, pp. 24–25.

30. Jürgen Gerhards, "Westeuropäische Integration und die Schwierigkeiten der Entstehung einer Europäischen Öffentlichkeit," *Festschrift für Soziologie* 22 (1993): 96–110.

31. See, for example, Maria Eduarda Gonçalves, "The Importance of Being European: The Science and Politics of BSE in Portugal," *Science, Technology, and Human Values* 25, 4 (2000): 417–448; Sheila Jasanoff, "Civilization and Madness: The Great BSE Scare of 1996," *Public Understanding of Science* 6: 221–232 (1997).

32. Klaus Eder, "Zur Transformation nationalstaatlicher Öffentlichkeit in Europa: von der Sprachgemeinschaft zur inspezifischen Kommunikationsgemeinschaft," *Berliner*

*Journal für Soziologie* 10, 2 (2000): 167–184; Klaus Eder and Cathleen Kantner, "Transnationale Resonanzstrukturen in Europa: Eine Kritik der Rede von Öffentlichkeitsdefizit," in Maurizio Bach, ed., *Die Europäisierung nationaler Gesellschaften* (Wiesbaden: Westdeutscher Verlag, 2000), pp. 306–331.

33. Gottweis, *Governing Molecules*, p. 174, Gottweis usefully reviews the early phases of European policy, pp. 166–181.

34. Ezrahi, *The Descent of Icarus.*

35. For discussion of the reception of the Apollo missions, see Wolfgang Sachs, *Planet Dialectics: Explorations in Environment and Development* (Halifax, Nova Scotia: Fernwood Publishing, 1999); Sheila Jasanoff, "Image and Imagination: The Formation of Global Environmental Consciousness," in Paul Edwards and Clark Miller, eds., *Changing the Atmosphere: Expert Knowledge and Environmental Governance* (Cambridge: MIT Press, 2001), pp. 309–337.

36. In the large literature on the rise of statistics and its role in "making" society and underwriting social policy, the following are especially illuminating for our purposes: Ian Hacking, *The Taming of Chance* (Cambridge: Cambridge University Press, 1990); Porter, *The Rise of Statistical Thinking*; Wagner, Wittrock, and Whitley, eds., *Discourses on Society*; Dietrich Rueschemeyer and Theda Skocpol, eds., *States, Social Knowledge, and the Origins of Modern Social Policies* (Princeton: Princeton University Press, 1996).

37. BSC operated with increasing lethargy over its short period of existence. It met five times in 1985, four in 1986, three in 1987, two in 1988, and one final time in July 1988. Interview with Mark Cantley, OECD, Paris, May 18, 1993.

38. DG XI is now just DG Environment.

39. The FAST Program was established by a July 1978 decision of the Council of Ministers.

40. Cantley interview, May 18, 1993 (source of this and subsequent MC quotes).

41. Commission of the European Communities, *Eurofutures: The Challenges of Innovation, The FAST Report* (London: Butterworths, 1984).

42. Gottweis, *Governing Molecules*, pp. 168–172.

43. See, for example, the discussion of modernity's "gardening instinct" in Bauman, *Modernity and Ambivalence*, pp. 18–52.

44. *Eurofutures*, note 41, p. 8.

45. Ibid., p. 9.

46. At the time of this quotation, there were seventeen commissioners from twelve member states: one from each of the seven smaller states, and two each from the five larger.

47. Les Levidow et al., "Bounding the Risk Assessment of a Herbicide-Tolerant Crop," in Ad van Dommelen, ed., *Coping with Deliberate Release: The Limits of Risk Assessment* (Tilburg, NL: International Centre for Human Rights, 1996), p. 83.

48. Interview, DG XI, Brussels, July 13, 1993.

49. MFC/CORR/NOT/jh, XII/87, Brussels, 17.08.1987.

50. A personal attack on Cantley in a German environmental magazine charged him with being a public relations front for the biotechnology industry and publishing their propaganda as EC documents. Michael Bullard, "*Unser Mann in der Kommission* (Our Man in the EC Commission)," *Natur* 8 (August 1991): 34–35. Cantley vigorously denied the charge, pointing to numerous inaccuracies in the report and noting

that many decisions attributed to him individually were official, peer-reviewed EC undertakings, such as the design of *Eurobarometer's* biotechnology questions. Personal communication, June 21, 1993.

51. See Directive 2001/18/EC of the European Parliament and of the Council on the deliberate release into the environment of genetically modified organisms and repealing Council Directive 90/220/EEC; and Regulation on Novel Foods and Novel Food Ingredients of 27 January 1997 (Regulation (EC) 258/97). The directive on research with GMOs was also revised in the late 1990s. See Council Directive 98/81/EC of 26 October 1998 amending Directive 90/219/EEC on the contained use of genetically modified micro-organisms.

52. These challenges arose in part under the safeguarding provision of article 16 of the 1990 directive on deliberate release.

53. Gordon Lake, "Scientific Uncertainty and Political Regulation: European Legislation on the Contained Use and Deliberate Release of Genetically Modified (Micro)organisms," *Project Appraisal* 6, 1 (March 1991): 7–15.

54. Commission of the European Communities, Directorate-General for Agriculture, Draft Proposal for a Council Regulation concerning the use of certain substances and techniques intended for administration or application to animals to simulate their productivity, VI/3670/90-REV.1 (1990). See also "Controversial Proposal on Fourth Criterion in Commission Pipeline," *European Report*, January 26, 1991, pp. 3–5.

55. For a provocative institutional analysis of the European Environment Agency's role in the formation of European identity, see Waterton and Wynne, "Knowledge and Political Order in the European Environment Agency."

56. Interview with Ulrike Riedel, former German Health Ministry official, Berlin, July 2002. See also Directive 2001/18/EC of 12 March 2001 on the deliberate release into the environment of genetically modified organisms and repealing Council Directive 90/220/EEC.

57. See, for example, European Commission, "Promoting the Competitive Environment for the Industrial Activities Based on Biotechnology within the Community," SEC(91)629 final, Brussels, April 19, 1991; "Innovation and Competitiveness in European Biotechnology," Enterprise Papers No. 7, Eur-Op catalogue no. NB-40-01-690-EN-C (2002).

58. European Commission, *The EC-US Task Force on Biotechnology Research—Mutual Understanding: A Decade of Collaboration 1990–2000* (Brussels: European Communities, 2000). Interestingly, the report contains an interview with Bruno Hansen, the Task Force's European cochair, asking whether U.S. collaboration is consistent with the EU's mission to increase the competitiveness of European R&D. Hansen replied that "we can best assist Europe's competitiveness by understanding biotechnology on a global scale" (p. 34).

59. Mark F. Cantley, "Democracy and Biotechnology: Popular Attitudes, Information, Trust and the Public Interest," *Swiss Biotech* 5, 5 (1987): 6.

60. http://www.gesis.org/en/data_service/eurobarometer/index.htm (visited April 2004). The surveys are conducted on behalf of former Directorate-General X of the European Commission, now the Directorate-General for Education and Culture. The surveys have included Greece since fall 1980, Portugal and Spain since fall 1985, East Germany since fall 1990, and Austria, Finland, and Sweden since spring 1995.

61. There is a large literature in science and technology studies on the ontological role of representation. Among the most influential are works by Ian Hacking; see particularly *Representing and Intervening: Introductory Topics in the Philosophy of Natural Science* (Cambridge: Cambridge University Press, 1983).

62. The surveys were, respectively, EB 35.1, EB 39.1, EB 46.1, EB 52.1, and EB 58.0.

63. European Commission, Quality of Life Programme, *Eurobarometer 52.1—The Europeans and Biotechnology* (2000), http://europa.eu.int/comm/research/quality-of-life/eurobarometer.html (visited April 2004).

64. Brian Wynne, "Creating Public Alienation: Expert Cultures of Risk and Ethics on GMOs," *Science as Culture* 10, 4 (2001): 445–481; see particularly the discussion of the *Eurobarometer* on pp. 463–464.

65. Irwin and Wynne, eds., *Misunderstanding Science?* Other authors have shown that conditions of trust and distrust are intimately bound up with culturally specific strategies of public justification. See, in particular, Porter, *Trust in Numbers*; Jasanoff, "Science, Politics, and the Renegotiation of Expertise at EPA."

66. Claire Marris et al., *Public Perceptions of Agricultural Biotechnologies in Europe (PABE)*, Final Report of the PABE Research Project, Contract number: FAIR CT98-3844 (DG12—SSMI), Lancaster University, December 2001.

67. Jasanoff, "In a Constitutional Moment: Science and Social Order at the Millennium."

68. Brian Wynne, "Public Understanding of Science," in Jasanoff et al., eds., *The Handbook of Science and Technology Studies*, pp. 361–388.

69. Irwin and Wynne, *Misunderstandings Science?*; Marris et al., *PABE Report*.

70. Daniel S. Greenberg, *Science, Money, and Politics: Political Triumph and Ethical Erosion* (Chicago: University of Chicago Press, 2001).

71. For more on this move, see Paul Rabinow, *French DNA: Trouble in Purgatory* (Chicago: University of Chicago Press, 1999), pp. 71–111. The French committee was relatively large as befitted its consensus-building role. Initially consisting of 37 members, it was later expanded to forty-one.

72. *Eurofutures*, note 41, p. 3.

73. Ibid., p. 56.

74. Interview with Adrian van der Meer, Commission of the European Communities, Brussels, July 13, 1993. The group began with six members but expanded to nine in its second term, from 1994 to 1997.

75. The Warnock Committee, as it came to be known, was established in July 1982 to consider the implications of the birth of Britain's first test-tube baby, Louise Brown, in 1978. See Mary Warnock, *A Question of Life: The Warnock Report on Human Fertilisation and Embryology* (Oxford: Blackwell, 1984).

76. The other members were Noëlle Lenoir, a member of the Constitutional Council (France); Margareta Mikkelsen, a medical geneticist (Denmark); Marcelino Oreja, a lawyer and member of parliament (Spain); and Marcello Siniscalco, a professor of genetics (Italy).

77. GAEIB, "The Ethical Aspects of the 5th Research Framework Programme," Opinion no. 10, December 11, 1997, http://europa.eu.int/comm/european_group_ethics/ gaieb/en/opinion10.pdf (visited April 2004).

78. Interview, Theodoros Karapiperis, Brussels, May 14, 2002.

79. European Commission, DG Research, "Minutes of the Fifth Meeting of Contact Persons for 'Ethics in Research,'" Brussels, April 26, 2002.

80. Joel A. Tickner, ed., *Precaution: Environmental Science and Preventive Public Policy* (Washington, DC: Island Press, 2003).

## CHAPTER 4: UNSETTLED SETTLEMENTS

1. The concept of "regulatory science" and the dynamics of its production in the United States were elaborated in Jasanoff, *The Fifth Branch.*

2. "A Timeline of Biotechnology," compiled by the Biotechnology Industry Organization, http://www.biospace.com/articles/timeline.cfm; see also http://www.bio.org/er/timeline.asp (both visited July 2002).

3. Beck, *Risk Society.* Originally published in Germany as *Risikogesellschaft: auf dem Weg in eine andere Moderne* (Frankfurt: Suhrkamp, 1986), the book created an immediate stir and sold more than 100,000 copies.

4. Hacking, *The Taming of Chance.*

5. Barry A. Palevitz and Ricki Lewis, "Perspective: Fears or Facts? A Viewpoint on GM Crops," *The Scientist* 13, 20 (October 11, 1999): p. 10.

6. *Foundation on Economic Trends v. Heckler,* 756 F.2d 143 (D.C. Cir. 1985); see chapter 2.

7. For a detailed account of the events surrounding the release, see Sheldon Krimsky and Alonzo Plough, *Environmental Hazards: Communicating Risks as a Social Process,* chapter 3, "The Release of Genetically Engineered Organisms into the Environment: The Case of Ice Minus" (Dover, MA: Auburn House Publishing Company, 1988), pp. 75–110.

8. Palevitz and Lewis, "Perspective: Fears or Facts?"

9. Brian Tokar, "Resisting the Engineering of Life," in Tokar, ed., *Redesigning Life? The Worldwide Challenge to Genetic Engineering* (London: Zed Books, 2001). For a picture of the AGS scientist, Julie Lindemann, spraying the field site, see Mark Crawford, "California Field Test Goes Forward," *Science* 236, 4801 (1987): 511.

10. Paul Slovic, "Beyond Numbers: A Broader Perspective on Risk Perception and Risk Communication," in Deborah G. Mayo and Rachelle D. Hollander, eds., *Acceptable Evidence: Science and Values in Risk Management* (New York: Oxford University Press, 1991), pp. 48–65; Slovic et al., "Characterizing Perceived Risks," in Robert W. Kates, Christoph Hohenemser and Jeanne X. Kasperson, eds., *Perilous Progress: Managing the Hazards of Technology* (Boulder, CO: Westview, 1985), pp. 91–125; Slovic et al., "Facts and Fears: Understanding Perceived Risk," in R. Schwing and W. A. Albers, Jr., eds., *Societal Risk Assessment: How Safe is Safe Enough?* (New York: Plenum, 1980), pp. 181–214.

11. Mark Crawford, "RAC Recommends Easing Some Recombinant DNA Guidelines," *Science* 235, 4790 (1987): 740–741.

12. Henry I. Miller, "The Big Fed Freeze."

13. See, for example, Gina Kolata, "How Safe Are Engineered Organisms?" *Science* 229, 4708 (1985): 34–35. See also Gottweis, *Governing Molecules,* pp. 235–236.

14. Tokar, *Redesigning Life?*

15. Microbial pesticide research continued, but the focus shifted from uses in ordinary agriculture to agents, often fungus-based, that were used to combat dangerous

plants, such as opium in Latin America. These new agents were also potential instruments of biological warfare. Protest over ice-minus may have played a role in driving the demand for research on microbial pesticides from civilian to military end-users.

16. Jonathon Porritt, "Down-to-Earth Agenda; Suggestions to Mrs. Thatcher," *The Times (London)*, September 27, 1988. The full text of the speech can be found at the home page of the Margaret Thatcher Foundation, http://www.margaretthatcher.org/default.htm (visited April 2004).

17. Interview, David Bishop, Institute of Virology and Environmental Microbiology, Oxford, July 5, 1990.

18. Les Levidow, "The Oxford Baculovirus Controversy—Safely Testing Safety?" *Bioscience* 8, 45 (1995): 545–551.

19. Interview, U.K. Advisory Committee on Releases to the Environment, London, July 16, 1990.

20. Levidow, "The Oxford Baculovirus Controversy" (quoting Alan Lees, a campaigner for Friends of the Earth, who characterized the enfeebled baculovirus as "a Trojan horse for the genetic engineering industry").

21. On this point, see particularly Donald MacKenzie, *Inventing Accuracy: A Historical Sociology of Nuclear Missile Guidance* (Cambridge: MIT Press, 1990); Pinch, "'Testing—One, Two, Three . . . Testing!'"

22. Steven Dickman, "New Law Needs Changes Made," *Nature* 343 (1990): 298.

23. *Bericht über die zurückliegende Amtsperiode der Zentralen Kommission fur die Biologische Sicherheit, (29.01.81 bis 30.06.88)*, Bonn, 1989.

24. Dorothy Nelkin and Michael Pollak, *The Atom Besieged: Extraparliamentary Dissent in France and Germany* (Cambridge: MIT Press, 1981).

25. Eva Kolinsky, ed., *The Greens in West Germany: Organisation and Policy Making* (Oxford: Berg, 1989); see also Gottweis, *Governing Molecules*, pp. 237–245.

26. Steven Dickman, "Germany Edges towards Law," *Nature* 339, 6223 (1989): 327.

27. Klaus Töpfer served as federal minister for the environment, nature conservation, and nuclear safety from May 1987 to November 1994.

28. *Gentechnikgesetz*, sections 18(1) and 18(2) (1990).

29. Marker genes make it possible to isolate transgenic plants on the basis of their acquired resistance. The widespread use of antibiotic resistance genes in such studies was not at this time a matter of great public concern.

30. Peter Meyer et al., "A New Petunia Flower Colour Generated by transformation of a Mutant with a Maize Gene," *Nature* 330 (1987): 667–668. Highly colored accounts of the study appeared in several English-language papers. See, for instance, Boyce Rensberger, "Making a Pink Petunia Turn Red," *Washington Post*, December 21, 1987, p. A3.

31. Peter Meyer, "Regulations for the Release of Transgenic Plants according to the German Gene Act and Their Consequences for Basic Research," *AgBiotech News and Information* 3, 6 (1991): 999–1001.

32. Peter Meyer et al., "Endogenous and Environmental Factors Influence 35S Promoter Methylation of a Maize A1 Gene Construct in Transgenic Petunia and Its Color Phenotype," *Molecular and General Genetics* 231 (1991): 345–352.

33. Interview with Peter Meyer, Max-Planck Institute for Plant Breeding Research, Köln, July 1993.

34. Ibid.

35. Irwin and Wynne, eds., *Misunderstanding Science?*

36. "A field test of genetically engineered petunias that were designed to produce one color wound up having wildly fluctuating results in the field." Richard Caplan and Ellen Hickey, "Weird Science: The Brave New World of Genetic Engineering," October 21, 2000, http://www.mindfully.org/GE/GE-Weird-Science.htm (visited April 2004).

37. "Elfter Bericht nach Inkrafttreten des Gentechnikgesetzes (GenTG) fur den Zeitraum 1.1.2000 bis 31.12.2000," *Bundesgesundheitsblatt-Gesundheitsforschung-Gesundheitsschutz* 9 (2001): 929–941.

38. Nuffield Council on Bioethics, *Genetically Modified Crops: The Ethical and Social Issues* (London: Nuffield Council on Bioethics, 1999), p. 31.

39. Jasanoff, *The Fifth Branch*, pp. 76–83.

40. Jasanoff, "Science, Politics, and the Renegotiation of Expertise at EPA; *Risk Management and Political Culture.*

41. John E. Losey, Linda S. Rayor, and Maureen E. Carter, "Transgenic Pollen Harms Monarch Larvae," *Nature* 399 (1999): 214.

42. According to the company's own promotional web site, "Monsanto is the world leader in biotechnology crops. Seeds with Monsanto traits accounted for more than 90 percent of the acres planted worldwide with herbicide-tolerant or insect-resistant traits in 2001." http://www.monsanto.com/monsanto/layout/about_us/ataglance.asp (visited August 2002).

43. Monsanto, "Bt Corn and the Monarch Butterfly," Biotech Knowledge Center, http://www.biotechknowledge.monsanto.com/biotech/knowcenter.nsf/f055f4dc645999 ad86256ac4000e6b68/0231086dd38f9a3d86256af6005433ae?OpenDocument (visited August 2002). An interesting rhetorical feature of the entries under this heading is the repeated reference to the Losey group's study as a "Cornell report," omitting any mention of its publication in *Nature*.

44. See, for instance, Donald MacKenzie's extended demonstration that the accuracy of the U.S. antiballistic missile system was "invented." Mackenzie, *Inventing Accuracy.*

45. David Quist and Ignacio H. Chapela, "Transgenic DNA Introgressed into Traditional Maize Landraces in Oaxaca, Mexico," *Nature* 414 (2001): 541–543.

46. See, for example, Marc Kaufman, "The Biotech Corn Debate Grows Hot in Mexico," *Washington Post*, March 25, 2002, p. A9.

47. Bizarre twists in the story included the charge that biotechnology companies had invented fake people to attack Chapela and Quist on the Internet. Industry representatives vehemently denied this accusation. George Monbiot, "The Fake Persuaders: Corporations are inventing people to rubbish their opponents on the internet," *The Guardian*, May 14, 2002, p. 15.

48. See Philip Campbell, "Editorial Note," *Nature* 416 (2002): 601.

49. Scientific journal editors often adjust the stringency of their peer review practices to take account of factors such as the novelty and possible political impact of research reports. See Jasanoff, *Fifth Branch*, pp. 66–68. The peculiarity in this case was the editor's delegation of interpretive discretion to the journal's readership.

50. Levidow, "The Oxford Baculovirus Controversy," p. 545.

51. On these points see the series of articles published from May to November 1994 by Susan Watts, science correspondent for the *Independent*, in particular, Watts,

"Genetics Row Fueled by Scorpion's Venom," *Independent*, May 17, 1994, p. 3; "Legal Fight Planned to Halt Scorpion Toxin Test," *Independent*, May 18, 1994, p. 3; "Warning: This Thing Isn't Natural," *Independent*, May 26, 1994, p. 20; "Safety Scare on Eve of Mutant Virus Test," *Independent*, June 26, 1994, p. See also Editorial, "Controversy in the Cabbage Patch," *Independent*, May 17, 1994, p. 15.

52. Susan Watts, "Genetic Riddle of 'Scorpion' Pesticide Virus, *Independent*, September 4, 1994, p. 2; Oliver Tickell, "Scorpion Gene Virus Experiment Abandoned," *Pesticides News*, no. 25 (September 1994), p. 21.

53. Watts, "Safety Scare."

54. Watts, "Legal Fight."

55. Steve Connor, "Gene Scientist 'Sacked without Warning,'" *Independent*, March 18, 1994, p. 5; Christian Tyler, "Private View: Professor with Killer Gene Blues," *Financial Times*, April 8, 1995, p. 18.

56. "Scorpion Has Sting in Tale," *The Splice of Life*, Bulletin of the Genetics Forum, 1, 8/9 (May 1995).

57. British Government Panel on Sustainable Development, *Second Report*, January 1996.

58. Interview, Sir Crispin Tickell, Warden, Green College, Oxford, July 9, 1996.

59. Government Response to the Second Annual Report of the Government's Panel on Sustainable Development, Department of the Environment, London, March 1996.

60. Agriculture and Environment Biotechnology Commission, *Crops on Trial* (September 2001), www.aebc.gov.uk/aebc/pdf/crops.pdf (visited July 2003).

61. Robin Grove-White, "New Wine, Old Bottles? Personal Reflections on the New Biotechnology Commissions," *Political Quarterly* 72, 4 (October 2001): 466–472.

62. Jasanoff, "Product, Process, or Programme."

63. Position paper of Hoechst AG, submitted to Hearing on Experiences with the Law for Regulating Questions of Gene Technology, January 31, 1992, pp. 16–17.

64. Position paper of Robert Koch Institute, Federal Health Office, submitted to Hearing on Experiences with the Law for Regulating Questions of Gene Technology, February 7, 1992, pp. 9–11.

65. "Die Tätigkeit in der ZKBS ist kein Geheimdienst, sondern Aktivierung gesellschaftlicher Sachkunde." Gerd Winter, position paper of ZERP, University of Bremen, submitted to Hearing on Experiences with the Law for Regulating Questions of Gene Technology, January 28, 1992, p. 15.

66. The Munich-based Max-von-Pettenkofer Institute, for instance, endorsed the views expressed on this score by the influential Max-Planck Institute for Biochemistry in Martinsried and appended to its own position paper a *Science* article rehearsing the obstacles that the new law posed to free inquiry and exchange; see Patricia Kahn, "Germany's Gene Law Begins to Bite," *Science* 255 (1992): 524–526.

67. Quirin Schiermeier, "German Transgenic Crop Trials Face Attack," *Nature* 394 (1998): 819.

68. Planet Ark, "German GM Wheat Trials Approved but Site Sabotaged," Hamburg, April 11, 2003, http://www.planetark.org/dailynewsstory.cfm/newsid/20444/ newsDate/ 11-Apr-2003/story.htm (visited July 2003).

69. Ned Stafford, "GM Crop Sites Stay Secret," *The Scientist*, 28 May 2004, http://www. biomedcentral.com/news/20040528/02 (visited June 2004).

### CHAPTER 5: FOOD FOR THOUGHT

1. http://news.bbc.co.uk/hi/english/uk_politics/newsid_282000/282376.stm (visited April 2004).

2. For a case study of the Alar controversy, see Jasanoff, *The Fifth Branch*, pp. 141–149.

3. For an account of prion-induced brain diseases, from Kuru in the New Guinea highlands to CJD, see Richard Rhodes, *Deadly Feasts* (New York: Simon and Schuster, 1997; paperback edition with new afterword, 1998).

4. Jasanoff, "Civilization and Madness."

5. In April 2000 the U.K. government estimated that the total cost of the BSE crisis to the public sector would be £3.7 billion by the end of the 2001–2002 fiscal year. *The Inquiry into BSE and Variant CJD in the United Kingdom* [hereafter cited as *The Phillips Inquiry*] (2000), vol. 10, Economic Impact and International Trade, http://www.bseinquiry.gov.uk/report/volume10/chapter1.htm#258548 (visited April 2004).

6. Ibid.

7. Department of Health, Ministry of Agriculture, Fisheries and Food, 1989, *Report of the Working Party on Bovine Spongiform Encephalopathy* (Southwood Committee Report), p. 22.

8. Martin Enserink, "Preliminary Data Touch Off Genetic Food Fight," *Science* 283 (1999): 1094–1095; also see Ehsan Masood, "Gag on Food Scientists Lifted as Gene Modification Row Hots Up . . . ," Nature 397 (1999): 547; and Editorial, "Food for Thought," *Nature* 397 (1999): 545, saying that James's actions against Pusztai had been "provocative" and hard to understand, and that the "lessons of BSE have yet to be fully absorbed."

9. Julian Barnes, *England, England* (London: Picador, 1998).

10. I refer here to the anthropological idea of nationhood conceived by Anderson, *Imagined Communities*.

11. David Starkey, *Elizabeth* (London: Vintage, 2001), p. 242.

12. Linda Colley, *Britons: Forging the Nation 1707–1807* (New Haven: Yale University Press, 1992), p. 233.

13. For a specimen of the prince's lampooning in the media, see Christopher Buckley, "Royal Pain: Further Adventures of Rick Renard," *The Atlantic Monthly* (April 2004): 94–106.

14. Prince Charles, "My 10 Fears for GM Food," *The Daily Mail*, June 1, 1999, pp. 10–11.

15. Professor Derek Burke was vice-chancellor of the University of East Anglia, chairman of the Advisory Committee for Novel Foods and Processes (1988–97), and member of the Nuffield Council on Bioethics working party on GM crops.

16. "Food for Our Future," Food and Drink Federation, *Feedback*, http://www.foodfuture.org.uk/answer.htm (visited July 2003).

17. The Reith 2000 series was titled "Respect for the Earth." The entire series is published at http://news.bbc.co.uk/hi/english/static/events/reith_2000/ (visited July 2003). The speakers were, in order, Chris Patten on governance; Tom Lovejoy on biodiversity; John Browne on business; Gro Harlem Brundtland on health and population; Vandana Shiva on poverty and globalization; and Prince Charles, speaking from his home at Highgrove.

18. Richard Dawkins, "Charles: Right or Wrong about Science?" Focus Special, *The Observer*, May 21, 2000, p. 21.

19. Ibid.

20. On this score, see particularly Brian Wynne, "The Prince and the GM Debate: Performing the Monarchy as Culture," http://domino.lancs.ac.uk/csec/bn.NSF/0/c3bbc73ca660b14f802569df005d6cdd?OpenDocument (visited July 2003).

21. On the messy mutual adjustment of ideas of nature and ideas of politics in contemporary societies, see Jasanoff and Martello, *Earthly Politics*.

22. Strategy Unit Study on the Costs and Benefits of GM Crops: Seminar on "Shocks and Surprises," April 3, 2003, http://www.number-10.gov.uk/output/ Page3673.asp (visited July 2003).

23. Interview, Dr. Doug Parr, chief scientist, Greenpeace UK, London, June 17, 2003.

24. http://www.number-10.gov.uk/output/Page3673.asp (visited July 2003).

25. See GM Science Review, http://www.gmsciencedebate.org.uk/default.htm (visited July 2003).

26. GM Open Meeting—Food Safety, Science Museum, London, January 23, 2003, Transcript, p. 9.

27. The government originally budgeted £250,000 for the exercise but doubled it to £500,000 on the recommendation of the Steering Board for the public debate. Eight major consumer and environmental groups criticized the process for inadequate preparation and unclear coordination with eventual policymaking. Personal communication, Sue Davies, principal policy adviser, Consumers' Association, July 23, 2003.

28. Interview, Sue Davies, Consumers' Association, London, June 17, 2003.

29. Consumers' Association Press Release, July 18, 2003, personal communication, Sue Davies.

30. Madeleine K. Albright, plenary address, AAAS Annual Meeting ("Science in an Uncertain Millennium"), Washington, DC, February 21, 2000, http://secretary.state.gov/www/statements/2000/000221.html (visited April 2004).

31. Senator Christopher Bond, AAAS Annual Meeting, Washington, DC, quoted in *Environment News Service*, February 23, 2000, http://www.wired.com/news/technology/ 0,1282,34507,00.html, visited July 2003.

32. There is a large literature on this point, but see particularly Jasanoff and Wynne, "Science and Decisionmaking," Cronon, ed., *Uncommon Ground*; Latour, *We Have Never Been Modern*.

33. On boundary work involving science and technology, see Gieryn, *Cultural Boundaries of Science*. On the notion of "situatedness" in knowledge production, see Haraway, *Simians, Cyborgs, and Women*, pp. 183–201.

34. The modern FDA is descended from a precursor agency formed in 1906 and is the oldest, and many would say most trusted, of the federal agencies responsible for health, safety, and environmental regulation.

35. Ann Gibbons, "Biotech Pipeline: Bottleneck Ahead," *Science* 254 (1991): 369–370.

36. Ann Gibbons, "Can David Kessler Revive the FDA?" *Science* 252 (1991): 200–201; "Kessler Gives FDA a Facelift," *Science* 255 (1992): 1350.

37. Statement of Policy: Foods Derived From New Plant Varieties, 57 *Fed. Reg.* 22,984 (proposed May 29, 1992).

38. David A. Kessler et al., "The Safety of Foods Developed by Biotechnology," *Science* 256 (1992): 1747–1749, 1832.

39. Richard Caplan and Skip Spitzer, "Regulation of Genetically Engineered Crops and Foods in the United States," March 2001, p. 3, http://www.gefoodalert.org/library/admin/uploadedfiles/Regulation_of_Genetically_Engineered_Crops_and.htm (visited April 2004). See also John Schwartz, "FDA Clears Tomato with Altered Genes," *Washington Post*, May 19, 1994, p. A1.

40. Emily Gersema, "FDA Opts against Further Biotech Review," *Associated Press Online*, June 17, 2003.

41. A well-known example of qualified labeling was that authorized for use by Ben & Jerry's ice cream company: "The family farmers who supply our milk and cream pledge not to treat their cows with recombinant Bovine Growth Hormone (rBGH). We oppose the use of rBGH even though the FDA has concluded that no significant difference has been shown and no test can now distinguish between milk from rBGH-treated and untreated cows."

42. On the propensity of the American regulatory system to deconstruct scientific claims, see Jasanoff, *Risk Management*; Brickman et al., *Controlling Chemicals*.

43. See Eric Brunner, *Bovine Somatotropin: A Product in Search of a Market*, Report to the London Food Commission's BST Working Party (London: London Food Commission, April 1988), p. 18. U.S. groups have also continued to worry about the human health effects of rBST. See, for example, Vermont Public Health Research Group, "rBGH, Monsanto, and the FDA," http://www.vpirg.org/campaigns/geneticEngineering/rBGHintro.html (visited July 2003).

44. SANET was named for the Sustainable Agriculture Network, a cooperative information sharing effort funded by the USDA's Sustainable Agriculture Research and Education program.

45. Pamela Andre, comments on bST/bGH postings, February 23, 1994.

46. Michele Gale-Sinex, BST & soul of the new machine, February 27, 1994.

47. For the Flavr Savr's regulatory history, see Secondary Bioengineering Protein Product Permitted as Food Additive, *Food Drug Cosmetic Law Reports*, para. 40301 (1994).

48. Sandra Sugawara, "For the Next Course, 'Engineered' Entrees? 'Genetic' Tomato May Launch an Industry," *Washington Post*, June 10, 1992, p. F1.

49. Maxine Singer, "Hot Tomato," *Washington Post*, August 10, 1993, p. A15.

50. Jane Rissler and Margaret Mellon, "A Real Hot Tomato," *Washington Post*, August 14, 1993, p. A19.

51. A different GM tomato variety produced by Zeneca was used in tomato puree and sold better than the Flavr Savr, but it was also withdrawn following the outbreak of the European GM controversy.

52. For more on the StarLink episode, see William Lin, Gregory K. Price, and Edward Allen, "StarLink™: Where No Cry9C Corn Should Have Gone Before," *Choices* (Winter 2001–2002): 31–34; Michael R. Taylor and Jody S. Tick, "The Star-Link Case: Issues for the Future," Pew Initiative on Food and Biotechnology and Resources for the Future (October 2001).

53. Comtex, "Greenpeace Dumps GM Corn at Whitman's EPA Door," February 8, 2001, Environmental News Network, http://www.enn.com/news/wire-stories/2001/02/02082001/greenpeace_41881.asp (visited July 2003).

54. Bill Hord, "The Road Back: Prodigene and Other Biotech Companies Are Moving Ahead in an Environment of Increasing Fear of Crop Contamination," *Omaha World Herald*, January 19, 2003, p. 1d.

55. Stephanie Simon, "The Food Industry Loves Engineered Crops, but Not When Plants Altered to 'Grow' Drugs and Chemicals Can Slip into Its Products," *Los Angeles Times*, December 23, 2002, p. 1.

56. Dan Glickman and Vin Weber, "Frankenfood Is Here to Stay. Let's Talk," *International Herald Tribune*, July 1, 2003, p. 9.

57. Charles Perrow, *Normal Accidents: Living with High-Risk Technologies* (New York: Basic Books, 1984).

58. Leaflets collected at Porter Square, Cambridge, MA, November 2002, author's files.

59. For a brief history, see the National Organic Program, Regulatory Impact Assessment for Proposed Rules Implementing the Organic Foods Production Act of 1990, http://www.ams.usda.gov/nop/archive/ProposedRule/RegImpAssess.html (visited July 2003).

60. Ben Lilliston and Ronnie Cummins, "Organic vs 'Organic': The Corruption of a Label," *The Ecologist* 28, 4 (July/Aug. 1998): 195–199.

61. Mikael Klintman, "Arguments Surrounding Organic and Genetically Modified Food Labelling: A Few Comparisons," *Journal of Environmental Policy and Planning* 4 (2002): 247–259.

62. See, for example, Rick Weiss, "'Organic' Label Ruled Out for Biotech, Irradiated Food," *Washington Post*, May 1, 1998, p. A2.

63. Marian Burros, "U.S. Imposes Standards for Organic-Food Labeling," *New York Times*, December 21, 2000, p. A22.

64. Klintman, "Arguments," elaborates on this point.

65. One such project, called LobsterTales, was started by an environmentally minded community organization called the Island Institute in Maine. It formed a partnership with eight lobstermen, who agreed to use special rubber bands for banding the claws of the lobsters they sell. The bands carry the words "Who Caught Me?" and supply the project's web address, www.lobstertales.org, and a four-digit number identifying the individual lobsterman. The hope is that the person eating the lobster will be able by these means to establish a personal connection with the person who caught it.

66. The issue of advertising organic foods can only be touched upon in this chapter, but the use of the first person appears to cut across the Anglophone world. The following signed message from an organic oatcakes package (viewed 2003, in author's files) is instructive: "For over 25 years, the Village Bakery Melmerby has been baking wholesome and distinctive bread and cakes in wood-fired ovens, using organic ingredients grown with a healthy respect for the delicate relationships between plants, animals and people. We'd be delighted if you paid us a visit or contacted us for further details."

67. As of this writing, Germany prohibited irradiated foods, another of the "Big Three" opposed by U.S. organics growers and consumers.

68. Novel foods are governed by Regulation (EC) No 258/97 of the European Parliament and of the Council of January 27, 1997 concerning novel foods and novel food ingredients. Labeling is subject to Commission Regulation (EC) No. 49/2000 and Commission Regulation (EC) No 50/2000 of January 10, 2000.

69. To market a GM novel food, the applicant first has to select a rapporteur member state, which conducts an in-depth evaluation of the risk assessment data submitted

by the applicant. The rapporteur has ninety days to forward an opinion on the application to the European Commission, plus any time the applicant requires to address questions. For the authorization procedure, the European Commission forwards the dossiers within 30 days to the other fourteen member states, whose competent authorities have 60 days to raise objections. If minor objections are resolved within forty-five days, the product is approved based on unanimous agreement with the positive rapporteur country opinion. If no decision is reached in the first stage, the dossier goes to the Regulatory Committee, which decides by qualified majority vote within 120 days. In the event of no decision, the process moves to a third stage, in which the Council of Ministers takes a qualified majority vote within 90 days. As a final recourse, the commission decides on the basis of scientific advice.

70. BfR was formerly known as the Federal Institute for Health Protection of Consumers and Veterinary Medicine (Bundesinstitut für gesundheitlichen Verbraucherschutz und Veterinärmedizin, BgVV).

71. European Commission, *Final Report of a Mission Carried out in Germany from 12 March 2001 to 16 March 2001 in Order to Evaluate Official Control Systems on Foods Consisting of or Produced from Genetically Modified Organisms (GMOs)*, DG(SANCO)/3233/2001-MR final (2001).

72. See, for example, Brickman et al., *Controlling Chemicals*.

73. The Novel Foods and Food Ingredients Regulation (Neuartige Lebensmittel- und Lebensmittelzutaten-Verordnung, NLV) of May 19, 1998.

74. Alison Abbott and Burkhardt Roeper, "Germany Seeks 'Non-modified' Food Label," *Nature* 391 (1998): 828.

75. Konrad Schuller, "Digging in the Dirt," *Frankfurter Allgemeine Zeitung* (English edition), April 6, 2002.

76. On the authoritarian impulses of the "gardening state," see Bauman, *Modernity and Ambivalence*, pp. 26–39.

77. EC Regulation nos. 1829/2003 and 1830/2003 of September 22, 2003, concerning the authorization, traceability, and labeling of GMOs and GMO derived products, http://europa.eu.int/eur-lex/pri/en/oj/dat/2003/l_268/l_26820031018en00010023.pdf and http://europa.eu.int/eur-lex/pri/en/oj/dat/2003/l_268/l_26820031018en00240028.pdf, respectively (visited March 2004).

78. Thomas Bernauer, *Genes, Trade, and Regulation: The Seeds of Conflict in Food Biotechnology* (Princeton: Princeton University Press, 2003).

79. Uwe Hessler, "Schroeder's Reluctant Cabinet to Allow GMO Foods," Deutsche Welle, Germany, http://www.gene.ch/genet/2004/Feb/msg00061.html (visited March 2004).

## CHAPTER 6: NATURAL MOTHERS AND OTHER KINDS

1. Gottweis, *Governing Molecules*.

2. The Abortion Act of 1967 made abortion legal to safeguard the physical or mental health of the mother and to prevent the birth of a severely disabled child.

3. Michael Mulkay, *The Embryo Research Debate: Science and the Politics of Reproduction* (Cambridge: Cambridge University Press, 1997), p. 11.

4. Report of the Commission of Inquiry into Human Fertilisation and Embryology (Cmnd. 9114) (London: HMSO, 1984). Republished with two added chapters by

Mary Warnock as *A Question of Life: The Warnock Report on Human Fertilisation and Embryology* (Oxford: Blackwell, 1985).

5. Interview, Baroness Warnock, U.K. House of Lords, June 29, 1993. Warnock noted that it had been extremely difficult to find a Roman Catholic appointee with whom she felt she could work, but that the eventual choice, a professor of neurology, was a wonderful colleague and an extraordinary draftsman, good at expressing even views he did not believe.

6. Jo Thomas, "British Debate Embryo Research," *New York Times*, October 16, 1984, p. 6.

7. For an influential statement of this concern, articulated during the debate on the Embryo Bill in 1990, see the editorial "Embryos Win Rights," Nature 343 (1990): 577.

8. Mulkay, *Embryo Research*, pp. 22–23.

9. David Dickson, "British Government Rekindles Debate on Embryo Research," *Science* 238 (1987): 1348.

10. Department of Health and Social Security, *Human Fertilisation and Embryology: A Framework for Legislation* (Cmnd. 259) (London: HMSO, 1987) [hereafter cited as White Paper].

11. White Paper, Prohibited Research, paras. 37–39, p. 7.

12. Lord Skelmersdale, Lords, January 15, 1988, col. 1451.

13. Mulkay, *Embryo Research*, p. 41.

14. Peter Aldhous, "Pressure Stepped Up on Embryo Research," *Nature* 344 (1990): 691.

15. Mulkay, *Embryo Research*, p. 104.

16. See, for example, Anne McLaren, "IVF: Regulation or Prohibition?" *Nature* 342 (1989): 469–470.

17. On policymaking by experts as public performance, see Hilgartner, *Science on Stage*.

18. For an interesting parallel example of the coproduction of rules for research and the object of research, see Jennifer Reardon, "The Human Genome Diversity Project: A Case Study in Coproduction," *Social Studies of Science* 31 (2001): 357–388. For a theoretical discussion of coproduction and exemplary essays, see Jasanoff, ed., *States of Knowledge*.

19. Michael Mulkay, "The Triumph of the Pre-Embryo: Interpretations of the Human Embryo in Parliamentary Debates over Embryo Research," *Social Studies of Science* 24 (1994): 611–639.

20. In their classic history of early modern science and politics in England, Steven Shapin and Simon Schaffer argued that the success of Robert Boyle's experimental method depended on the creation of a culture of actual and virtual witnessing, in which people were prepared to acknowledge the facticity of the experimentalist's observations. See Shapin and Schaffer, *Leviathan and the Air-Pump*, pp. 225–226.

21. Interview, Warnock, House of Lords, June 29, 1993.

22. Lord Kennet, Lords, January 15, 1988, col. 1497.

23. Archbishop of York, Lords, January 15, 1988, col. 1461–1462.

24. Lord Henderson of Brompton, Lords, January 15, 1988, col. 1496–1497.

25. Baroness Warnock, Lords, January 15, 1988, col. 1470.

26. I am indebted for this insight and for the details of Warnock's biography to Professor Anna Morpurgo Davies of Oxford University.

27. Dr. Malcolm Guite of Girton College was kind enough to confirm to me that this prayer was indeed often spoken in chapel during Warnock's tenure as Mistress. He wrote: "It's very encouraging to have these little signs of how the largely hidden life of a college chapel can occasionally surface in the midst of everyday life or academic endeavor, and particularly moving to know that college prayers were helping our former Mistress in the often-difficult task of discernment." Personal communication, March 30, 2004.

28. Earl Jellicoe, Lords, January 15, 1988, col. 1464.

29. Peter Aldhous, "Pro-life Actions Backfire," *Nature* 345 (1990): 7.

30. Earl Jellicoe, Lords, January 15, 1988, col. 1464. Jellicoe was rebuked for this analogy by the Earl of Lauderdale, who exclaimed: "Some noble Lords have put it even more crudely in trying to dismiss the pre-embryo as something no larger than a pinhead or a full stop at the end of a sentence. As if its size had anything to do with the issue!" Lords, January 15, 1988, col. 1485.

31. One reason may be that a strong professional consensus had grown among physicians that twenty-eight weeks was unacceptably late. Most physicians found such late abortions distasteful. Interview, Stephen Lock, British Medical Association, London, 1993.

32. Mulkay, *Embryo Research*, p. 205, n. 52.

33. Baroness Warnock, Lords, January 15, 1988, col. 1471.

34. McLaren, "IVF: Regulation or Prohibition," p. 470.

35. Editorial, "Embryo Research," *Nature* 344 (1990): 690.

36. Aldhous, "Pro-Life Actions."

37. Interview, Ruth Deech, chair, HFEA, and principal, St. Anne's College, Oxford, June 16, 1996.

38. Human Embryology and Fertilisation Act of 1990, schedule 3, paragraph 5.

39. Kathy Marks, "Widow Wins Fight to Bear Child of Dead Husband," *The Daily Telegraph*, February 7, 1997, p. 6; Associated Press, "In U.K. Case, Widow Wins Right to Use Spouse's Sperm," *International Herald Tribune*, February 7, 1997, p. 5.

40. In the United States the term is particularly associated with Leon Kass, head of President George W. Bush's National Bioethics Council, who wrote in favor of taking intuitively negative reactions to new technologies seriously as a basis for moral judgment. See Kass, "The Wisdom of Repugnance," *New Republic* (June 2, 1997): 17–26.

41. Interview, Deech, Oxford, June 16, 1996. See also Linda Grant, "Yuk Factor or SPUC Factor?" *Independent on Sunday*, July 17, 1994, p. 22.

42. *Quintavalle, R (on the application of) v Human Fertilisation and Embryology Authority* [2003] EWCA Civ 667 (May 16, 2003).

43. "A Chancellor's Dilemma," *The Economist*, January 31, 1998, pp. 54–55.

44. The Penal Code of the German Empire adopted in 1871 contained a provision (section 218) forbidding all abortions and subjecting them to criminal sanctions. See Walter F. Murphy and Joseph Tanenhaus, *Comparative Constitutional Law Cases and Commentaries* (New York: St. Martin's Press, 1977).

45. Federal Constitutional Court (Bundesverfassungsgericht), Abortion Reform Law Case, 39 BverfGE 1 (1975).

46. Interview, Birgitta Porz-Krämer, Federal Ministry of Justice, Bonn, Germany, July 15, 1993.

47. Andrea Wuerth, "National Politics/Local Identities: Abortion Rights Activism in Post-Wall Berlin," *Feminist Studies* 25, 3 (1999): 601–632. Wuerth characterizes the national unity discourse as hegemonic and patriarchal in its support for family values. It seems clear that unity was the hegemonic theme, and anything detracting from the drive for unity, including the Social Democrats' commitment to women's issues, was shunted aside.

48. Interview, Porz-Krämer, July 15, 1993.

49. Pro Familia, Theme Abortion, http://www.profamilia.de/article/show/933.html (visited July 2003).

50. For a discussion of this history, see Rolf Keller, Hans-Ludwig Günther, and Peter Kaiser, *Embryonenschutzgesetz* (Stuttgart: Kohlhammer, 1992), pp. 65–81.

51. Formal guidelines of the Bundesärztekammer may be found at its website, http://www.bundesaerztekammer.de/index.html (visited July 2003); relevant background is available under the heading Ethics and Science (*Ethik und Wissenschaft*).

52. 73rd Session of the Justice Committee, March 9, 1990, Bonn, Bundeshaus (chair: Rep. Helmrich, CDU/CSU). Presenters included experts on criminal, civil, and family law, representatives of major organizations like BÄK and Pro Familia, and spokespersons for the Protestant and Catholic churches.

53. Ibid., p. 80.

54. Ibid., pp. 39–40.

55. Ibid., pp. 58–59.

56. Ibid., pp. 95–96.

57. Ibid., p. 164.

58. Benno Müller-Hill, *Murderous Science: Elimination by Scientific Selection of Jews, Gypsies, and Others in Germany, 1933–1945* (Cold Spring Harbor: Cold Spring Harbor Press, 1998).

59. Michael Simm, "Violence Study Hits a Nerve in Germany," *Science* 264 (1994): 653.

60. Interview, Benno Müller-Hill, Institute for Genetics, University of Köln, July 14, 1993.

61. *Roe v. Wade*, 410 U.S. 113 (1973).

62. *Planned Parenthood of Southeastern Pennsylvania v. Casey*, 505 U.S. 833 (1992).

63. Ibid. at 846.

64. Ivy League student newspapers often run ads soliciting sperm and egg donations at varying prices. One famous ad offered $50,000 for eggs from a donor at least 5 feet 10, inches tall, with an SAT score of at least 1400, and various other characteristics.

65. 61 Cal. App. 4th 1410.

66. Jasanoff, *Science at the Bar*, pp. 160–182.

67. This remains the position under British law. In the United States, see *In the Matter of Baby M*, 109 N.J. 396 (1988). See also Valerie Hartouni, "Breached Birth: Anna Johnson and the Reproduction of Raced Bodies,' in *Cultural Conceptions: On Reproductive Technologies and the Remaking of Life* (Minneapolis: University of Minnesota Press, 1997), pp. 85–98.

68. 5 Cal. 4th at 93.

69. Ibid. at 97.

70. Genetics and IVF Institute advertisement, *New York Times Magazine*, March 14, 2004, p. 16.

71. Genetics and IVF Institute advertisement, *Attaché*, US Airways magazine, December 2003, p. 73.

## CHAPTER 7: ETHICAL SENSE AND SENSIBILITY

1. Hacking, "World-making by Kind-making."

2. Kass, "The Wisdom of Repugnance."

3. For a U.S. rejection of the "yuk factor" as a valid basis for policy argument, see Laurence Tribe, "Clone as Outlaw? Reasons *Not* to Ban 'Unnatural' Ways of Making Babies," in Martha C. Nussbaum and Cass R. Sunstein, eds., *Clones and Clones* (New York: Norton, 1998), pp. 223–234.

4. An early example was a comparative study of bioethics commissioned by the European Parliament in 1992. Scientific and Technological Options Assessment (STOA), *Bioethics in Europe*, PE 158.453, Luxembourg, September 8, 1992 (hereafter cited as STOA Report).

5. Ibid., p. 1.

6. Ibid., executive summary.

7. Ruth E. Bulger, Elizabeth M. Bobby and Harvey V. Fineberg, eds., *Society's Choices: Social and Ethical Decision Making in Biomedicine* (Washington, DC: National Academies Press, 1995), p. 14.

8. Foreshadowing the 1948 United Nations Universal Declaration of Human Rights, which guarantees the security of the person, the Geneva Declaration demands that medical knowledge not be used "contrary to the laws of humanity."

9. Beecher's name at birth was Harry Unangst (German for *unfear*), but he changed it for unknown reasons in his twenties. Vincent J. Kopp, "Henry K. Beecher, M.D.: Contrarian (1904–1976)," *American Society of Anesthesiologists Newsletter* 63, 9 (September 1999).

10. Henry K. Beecher, "Ethics and Clinical Research," *New England Journal of Medicine* 274 (1966): 1354–1360.

11. For a useful summary of the major developments, see IOM, *Society's Choices*, pp. 87–131.

12. President's Commission for the Study of Ethical Problems in Medicine and Biomedical and Behavioral Research, *Splicing Life: A Report on the Social and Ethical Issues of Genetic Engineering with Human Beings* (Washington, DC: The Commission, 1982).

13. See for example the collection of essays entitled "Bioethics and Beyond," edited by Arthur Kleinman, Renee C. Fox and Allan Brandt, *Daedalus* 128, 4 (1999); also Kleinman, *Writing at the Margin: Discourse between Anthropology and Medicine* (Berkeley: University of California Press, 1996).

14. By 1997 ELSI's budget reached an estimated ten million dollars, of which about three-quarters were spent on research. Eliot Marshall, "The Genome Program's Conscience," *Science* 274 (1996): 488–490.

15. Robert M. Cook-Deegan, "The Human Genome Project," in Kathi E. Hanna, ed., *Biomedical Politics* (Washington, DC: National Academy Press, 1991), pp. 148–149.

16. On this theme, see particularly Daniel S. Greenberg, *Science, Money, and Politics: Political Triumph and Ethical Erosion* (Chicago: University of Chicago Press, 2001).

17. Interview, Troy Duster and Stephen Hilgartner, Phoenix, AZ, May 31, 1996.

18. Marshall, "Genome Program's Conscience," p. 488.

19. Report of the Joint NIH/DOE Committee to Evaluate the Ethical, Legal, and Social Implications Program of the Human Genome Project, December 21, 1996.

20. An ELSI Research, Planning and Evaluation Group was established in May 1997 to provide expert advice to the ELSI grant programs and assist in developing a strategic plan for future ELSI activities. In July 1997 the NIH director established the Trans-NIH Ethical, Legal and Social Implications Coordinating Committee. This committee is charged with NIH-wide coordination of bioethical activities—not only those related to genetics, but also other emerging and contentious issues such as xenotransplantation and research involving the cognitively impaired.

21. See, for example, Erik Parens, "Respect for the RAC," Letter, *Science* 272 (1996): 1569–1570.

22. Sheryl Gay Stolberg, "The Biotech Death of Jesse Gelsinger," *New York Times*, Sunday Magazine, November 28, 1999, p. 137.

23. Detailed documentation would unnecessarily complicate the text, but for evidence one needs only to read the recurrent articles on ethics in *Science*, one of the premier U.S. journals for biomedical research. Examples include Charles Weijer and Ezekiel J. Emanuel, "Protecting Communities in Biomedical Research," *Science* 289 (2000): 1142–1144; Gretchen Vogel, "Study of HIV Transmission Sparks Ethics Debate," *Science* 288 (2000): 22–23; Jeremy Sugarman, "Ethical Considerations in Leaping from Bench to Bedside," *Science* 285: 2071–2072; Eliot Marshall, "NIMH to Screen Studies for Science and Human Risks," *Science* 283 (1999): 464–465 [noting the NIMH director's opinion that more stringent review of risks was needed to ensure the "beneficence" of the studies].

24. See IOM, *Society's Choices*, pp. 93–94.

25. Harold T. Shapiro, Editorial, "Ethical Dilemmas and Stem Cell Research," *Science* 285 (1999): 2065.

26. Remarks by the President in Meeting with Bioethics Committee, January 17, 2002 (on PCBE website), http://www.bioethics.gov/about/ (visited August 2003).

27. See, for example, Jürgen Habermas. *The Future of Human Nature* (Cambridge: Polity, 2003).

28. Michael Specter, "The Dangerous Philosopher," *New Yorker*, (September 6, 1999): 46–55.

29. Peter Singer, *Practical Ethics* (Cambridge: Cambridge University Press, 1979).

30. Peter Singer, "On Being Silenced in Germany," *New York Review of Books* (August 15, 1991): pp. 36–42.

31. Text of interview with Peter Singer (in German) in author's files, June 6, 1989.

32. Manfred D. Laubichler, "Frankenstein in the Land of *Dichter* and *Denker*," *Science* 286 (1999): 1859–1860.

33. Charles S. Maier, *The Unmasterable Past: History, Holocaust, and German National Identity* (Cambridge: Harvard University Press, 1988).

34. This is not the place to engage in a full-scale analysis of German national self-examination at the moment of reunification, but it should be noted that there were many controversies and writings suggesting that the Nazi phenomenon was both more pervasively distributed across German society and more deeply embedded in historical traditions than the exceptionalists wished to believe. Consider, for example, the

controversial exhibition *"Crimes of the German Wehrmacht: Dimensions of a War of Annihilation, 1941–1944"* mounted in Germany and Austria in the 1990s and again in 2002, challenging the myth of the "clean" Wehrmacht and showing it rather to have been more complicit in Nazi crimes than previously assumed. A similar story of unsuspectedly wide complicity was told in Daniel J. Goldhagen, *Hitler's Willing Executioners: Ordinary Germans and the Holocaust* (New York: Vintage, 1997); that book also evoked intense resonance in Germany. Work in the history of science also stressed the continuity of Nazi biological thought with earlier German research. See, for example, Susan Heim, "Research for Autarky: The Contribution of Scientists to Nazi Rule in Germany," Max-Planck Institute for the History of Science, Berlin (2001); George Stein, "Biological Science and the Roots of Nazism," *American Scientist*, 76 (1988): 50–58.

35. As a parliamentary body, the bioethics Enquete Commission had both political and professional members. The formula for appointing members tracked the Bundestag's electoral composition, so that the ruling SPD nominated six members in each category, the CDU/CSU picked four members in each, and the three minor parties (Greens, FDP, PDS) each chose one representative and one expert.

36. Robert Leicht, "Ein Rat der Anständigen," *Die Zeit*, May 2, 2001.

37. Tom Wilkie, "Whose Genes Are They Anyway?" *Independent*, May 6, 1991, p. 19. Wilkie notes that a Royal Commission would have been the appropriate constitutional device for such an undertaking.

38. Nuffield Council on Bioethics, *Annual Report 1991–1992*, Nuffield Foundation, London, 1992, p. 4.

39. "The Need for a New National Bioethics Body: A Consultation Document," Nuffield Foundation Conference on Bioethics, Cumberland Lodge, April 20–22, 1990.

40. Report of the Steering Group on the Nuffield Foundation Bioethics Initiative to the Trustees of the Nuffield Foundation, December 1990, p. 6.

41. Ibid., p. 13.

42. On the instrumental and self-legitimating uses of science by the liberal democratic state, see Ezrahi, *The Descent of Icarus*.

43. Richard Dashefsky, "The High Road to Success: How Investing in Ethics Enhances Corporate Objectives," *Journal of Biolaw and Business* 6, 3 (2003): 3–7.

44. History of BIO, http://www.bio.org/aboutbio/history.asp (visited August 2003).

45. Interview, Sheldon Krimsky, Cambridge, MA, July 26, 2004.

46. Gen-Ethischer Informationsdienst, *Collected Edition*, issues 0–30, Berlin (1987).

47. A feminist orientation is graphically signaled in German by using a single term with a capitalized internal "I" to designate both masculine and feminine variants of the same agent noun: thus, *WissenschaftlerInnen, BiologInnen, InformatikerInnen*, referring to male *and* female scientists, biologists, information experts. GID newsletters adopted this practice.

48. STOA Report, p. 104.

49. On the GM foods campaign, see GeN campaign leaflet ("Essen aus dem Genlabor, natürlich nicht!"); on DNA fingerprinting, see open letter to the justice minister for Lower Saxony, April 22, 1991 (both documents in author's files).

50. Wynne, "Public Understanding of Science"; "Public Uptake of Science."

51. Brian Wynne, "Creating Public Alienation: Expert Cultures of Risk and Ethics on GMOs," *Science as Culture* 10, 4 (2001): 445–481.

52. Phil McNaghten, "Animals in Their Nature," *Sociology* 38, 3 (2004): 533–551.

53. Stem cells were presented as the breakthrough of the year in 1999 in the prestigious journal *Science*. Gretchen Vogel, "Capturing the Promise of Youth," *Science* 286 (1999): 2238–2240.

54. Shirley J. Wright, "Human Embryonic Stem-Cell Research: Science and Ethics," *American Scientist* 87 (1999): 352–361.

55. One of many sources in which these processes are described in accessible terms is NBAC's report, *Ethical Issues in Stem Cell Research*, Washington, DC, September 1999.

56. Statement by the President to NBAC, September 13, 1999.

57. James Fallows, "The Political Scientist," *New Yorker* (June 7, 1999): 66–75.

58. Gretchen Vogel, "NIH Sets Rules for Funding Embryonic Stem Cell Research," *Science* 286 (1999): 2050–2051.

59. PCBE, executive summary, *Human Cloning and Human Dignity: An Ethical Inquiry*, July 2002.

60. G. Nigel Gilbert and Michael Mulkay, *Opening Pandora's Box: A Sociological Analysis of Scientists' Discourse* (Cambridge: Cambridge University Press, 1984).

61. Habermas, *Future of Human Nature*, p. 95.

62. Interestingly, this debate paralleled in key respects a contemporaneous discussion of German immigration policy, which also flagged basic questions about what kind of state Germany was or wanted to be. See Stefan Sperling, "Managing Potential Selves: Stem Cells, Immigrants, and German Identity," *Science and Public Policy* 39, 2 (2004): 139–149.

63. Gretchen Vogel, "U.K. Backs Use of Embryos, Sets Vote," *Science* 289 (2000): 1269–1273.

64. House of Lords, *Stem Cell Research—Report*, February 13, 2002

65. *Quintavalle, R (on the application of) v. Secretary of State for Health* [2003] UKHL 13.

66. *Diamond v. Chakrabarty*, 447 U.S. 303 (1980).

67. The Law Lords took as their controlling authority on the matter of interpretation. Lord Wilberforce's dissenting opinion in *Royal College of Nursing of the United Kingdom v. Department of Health and Social Security* [1981] AC 800.

68. On this point, see particularly the opinions of Lord Bingham of Cornhill and Lord Steyn.

69. I have discussed judicial empiricism and the role of judicial vision in "The Eye of Everyman: Witnessing DNA in the Simpson Trial," *Social Studies of Science* 28, 5–6 (1998): 713–740; and "Science and the Statistical Victim: Modernizing Knowledge in Breast Implant Litigation," *Social Studies of Science* 32, 1 (2002): 37–70.

70. [2003] UKHL 13 (Opinion of Lord Bingham, para. 15).

### CHAPTER 8: MAKING SOMETHING OF LIFE

1. See, for example, Reid G. Adler, "Biotechnology as an Intellectual Property," *Science* 224 (1984): 357–363.

2. On the idea of romantic authorship as the basis for awarding intellectual property rights in U.S. law, see James Boyle, *Shamans, Software, and Spleens: Law and the Constitution of the Information Society* (Cambridge: Harvard University Press, 1996).

3. For a fascinating account of biotechnology as both capital and a cultural formation, see Kaushik Sunder Rajan, "Genomic Capital: Public Cultures and Market Logics of Corporate Biotechnology," *Science as Culture* 12, 1 (2003): 87–121.

4. Winner, "Do Artifacts Have Politics," in *The Whale and the Reactor*, pp. 19–39.

5. Bijker, Hughes, and Pinch, eds., *The Social Construction of Technological Systems*.

6. Callon, "Some Elements of a Sociology of Translation."

7. Jasanoff, ed., *Learning from Disaster*; Brian Wynne, "Unruly Technology," *Social Studies of Science* 18 (1988): 147–167.

8. For an elegant anthropological demonstration of this point, calling attention to the distributed sources of invention in biotechnology, see Paul Rabinow, *Making PCR* (Chicago: University of Chicago Press, 1996).

9. *Diamond v. Chakrabarty*, 447 U.S. 303 (1980).

10. Richard Stone, "Religious Leaders Oppose Patenting Genes and Animals," *Science* 268 (1995): 1126.

11. United States Code, Title 35, sections 101–103. The law defines what can be patented as "any new and useful process, machine, manufacture, or composition of matter, or any new and useful improvement thereof."

12. Editorial, "Yes, Patent Life," *New York Times*, April 21, 1987, p. A30.

13. Shelley Rowland and Jared Scarlett, "The World's Most Litigated Mouse," *NZ Bio Science* 13 (February 2003), http://www.bsw.co.nz/articles/xfactor13.html (visited August 2003).

14. 447 U.S. at 317.

15. For more on this point, see Jasanoff, *Science at the Bar*, pp. 144–145.

16. 100 S. Ct. at 2211.

17. Judicial dissents are rare in the civil law world. The German constitutional court, for example, typically does not issue dissenting opinions.

18. *President and Fellows of Harvard College v. Canada (Commissioner of Patents)*, 2002 SCC 76.

19. See Factum of the Interveners, Canadian Environmental Law Association et al., Court File no. 28155, May 2002.

20. *Harvard College v. Canada*, para. 163.

21. Ibid., para. 78.

22. Ibid., para. 181.

23. *Moore v. Regents of the University of California*, 51 Cal. 3d 134 (1990).

24. See, for example, Boyle, *Shamans, Software, and Spleens*.

25. *Moore v. Regents*, 249 Cal. Rptr. 494 (Cal. App. 2nd Dist. 1988), at 510.

26. "Because of the novelty of Moore's claim to own the biological materials at issue, to apply the theory of conversion in this context would frankly have to be recognized as an extension of the theory. Therefore, we consider next whether it is advisable to extend the tort to this context." 51 Cal. 3d at 136.

27. Ibid., at 139.

28. Boyle, *Shamans, Software, and Spleens*, p. 107.

29. Leslie Roberts, "NIH Gene Patents, Round Two," *Science* 255 (1992): 912–913.

30. Rebecca Eisenberg, "Genes, Patents, and Product Development," *Science* 257 (1992): 903–908; Michael A. Heller and Rebecca Eisenberg, "Can Patents Deter Innovation? The Anticommons in Biomedical Research," *Science* 280 (1998): 698–791.

31. USPTO Utility Examination Guidelines, Federal Register xxxx, January 5, 2001.

32. *European Directive on the Legal Protection of Biotechnological Inventions* (Directive 98/44/EC).

33. Alison Abbott, "Euro-vote Lifts Block on Biotech Patents," *Nature* 388 (1997): 314–315.

34. GenEthics News, issue 5 (1995): 3

35. European Parliament, Scientific and Technological Options Assessment, Bioethics in Europe, Final Report, Luxembourg, October 1992, pp. 100–106.

36. Directive EC 98/44/EC on the legal protection of biotechnological inventions of July 6, 1998.

37. Nuffield Council on Bioethics, *The Ethics of Patenting DNA*, July 23, 2002, p. 27 (para. 3.20).

38. The process of isolating DNA sequences is unproblematically patentable. The confusions arise principally around the patenting of the products, in this case the isolated DNA sequences.

39. Nuffield Council, *The Ethics of Patenting DNA*, pp. 35–36.

## CHAPTER 9: THE NEW SOCIAL CONTRACT

1. For descriptions of what the so-called contract entailed, see David H. Guston, *Between Politics and Science: Assuring the Integrity and Productivity of Research* (New York: Cambridge University Press, 2000), pp. 37–63; Donald E. Stokes, *Pasteur's Quadrant: Basic Science and Technological Innovation* (Washington, DC: Brookings Institution, 1997).

2. Vannevar Bush, *Science—The Endless Frontier* (Washington, DC: US Government Printing Office, 1945).

3. For an overview of some commonly used measures of national performance in science, see Robert M. May, "The Scientific Wealth of Nations," *Science* 275 (1997): 793–796.

4. Derek de Solla Price, *Little Science, Big Science* (New York: Columbia University Press, 1963).

5. Ezrahi, "Science and Utopia in Late 20th Century Pluralist Democracy."

6. Greenberg, *Science, Money, and Politics*.

7. Merton, "The Normative Structure of Science."

8. The creation of the National Institutes of Health to sponsor biomedical research divided US science policy in a way not contemplated in Bush's original design. In the recent politics of science, NIH budgets have proved consistently easier to justify than appropriations for other branches of science. See Michael Dennis, "Reconstructing Sociotechnical Order: Vannevar Bush and US Science Policy," in Jasanoff, ed., *States of Knowledge*, pp. 225–253; also see Greenberg, *Science, Money, and Politics*.

9. David M. Hart, *Forged Consensus: Science, Technology, and Economic Policy in the United States, 1921–1953* (Princeton: Princeton University Press, 1998).

10. Daniel J. Kevles, *The Baltimore Case: A Trial of Politics, Science, and Character* (New York: W. W. Norton, 1998).

11. Guston, *Between Politics and Science*.

12. Jasanoff, *The Fifth Branch*.

13. H. M. Collins, "The Place of the 'Core Set' in Modern Science: Social Contingency with Methodological Propriety in Science," *History of Science* 19 (1981): 6–19.

14. *Daubert v. Merrell Dow*, 509 U.S. 579 (1993).

15. The Office of Management and Budget's Proposed Bulletin on Peer Review and Information Quality, together with public comments on the proposal, were posted at http://www.thecre.com/ (visited January 2004).

16. Kuhn noted that most scientists much of the time are engaged in work that is unproblematically bounded by a governing paradigm. Their work is "normal science" because it is ordinary in this sense. It does not question its own foundations. See Kuhn, *The Structure of Scientific Revolutions*.

17. Silvio O. Funtowicz and Jerome R. Ravetz, "Three Types of Risk Assessment and the Emergence of Post Normal Science," in Sheldon Krimsky and David Golding, eds., *Social Theories of Risk* (London: Praeger, 1992), pp. 251–273.

18. Michael Gibbons et al., *The New Production of Knowledge: The Dynamics of Science and Research in Contemporary Societies* (London: Sage, 1994).

19. Ibid., p. 8.

20. Anderson, *Imagined Communities*.

21. Cover illustration, *Science* 291 (February 16, 2001).

22. The act was sponsored by senators Birch Bayh (D-Indiana) and Robert Dole (R-Kansas). For its legislative history, see Jennifer A. Henderson and John J. Smith, "Academia, Industry, and the Bayh-Dole Act: An Implied Duty to Commercialize," http://www.cimit.org/coi_part3.pdf (visited October 2003).

23. Arti K. Rai and Rebecca Eisenberg, "Bayh-Dole Reform and the Progress of Biomedicine," *American Scientist* 91 (2003): 53.

24. National Science Foundation, *Science and Engineering Indicators*, chap. 5 (Academic Research and Development), http://www.nsf.gov/sbe/srs/seind02/c5/c5h.htm (visited October 2003).

25. Eyal Press and Jennifer Washburn, "The Kept University," *Atlantic Monthly* (March 2000): 39–54.

26. Wil Lepkowski, "Biotech's OK Corral," *Science and Policy Perspectives*, no. 13 (July 9, 2002), http://www.cspo.org/s&pp/060902printer.html (visited October 2003).

27. Personal communication, Lawrence Busch, Michigan State University, October 5, 2003.

28. Sheldon Krimsky, *Science in the Private Interest: How the Lure of Profits Has Corrupted the Virtue of Biomedical Research* (Lanham, MD: Rowman-Littlefield, 2003). See also Melody Petersen, "Uncoupling Campus and Company," *New York Times*, September 23, 2003, p. F2.

29. For an interesting discussion of this phenomenon, see Steve Fuller, *The Governance of Science* (Buckingham, UK: Open University Press, 2000), pp. 22–25.

30. See, for example, Daniel Zalewski, "Ties That Bind," *Lingua Franca* (June/July 1997): 51–59.

31. The humorous designation of British science as a "private limited company (plc)" calls attention to its corporate character as well as its close ties to the state.

32. Peter Aldhous, "The Biggest Shake-Up for British Science in 30 Years," *Science* 260 (1993): 1419–1420.

33. Georges Köhler and César Milstein, "Continuous Cultures of Fused Cells Secreting Antibody of Predefined Specificity," *Nature* 256 (1975): 495–497.

34. NRDC enjoyed this right under a Treasury Department circular, TC5/50, of 1950. The right was later transferred to the successor body, the British Technology Group.

35. Letter dated October 7, 1976, from EJT (Eric), National Research Development Corporation to L. D. Hamlyn at the Medical Research Council, http://www.path.cam.ac.uk/~mrc7/mab25yrs/ (visited October 2003).

36. SET Forum, "Shaping the Future: A Policy for Science, Engineering and Technology," Discussion Document for the Labour Party (1995), p. 10.

37. Tony Blair, "Science Matters," speech delivered at the Royal Society, London, May 23, 2002.

38. Aldhous, "The Biggest Shake-Up."

39. Astex Technology website, http://www.astex-technology.com/index.html (visited October 2003).

40. Interview with Tom Blundell, Department of Biochemistry, University of Cambridge, Cambridge, U.K., April 2002.

41. Simon Jenkins, "Face It, the Last Thing We Need Is More Scientists," *Times Higher Supplement*, September 11, 1998, pp. 19–20.

42. Interview with Peter Hiscocks, Cambridge Entrepreneurship Centre, Cambridge, U.K., April 2002.

43. May, "The Scientific Wealth of Nations," p. 796. Of course, the citation game can be played in many ways. A 1996 article by a leading German biochemist observed that, between 1981 and 1990, eight of the fifty most cited articles in molecular biology had stemmed from Germany. Peter-Hans Hofschneider, "Grundlagenforschung und Industrie in Deutschland—warnendes Beispiel Gentechnologie," *Futura* (February 1996): 104–109.

44. David Dickson, "German Firms Move into Biotechnology," *Science* 218 (1982): 1287–1289.

45. Michael Wortmann, "Multinationals and the Internationalization of R&D: New Developments in German Companies," *Research Policy* 19 (1990): 175–183.

46. Ronald Bailey, "Brain Drain," *Forbes* (November 27, 1989): 261–262.

47. Hofschneider, "Grundlagenforschung," pp. 105–106.

48. For historical writing on these topics, see Fritz Stern, *Einstein's German World* (Princeton: Princeton University Press, 2001); Robert N. Proctor, *The Nazi War on Cancer* (Princeton: Princeton University Press, 1999); *Racial Hygiene*.

49. Federal Ministry for Education and Research web site, http://www.bioregio.com/ (visited October 2003).

50. See, for instance, Federal Ministry for Education and Research, *Rahmenprogramm Biotechnologie—Chancen, Nutzen und Gestalten* (Framework Program Biotechnology—Opportunities, Uses and Structures), Bonn, April 2001.

51. Richard Bernstein, "Letter from Europe: Listen to the Germans: Oh, What a Sorry State We're In," *New York Times*, March 24, 2004, p. A4.

52. Jasanoff, "Technological Innovation in a Corporatist State: The Case of Biotechnology in the Federal Republic of Germany," pp. 23–38.

53. I am indebted to Stefan Sperling for pointing out the importance of these concurrent debates for German identity formation.

54. Polanyi, "The Republic of Science," pp. 54–73.

## CHAPTER 10: CIVIC EPISTEMOLOGY

1. Michel Foucault, *Madness and Civilization: A History of Insanity in the Age of Reason* (New York: Vintage Books, 1973); *The History of Sexuality*; *Discipline and Punish: The Birth of the Prison* (New York: Random House, 1979).

2. On the looping effects of human kinds, see Hacking, *The Social Construction of What?*; *Rewriting the Soul*. On the rationalizing impacts of classification, see Bauman, *Modernity and Ambivalence*; Bowker and Star, *Sorting Things Out*.

3. Ezrahi, *The Descent of Icarus*.

4. John Ziman, *Public Knowledge* (Cambridge: Cambridge University Press, 1968), p. 33.

5. Significantly, the three sponsoring organizations of COPUS announced in December 2002 that they were disbanding the body because "We have reached the conclusion that the top-down approach which Copus currently exemplifies is no longer appropriate to the wider agenda that the science communication community is now addressing. We believe it will be more effective to allow organisations to seek their own partnerships and develop their own activities." http://www.copus.org.uk/news_detail_091202.html (visited April 2003).

6. The science journalist Daniel Greenberg has written scathingly of U.S. scientists' persistent blaming of a scientifically illiterate public for their imagined woes. Greenberg finds no association between PUS and science funding. See Greenberg, *Science, Money, and Politics*, pp. 205–233.

7. National Science Foundation, *Science and Engineering Indicators 2002*, chapter 7, http://www.nsf.gov/sbe/srs/seind02/c7/c7s2.htm#attb (visited April 2004) (hereafter, *2002 Indicators*).

8. Ibid., http://www.nsf.gov/sbe/srs/seind02/c7/c7s1.htm (visited April 2004).

9. Jon Turney, "Public Understanding of Science," *Lancet* 347 (1996): 1087–1090.

10. See particularly Irwin and Wynne, eds., *Misunderstanding Science?*

11. Jasanoff, "Civilization and Madness."

12. *Phillips Inquiry Report*, vol. 6: Human Health, 1989–1996, para. 4.564, http://www.bseinquiry.gov.uk/report/volume6/chapt413.htm (visited April 2004).

13. *Guardian Unlimited*, October 22, 1998.

14. Edmund Burke, *Thoughts on the Present Discontents* (1770). I am indebted to Maya Jasanoff for calling my attention to this passage and to the Gillray cartoons discussed above.

15. Ezrahi, *The Descent of Icarus*.

16. Brickman et al., *Controlling Chemicals*.

17. In a September 18, 2003, survey released by the Pew Initiative on Food and Biotechnology, 58 percent of Americans said they did not believe they had ever eaten GM foods. It is estimated that 70–75 percent of processed foods in grocery stores contain GM ingredients. See http://pewagbiotech.org/research/2003update/ (visited September 2003).

18. Ezrahi, *The Descent of Icarus*.

19. For contrasts between the United States and Europe with regard to these assumptions, see, Sheila Jasanoff, "Citizens at Risk: Cultures of Modernity in Europe and the U.S.," *Science as Culture* 11, 3 (2002): 363–380.

20. For an account of the fluid and diverse accommodations between science and Western democracy, see Roy MacLeod, "Science and Democracy: Historical Reflections on Present Discontents," *Minerva* 35 (1997): 369–384.

21. Brian Wynne and Peter Simmons (with Claire Waterton, Peter Hughes, and Simon Shackley), "Institutional Cultures and the Management of Global Environmental Risks in the United Kingdom," in The Social Learning Group, *Learning to Manage Global Environmental Risks*, pp. 93–113.

22. Brickman et al., *Controlling Chemicals*. See also Sheila Jasanoff, "Acceptable Evidence in a Pluralistic Society," in Rachelle Hollander and Deborah Mayo, eds., *Acceptable Evidence: Science and Values in Hazard Management* (New York: Oxford University Press, 1991), pp. 29–47; "Cultural Aspects of Risk Assessment in Britain and the United States," in Johnson and Covello, eds., *The Social and Cultural Construction of Risk*, pp. 359–397.

23. Skocpol, *Protecting Soldiers and Mothers*.

24. Porter, *Trust in Numbers*.

25. For more on the politics of numbers in U.S. environmental regulation, see Jasanoff, *Risk Management and Political Culture*; also see Brickman et al., *Controlling Chemicals*; Jasanoff, "Acceptable Evidence" and "Cultural Aspects of Risk Assessment."

26. See Brickman et al., *Controlling Chemicals*, pp. 122–126.

27. There is a large literature on this topic. Important examples include Paul Stern and Harvey Fineberg, eds., *Understanding Risk* (Washington, DC: National Academy Press, 1996); Beck, *Risk Society*; Langdon Winner, "On Not Hitting the Tar-Baby," in *The Whale and the Reactor*, pp. 138–154. For a review of the implications of this work for comparative politics, see Sheila Jasanoff, "Technological Risk and Cultures of Rationality," in National Research Council, *Incorporating Science, Economics, and Sociology in Developing Sanitary and Phytosanitary Standards in International Trade* (Washington, DC: National Academy Press, 2000), pp. 65–84.

28. Harried administrators were not the only ones who jumped on the risk assessment bandwagon. The movement received powerful support from legal scholars and the courts. On the role of the courts, see Jasanoff, *Science at the Bar*. On the attitudes of legal academics, see Cass Sunstein, *Risk and Reason* (Cambridge: Cambridge University Press, 2002); Breyer, *Breaking the Vicious Circle*.

29. U.S. House of Representatives, Committee on Government Reform (Minority Report), *Politics and Science in the Bush Administration*.

30. In the rich literature on the sociology of testing, the following works are especially illuminating: Pinch, "Testing—One, Two, Three ... Testing!'" and MacKenzie, *Inventing Accuracy*.

31. John Carson, "The Merit of Science and the Science of Merit," in Jasanoff, ed., *States of Knowledge*, pp. 181–205.

32. Press and Washburn, "The Kept University."

## CHAPTER 11: REPUBLICS OF SCIENCE

1. See Press Release, United States Trade Representative, Executive Office of the President, Washington, DC, May 13, 2003, http://www.ustr.gov/releases/2003/05/03-31.htm (visited October 2003).

2. "European Commission regrets the request for a WTO panel on GMOs," EU Institutions Press Releases, Brussels, August 18, 2003.

3. Bernauer, *Genes, Trade, and Regulation*.

4. Jasanoff, ed., *States of Knowledge*.

5. These observations are similar to what Peter Hall and others term "policy paradigms" but the coproduction framework brings the cognitive dimension of framing to the fore.

6. Jasanoff, "Product, Process, or Programme: Three Cultures and the Regulation of Biotechnology."

7. Francis Fukuyama, *The End of History and the Last Man* (New York: Penguin, 1992).

8. *Diamond v. Chakrabarty*, 447 U.S. 303 (1980).

9. Consider, for example, the crucial role of venture capital in the early growth of biotechnology in the United States. United States Congress, Office of Technology Assessment, *Commercial Biotechnology: An International Analysis* (Washington, DC: US GPO, 1984).

10. See, for instance, Smith and Marx, eds., *Does Technology Drive History?*

11. *Moore v. Regents of the University of California*, 51 Cal. 3d 134 (1990).

12. *Johnson v. Calvert*, 5 Cal. 4th 84 (1993). See also Jasanoff, "Ordering Life: Law and the Normalization of Biotechnology."

13. See Habermas, *The Future of Human Nature*.

14. John Krebs had successfully served as the head of the Natural Environment Research Council and, in that capacity, had dealt with the dismissal of David Bishop over the release of a viral pesticide modified with a scorpion gene (see chapter 4).

15. On Germany's wider struggles with historical memory in this period, see Maier, *The Unmasterable Past*; Ian Buruma, *The Wages of Guilt: Memories of War in Germany and Japan* (London: Vintage, 1995).

16. Interview with Wolfgang van den Daele, Berlin, July 2002.

17. This point builds on the work of Michel Foucault and his followers. See particularly Foucault, *The Foucault Reader*, ed. Paul Rabinow (New York: Random House, 1984); Paul Rabinow, *Essays on the Anthropology of Reason* (Princeton: Princeton University Press, 1996); Hardt and Negri, *Empire*.

18. For a comprehensive history of this phase of U.K. policymaking, see Wright, *Molecular Politics*.

19. Jasanoff, "Civilization and Madness."

20. Gottweis, *Governing Molecules*.

21. The data were based on a survey of one thousand American consumers conducted on August 5–10, 2003. Pew Initiative on Food and Biotechnology, "Public Sentiment about Genetically Modified Food," http://pewagbiotech.org/research/2003update (visited April 2004).

22. Jasanoff, "Ordering Life."

23. Walter Lippmann, *The Phantom Public* (New Brunswick, NJ: Transaction Publishers, 1993 [1925]).

24. Kuhn, *The Structure of Scientific Revolutions*.

25. See particularly Bruno Latour, *Science in Action: How to Follow Scientists and Engineers through Society* (Cambridge: Harvard University Press, 1987).

26. For an argument that this confluence of scientific with political culture in the West dates back to the time of the scientific revolution, see the magisterial account of Robert Boyle's struggles with Thomas Hobbes in Shapin and Schaffer, *Leviathan and the Air-Pump*.

27. L. Frank Baum, *The Wonderful Wizard of Oz* (Chicago: G. M. Hill, 1900).

28. Baroness Warnock, Lords, January 15, 1988, col. 1470.

# References

Abbott, Allison. "Euro-vote Lifts Block on Biotech Patents." *Nature* 388 (1997): 314–315.

Abbott, Alison, and Burkhardt Roeper. "Germany Seeks 'Non-modified' Food Label." *Nature* 391 (1998): 828.

Adler, Reid G. "Biotechnology as an Intellectual Property." *Science* 224 (1984): 357–363.

Agamben, Giorgio. *Homo Sacer: Sovereign Power and Bare Life*. Stanford: Stanford University Press, 1998.

Agriculture and Environment Biotechnology Commission, *Crops on Trial*. September 2001. www.aebc.gov.uk/aebc/pdf/crops.pdf (visited July 2003).

Aldhous, Peter. "The Biggest Shake-Up for British Science in 30 Years." *Science* 260 (1993): 1419–1420.

———. "Pressure Stepped up on Embryo Research." *Nature* 344 (1990): 691.

———. "Pro-life Actions Backfire." *Nature* 345 (1990): 7.

Allen, Garland E. "Modern Biological Determinism: The Violence Initiative." In *The Practices of Human Genetics*, edited by M. Fortun and Everett Mendelsohn, pp. 1–23. Dordrecht: Kluwer, 1999.

Alpers, Svetlana. *The Art of Describing: Dutch Art in the Seventeenth Century*. Chicago: University of Chicago Press, 1983.

Anderson, Benedict. *Imagined Communities*. 2d ed. London: Verso, 1991.

Associated Press. "In U.K. Court Case, Widow Wins Right to Use Spouse's Sperm." *International Herald Tribune*, February 7, 1997, p. 5.

Badaracco, Joseph. *Loading the Dice: A Five Country Study of Vinyl Chloride Regulation*. Cambridge: Harvard University Press, 1985.

Bailey, Ronald. "Brain Drain." *Forbes*, November 27, 1989: 261–262.

Balogh, Brian. *Chain Reaction: Expert Debate and Public Participation in American Commercial Nuclear Power, 1945–1975*. New York: Cambridge University Press, 1991.

Barnes, Barry. *T. S. Kuhn and Social Science*. London: Macmillan, 1982.

Barnes, Julian. *England, England*. London: Picador, 1998.

Bartelson, Jens. *A Genealogy of Sovereignty*. Cambridge: Cambridge University Press, 1995.

REFERENCES

Baum, L. Frank. *The Wonderful Wizard of Oz*. Chicago: G. M. Hill, 1900.
Bauman, Zygmunt . *Modernity and Ambivalence*. Ithaca: Cornell University Press, 1991.
Beck, Ulrich. *Risk Society: Towards a New Modernity*. London: Sage, 1992.
Beecher, Henry K. "Ethics and Clinical Research." *New England Journal of Medicine* 274 (1966): 1354–1368.
Bell, Daniel. *The Coming of Post-Industrial Society*. London: Heinemann, 1973.
Berenson, Alex, and Nicholas Wade. "A Call for Sharing of Research Causes Gene Stocks to Plunge." *New York Times*, March 15, 2000, p. A1.
Berg, Paul, et al. "Potential Biohazards of Recombinant DNA Molecules." *Science* 185 (1974): 303.
Bernauer, Thomas. *Genes, Trade, and Regulation: The Seeds of Conflict in Food Biotechnology*. Princeton: Princeton University Press, 2003.
Bernstein, Richard. "Europe's Lofty Vision of Unity Meets Headwinds." *New York Times*, December 4, 2003, p. A1.
———. "Letter from Europe: Listen to the Germans: Oh, What a Sorry State We're In." *New York Times*, March 24, 2004, p. A4.
Bierstecker, Thomas, and Cynthia Weber, eds. *State Sovereignty as Social Construct*. Cambridge: Cambridge University Press, 1996.
Bijker, Wiebe E., Thomas P. Hughes, and Trevor Pinch, eds. *The Social Construction of Technological Systems: New Directions in the Sociology and History of Technology*. Cambridge: MIT Press, 1987.
Bloor, David. *Knowledge and Social Imagery*. Chicago: University of Chicago Press, 1976.
Bohme, Gernot, and Nico Stehr. *The Knowledge Society: The Growing Impact of Scientific Knowledge on Social Relations*. Dordrecht, NL: Reidel, 1986.
Botkin, Daniel. *Discordant Harmonies: A New Ecology for the Twenty-First Century*. Oxford: Oxford University Press, 1990.
Bowker, Geoffrey C., and Susan Leigh Star. *Sorting Things Out: Classification and Its Consequences*. Cambridge: MIT Press, 1999.
Boyle, James. *Shamans, Software, and Spleens: Law and the Constitution of the Information Society*. Cambridge: Harvard University Press, 1996.
Breyer, Stephen G. *Breaking the Vicious Circle: Toward Effective Risk Regulation*. Cambridge: Harvard University Press, 1993.
Brickman, Ronald, Sheila Jasanoff, and Thomas Ilgen. *Controlling Chemicals: The Politics of Regulation in Europe and the United States*. Ithaca: Cornell University Press, 1985.
British Government Panel on Sustainable Development. *Second Report*, January 1996.
Brunner, Eric. "Bovine Somatotropin: A Product in Search of a Market." Report to the London Food Commission's BST Working Party. London: London Food Commission, April 1988.
Buckley, Christopher. "Royal Pain: Further Adventures of Rick Renard." *The Atlantic Monthly*, April 2004, pp. 94–106.
Bud, Robert. *The Uses of Life: A History of Biotechnology*. Cambridge: Cambridge University Press, 1993.

Bulger, Ruth E., Elizabeth M. Bobby, and Harvey V. Fineberg, eds., *Society's Choices: Social and Ethical Decision Making in Biomedicine* (Washington, DC: National Academics Press, 1995).

Bullard, Michael. "*Unser Mann in der Kommission* (Our Man in the EC Commission)." *Natur* 8 (August 1991): 34–35.

Burke, Edmund. *Thoughts on the Present Discontents* (1770).

Burros, Marian. "U.S. Imposes Standards for Organic-Food Labeling." *New York Times*, December 21, 2000, p. A22.

Buruma, Ian. *Wages of Guilt: Memories of War in Germany and Japan.* London: Vintage, 1994.

Busch, Lawrence, et al. *Plants, Power, and Profit: Social, Economic, and Ethical Consequences of the New Biotechnologies.* Oxford: Blackwell, 1991.

Bush, Vannevar. *Science—The Endless Frontier.* Washington, DC: U.S. Government Printing Office, 1945.

Callon, Michel. "Some Elements of a Sociology of Translation: Domestication of the Scallops and Fishermen of St. Brieuc Bay." In *Power, Action, and Belief: A New Sociology of Knowledge?* edited by John Law, pp. 196–233. London: Routledge and Kegan Paul, 1986.

Cambridge Biomedical Consultants, *The Impact of New and Impending Regulations on UK Biotechnology.* Cambridge, 1990.

Camilleri, Joseph, and Jim Falk. *The End of Sovereignty? The Politics of a Shrinking and Fragmenting World.* Hants, England: Edward Elgar, 1992.

Campbell, Philip. Editorial Note. *Nature* 416 (2002): 601.

Cantley, Mark F. "Democracy and Biotechnology: Popular Attitudes, Information, Trust and the Public Interest." *Swiss Biotech* 5(5) (1987): 5–15.

———. "The Regulation of Modern Biotechnology: A Historical and European Perspective." In *Biotechnology*, vol. 12 (Legal, Economic and Ethical Dimensions). New York: VCH Weinheim, 1995, chap. 18.

Caplan, Richard, and Ellen Hickey. "Weird Science: The Brave New World of Genetic Engineering." October 21, 2000. http://www.mindfully.org/GE/GE-Weird-Science.htm.

Caplan, Richard, and Skip Spitzer. "Regulation of Genetically Engineered Crops and Foods in the United States," March 2001, p. 3. http://www.gefoodalert.org/library/admin/uploadedfiles/Regulation_of_Genetically_Engineered_Crops_and.htm.

Carson, John. "The Merit of Science and the Science of Merit." In *States of Knowledge: The Co-Production of Science and Social Order*, edited by Sheila Jasanoff, pp. 181–205. London: Routledge, 2004.

Carson, Rachel. *Silent Spring.* New York: Houghton Mifflin, 1962.

"A Chancellor's Dilemma." *The Economist*, January 31, 1998: 54–55.

Christopher, John. *The Death of Grass.* London: Michael Joseph, 1956.

Cobb, Roger W., and Charles D. Elder. *Participation in American Politics: The Dynamics of Agenda-Building.* Baltimore: Johns Hopkins University Press, 1972.

Cohen, Stanley N., et al. "Construction of Biologically Functional Bacterial Plasmids in Vitro." *Proceedings of the National Academy of Sciences* 70 (1973): 3240–3244.

Colley, Linda. *Britons: Forging the Nation 1707–1807.* New Haven: Yale University Press, 1992.

Collins, H. M. "The Place of the 'Core Set' in Modern Science: Social Contingency with Methodological Propriety in Science." *History of Science* 19 (1981): 6–19.

———. *Changing Order: Replication and Induction in Scientific Practice*. London: Sage Publications, 1985.

Commission of the European Communities. *Eurofutures: The Challenges of Innovation, The FAST Report*. London: Butterworths, 1984.

———. *European Governance: A White Paper*, COM (2001) 428, Brussels, July 27, 2001. http://europa.eu.int/eur-lex/en/com/cnc/2001/com2001_0428en01.pdf.

Comtex. "Greenpeace Dumps GM Corn at Whitman's EPA Door," February 8, 2001, Environmental News Network. http://www.enn.com/news/wirestories/2001/02/02082001/greenpeace_41881.asp.

Connor, Steve. "Gene Scientist 'Sacked without Warning.'" *Independent*, March 18, 1994, p. 5.

"Controversial Proposal on Fourth Criterion in Commission Pipeline." *European Report*, January 26, 1991: 3–5.

Cook-Deegan, Robert M. "The Human Genome Project." In *Biomedical Politics*, edited by Kathi E. Hanna, pp. 148–149. Washington, DC: National Academy Press, 1991.

Crawford, Mark. "California Field Test Goes Forward." *Science* 236, 4801 (1987): 511.

———. "RAC Recommends Easing Some Recombinant DNA Guidelines." *Science* 235, 4790 (1987): 740–741.

Crichton, Michael. *Jurassic Park: A Novel*. New York: Knopf, 1990.

Crick, Francis H.C. *What Mad Pursuit: A Personal View of Scientific Discovery*. London: Weidenfeld and Nicolson, 1989.

Cronon, William, ed.. *Uncommon Ground: Rethinking the Human Place in Nature*. New York: Norton, 1996.

Cussins, Charis. "Ontological Choreography: Agency through Objectification in Infertility Clinics." *Social Studies of Science* 26 (1996): 575–610.

Cvetkovich, Ann, and Douglas Kellner, eds. *Articulating the Global and Local: Globalization and Cultural Studies*, Boulder: Westview Press, 1997.

Daimler-Chrysler Special Report. "Moral Responsibility: Confronting the Past." http://www.daimlerchrysler.com/index_e.htm?/specials/zwangs/zwarb3_e.htm.

Dashefsky, Richard. "The High Road to Success: How Investing in Ethics Enhances Corporate Objectives." *Journal of Biolaw and Business* 6, 3 (2003): 3–7.

Dawkins, Richard. "Charles: Right or Wrong about Science?" Focus Special, *The Observer*, May 21, 2000, p. 21.

de Solla Price, Derek. *Little Science, Big Science*. New York: Columbia University Press, 1963.

Dennis, Michael A. "Reconstructing Sociotechnical Order: Vannevar Bush and US Science Policy." In *States of Knowledge: The Co-production of Science and Social Order*, edited by Sheila Jasanoff. 225–253. London: Routledge, 2004.

Department of Health, Ministry of Agriculture, Fisheries and Food, 1989. *Report of the Working Party on Bovine Spongiform Encephalopathy* (Southwood Committee Report).

Department of Health and Social Security. *Human Fertilisation and Embryology: A Framework for Legislation* (Cmnd. 259). London: HMSO, 1987.

Dickman, Steven. "Germany Edges Towards Law." *Nature* 339, 6223 (1989): 327.

———. "New Law Needs Changes Made." *Nature* 343 (1990): 298.

Dickson, David. "British Government Rekindles Debate on Embryo Research." *Science* 238 (1987): 1348.

———. "German Firms Move into Biotechnology." *Science* 218 (1982): 1287–1289.

Dixon, Bernard. "Who's Who in European Antibiotech." *Bio/Technology* 11 (1993): 44–48.

Douglas, Mary. *Purity and Danger: An Analysis of Concepts of Pollution and Taboo*. London: Routledge and Kegan Paul, 1966.

———. *How Institutions Think*. Syracuse, NY: Syracuse University Press, 1986.

Drayton, Richard. *Nature's Government: Science, Imperial Britain, and the Improvement of the World*. New Haven: Yale University Press, 2000.

Duster, Troy. *Backdoor to Eugenics*. New York: Routledge, 1990.

Earnshaw, David, and David Judge. *The European Parliament*. Houndmills, Hamps.: Palgrave Macmillan, 2003.

Eder, Klaus. "Sustainability as a Discursive Device for Mobilizing European Publics." In *Environmental Politics in Southern Europe*, edited by Klaus Eder and Maria Kousis, pp. 25–52. Dordrecht: Kluwer, 2001.

———. "Zur Transformation nationalstaatlicher Öffentlichkeit in Europa: von der Sprachgemeinschaft zur inspezifischen Kommunikationsgemeinschaft." *Berliner Journal für Soziologie* 10, 2 (2000):167–184.

Eder, Klaus, and Cathleen Kantner. "Transnationale Resonanzstrukturen in Europa: Eine Kritik der Rede von Öffentlichkeitsdefizit." In *Die Europäisierung nationaler Gesellschaften*, edited by Maurizio Bach., pp. 306–331. Wiesbaden: Westdeutscher Verlag, 2000.

Eder, Klaus, and Maria Kousis, eds. *Environmental Politics in Southern Europe*. Dordrecht: Kluwer, 2001.

Eisenberg, Rebecca. "Genes, Patents, and Product Development." *Science* 257 (1992): 903–908.

"Elfter Bericht nach Inkrafttreten des Gentechnikgesetzes (GenTG) fur den Zeitraum 1.1.2000 bis 31.12.2000." *Bundesgesundheitsblatt-Gesundheitsforschung-Gesundheitsschutz* 9 (2001): 929–941.

Ellul, Jacques. *The Technological Society*. New York: Vintage Books, 1964.

Enserink, Martin. "Preliminary Data Touch off Genetic Food Fight." *Science* 283 (1999): 1094–1095.

Epstein, Steven. *Impure Science: AIDS, Activism, and the Politics of Knowledge*. Berkeley: University of California Press, 1996.

Ernst and Young. *Eighth Annual European Life Sciences Report 2001*. London: Ernst and Young International, 2001.

Escudier, Alexandre, Brigitte Sauzay, and Rudolf von Thadden, eds. *Gedenken im Zwiespalt: Konfliktlinien europäischen Erinnerns*. Göttingen: Wallstein, 2001.

European Commission. *The EC-US Task Force on Biotechnology Research—Mutual Understanding: A Decade of Collaboration 1990–2000*. Brussels: European Communities, 2000.

———. *Final Report of a Mission Carried out in Germany from 12 March 2001 to 16 March 2001 in Order to Evaluate Official Control Systems on Foods Consisting of or Produced from Genetically Modified Organisms* (GMOs., DG(SANCO)/3233/ 2001-MR final (2001).

————. "Promoting the Competitive Environment for the Industrial Activities Based on Biotechnology within the Community." SEC(91)629 final, Brussels, April 19, 1991.

————. Quality of Life Programme, *Eurobarometer 52.1—The Europeans and Biotechnology*. http://europa.eu.int/comm/research/quality-of-life/eurobarometer.html.

European Parliament. Scientific and Tecnological Options Assessment, Bioethics in Europe. Final Report, Luxembourg, October 1992, pp. 100–106.

Evans, Peter, Dietrich Rueschemeyer, and Theda Skocpol, eds. *Bringing the State Back In*. Cambridge: Cambridge University Press, 1985.

Ezrahi, Yaron. *The Descent of Icarus: Science and the Transformation of Contemporary Democracy*. Cambridge: Harvard University Press, 1990.

————. "Science and Political Imagination in Contemporary Democracies." In *States of Knowledge: The Co-Production of Science and Social Order*, edited by Sheila Jasanoff. London: Routledge, 2004.

————."Science and Utopia in Late 20th Century Pluralist Democracy." In *Nineteen Eighty-Four: Science between Utopia and Dystopia. Sociology of the Sciences Yearbook*. Vol. 8, edited by Everett Mendelsohn and Helga Nowotny, pp. 273–290. Dordrecht: Reidel, 1984.

Ezrahi, Yaron, Everett Mendelsohn, and Howard Segal, eds. *Technology, Pessimism, and Postmodernism*. Dordrecht: Kluwer, 1994.

Falk, Richard. *This Endangered Planet: Prospects and Proposals for Human Survival*. New York: Vintage Books, 1971.

Fallows, James. "The Political Scientist." *New Yorker*, June 7, 1999, pp. 66–75.

Featherstone, Mike. *Undoing Culture: Globalization, Postmodernism and Identity*. London: Sage Publications, 1995.

Food and Drink Federation. "Food for Our Future." *Feedback*. http://www.foodfuture.org.uk/answer.htm.

Foucault, Michel. *The Archaeology of Knowledge*. New York: Harper and Row, 1976.

————. *Discipline and Punish: The Birth of the Prison*. New York: Random House, 1979.

————. *The History of Sexuality*. New York: Pantheon, 1978.

————. *Madness and Civilization: A History of Insanity in the Age of Reason*. New York: Vintage Books, 1973.

————. *The Order of Things: An Archaeology of the Human Sciences*. New York: Pantheon, 1970.

————. *Power/Knowledge: Selected Interviews and Other Writings 1972–1977*. New York: Pantheon, 1980.

Frederickson, Donald S. "Asilomar and Recombinant DNA: The End of the Beginning." In *Biomedical Politics*, edited by Kathi E. Hanna, pp. 258–292. Washington, DC: National Academy Press, 1991.

————. *The Recombinant DNA Controversy: A Memoir: Science, Politics, and the Public Interest 1974–1981*. Washington, DC: ASM Press, 2001.

Friedland, Roger, and Robert A. Alford. "Bringing Society Back In: Symbols, Practices, and Institutional Contradictions." In *The New Institutionalism in Organizational Analysis*, edited by Walter W. Powell and Paul J. DiMaggio, p. 251. Chicago: University of Chicago Press, 1991.

Fukuyama, Francis. *The End of History and the Last Man*. New York: Penguin, 1992.

―――. *Our Posthuman Future: Consequences of the Biotechnology Revolution.* New York: Picador, 2002.

Fuller, Steve. *The Governance of Science.* Buckingham, UK: Open University Press, 2000.

Fuller, Thomas. "A Blunt Appraisal of EU's Laggards." *International Herald Tribune,* January 13, 2004, p. 1.

Funtowicz, Silvio O., and Jerome R. Ravetz. "Three Types of Risk Assessment and the Emergence of Post Normal Science." In *Social Theories of Risk,* edited by Sheldon Krimsky and David Golding, pp. 251–273. London: Praeger, 1992.

GAEIB. "The Ethical Aspects of the 5$^{th}$ Research Framework Programme." Opinion No. 10, December 11, 1997. http://europa.eu.int/comm/european_group_ethics/gaieb/en/opinion10.pdf.

Gaskell, George, Nick Allum, and Sally Stares, "Europeans and Biotechnology in 2002." Report to the EC Directorate General for Research, 2nd edition, March 21, 2003.

Geertz, Clifford. *Available Light: Anthropological Reflections on Philosophical Topics.* Princeton: Princeton University Press, 2000.

―――. *The Interpretation of Cultures: Selected Essays.* New York: Basic Books, 1973.

Gen-Ethischer Informationsdienst. *Collected Edition.* Issues 0–30, Berlin (1987).

Gerhards, Jürgen. "Westeuropäische Integration und die Schwierigkeiten der Entstehung einer Europäischen Öffentlichkeit." *Festschrift für Soziologie* 22 (1993): 96–110.

"German GM Wheat Trials Approved but Site Sabotaged," Hamburg, April 11, 2003. http://www.planetark.org/dailynewsstory.cfm/newsid/20444/newsDate/11-Apr-2003/story.htm.

Gersema, Emily. "FDA Opts Against Further Biotech Review." *Associated Press Online,* June 17, 2003.

Gerth, Hans H., and C. Wright Mills, eds. *From Max Weber: Essays in Sociology.* New York: Oxford University Press, 1946.

Gibbons, Ann. "Biotech Pipeline: Bottleneck Ahead." *Science* 254 (1991): 369–370.

―――. "Can David Kessler Revive the FDA?" *Science* 252 (1991): 200–201.

―――. "Kessler Gives FDA a Facelift." *Science* 255 (1992): 1350.

Gibbons, Michael, et al. *The New Production of Knowledge: The Dynamics of Science and Research in Contemporary Societies.* London: Sage, 1994.

Gieryn, Thomas. "Boundaries of Science." In *The Handbook of Science and Technology Studies,* edited by Sheila Jasanoff et al., pp. 393–456. Thousand Oaks, CA: Sage, 1995.

―――. *Cultural Boundaries of Science: Credibility on the Line.* Chicago: University of Chicago Press, 1999.

Gilbert, G. Nigel, and Michael Mulkay. *Opening Pandora's Box: A Sociological Analysis of Scientists' Discourse.* Cambridge: Cambridge University Press, 1984.

Gillespie, Brendan, Dave Eva, and Ron Johnston. "Carcinogenic Risk Assessment in the United States and Great Britain." *Social Studies of Science* 9 (1979): 265–301.

Glickman, Dan, and Vin Weber. "Frankenfood Is Here to Stay. Let's Talk." *International Herald Tribune,* July 1, 2003, p. 9.

Glover, Jonathan. *What Sort of People Should There Be? Genetic Engineering, Brain Control and Their Impact on Our Future World.* Middlesex, UK: Penguin, 1984.

Goffman, Erving. *Frame Analysis: An Essay on the Organization of Experience*. Cambridge: Harvard University Press, 1974.

Goldhagen, Daniel J. *Hitler's Willing Executioners: Ordinary Germans and the Holocaust*. New York: Vintage, 1997.

Goldmann, Kjell. *Transforming the European Nation-State*. London: Sage, 2001.

Gonçalves, Maria Eduarda. "The Importance of Being European: The Science and Politics of BSE in Portugal." *Science, Technology, and Human Values* 25, 4 (2000): 417–448.

Gottweis, Herbert. *Governing Molecules: The Discursive Politics of Genetic Engineering in Europe and the United States*. Cambridge: MIT Press, 1998.

Grant, Linda. "Yuk Factor or SPUC Factor?" *Independent on Sunday*, July 17, 1994, p. 22.

Greenberg, Daniel S. *Science, Money, and Politics: Political Triumph and Ethical Erosion*. Chicago: University of Chicago Press, 2001.

Grobstein, Clifford. *A Double Image of the Double Helix: The Recombinant DNA Debate*. San Francisco: Freeman, 1979.

Grove-White, Robin. "New Wine, Old Bottles? Personal Reflections on the New Biotechnology Commissions." *Political Quarterly* 72, 4 (October 2001): 466–472.

Grünen, Die. *Erklärung zur Gentechnologie und zur Fortpflanzungs- und Gentechnik am Menschen*. Hagen, February 15–16, 1986.

Gusfield, Joseph. *The Culture of Public Problems: Drinking-Driving and the Symbolic Order*. Chicago: University of Chicago Press, 1981.

Guston, David. *Between Politics and Science: Assuring the Integrity and Productivity of Research*. New York: Cambridge University Press, 2000.

———. "Stabilizing the Boundary between U.S. Politics and Science." *Social Studies of Science* 29 (1999): 87–112.

Haas, Peter M., Robert O. Keohane, and Marc A. Levy, eds. *Institutions for the Earth: Sources of Effective International Environmental Protection*. Cambridge: MIT Press, 1993.

Habermas, Jürgen. *The Future of Human Nature*. Cambridge: Polity, 2003.

———. *Legitimation Crisis*. Boston: Beacon Press, 1975.

Hacking, Ian. *Representing and Intervening: Introductory Topics in the Philosophy of Natural Science*. Cambridge: Cambridge University Press, 1983.

———. *Rewriting the Soul: Multiple Personality and the Sciences of Memory*. Princeton: Princeton University Press, 1995.

———. *The Social Construction of What?* Cambridge: Harvard University Press, 1999.

———. *The Taming of Chance*. Cambridge: Cambridge University Press, 1990.

———. "World-Making by Kind-Making: Child Abuse for Example." In *How Classification Works: Nelson Goodman among the Social Sciences*, edited by Mary Douglas and David Hull, pp. 180–238. Edinburgh: Edinburgh University Press, 1992.

Hajer, Maarten. *The Politics of Environmental Discourse*. Oxford: Oxford University Press, 1995.

Haraway, Donna. *Primate Visions: Gender, Race, and Nature in the World of Modern Science*. New York: Routledge, 1989.

———. *Simians, Cyborgs, and Women: The Reinvention of Nature*. New York: Routledge, Chapman, and Hall, 1991.

Hardt, Michael, and Antonio Negri. *Empire*. Cambridge: Harvard University Press, 2000.

Hart, David M. *Forged Consensus: Science, Technology, and Economic Policy in the United States, 1921–1953*. Princeton: Princeton University Press, 1998.

Hartouni, Valerie. "Breached Birth: Anna Johnson and the Reproduction of Raced Bodies." In *Cultural Conceptions: On Reproductive Technologies and the Remaking of Life*, pp. 85–98. Minneapolis: University of Minnesota Press, 1997.

Heim, Susan. "Research for Autarky: The Contribution of Scientists to Nazi Rule in Germany." Max-Planck Institute for the History of Science, Berlin, 2001.

Heller, Michael A., and Rebecca Eisenberg. "Can Patents Deter Innovation? The Anticommons in Biomedical Research." *Science* 280, 5364 (1998): 698–791.

Henderson, Jennifer A., and John J. Smith. "Academia, Industry, and the Bayh-Dole Act: An Implied Duty to Commercialize." http://www.cimit.org/coi_part3.pdf.

Hessler, Uwe. "Schroeder's Reluctant Cabinet to Allow GMO Foods." Deutsche Welle, Germany. http://www.gene.ch/genet/2004/Feb/msg00061.html.

Hilgartner, Stephen. *Science on Stage: Expert Advice as Public Drama*. Stanford: Stanford University Press, 2000.

Hofschneider, Peter-Hans. "Grundlagenforschung und Industrie in Deutschland— warnendes Beispiel Gentechnologie." *Futura* (February 1996): 104–109.

Hord, Bill. "The Road Back: Prodigene and Other Biotech Companies Are Moving Ahead in an Environment of Increasing Fear of Crop Contamination." *Omaha World Herald*, January 19, 2003, p. 1d.

House of Lords. *Stem Cell Research—Report*. http://www.parliament.the stationery-office.co.uk/pa/ld200102/ldselect/ldstem/83/8301.htm.

Hunt, Bruce J. "Michael Faraday, Cable Telegraphy, and the Rise of Field Theory." *History of Technology* 13 (1991):1–19.

Hutton, Will, and Anthony Giddens, eds. *Global Capitalism*. New York: The New York Press, 2000.

Huxley, Aldous. *Brave New World: A Novel*. London: Chatto and Windus, 1932.

"Innovation and Competitiveness in European Biotechnology." Enterprise Papers no. 7, Eur-Op catalogue no. NB-40-01-690-EN-C. (2002).

Irwin, Alan, and Brian Wynne, eds. *Misunderstanding Science? The Public Reconstruction of Science and Technology*. Cambridge: Cambridge University Press, 1996.

Jasanoff, Sheila. "Acceptable Evidence in a Pluralistic Society." In *Acceptable Evidence: Science and Values in Hazard Management*, edited by Rachelle Hollander and Deborah Mayo, pp. 29–47. New York: Oxford University Press, 1991.

———. "Beyond Epistemology: Relativism and Engagement in the Politics of Science." *Social Studies of Science* 26, 2 (1996): 393–418.

———. "Citizens at Risk: Cultures of Modernity in Europe and the U.S." *Science as Culture*, 11, 3 (2002): 363–380.

———. "Civilization and Madness: The Great BSE Scare of 1996." *Public Understanding of Science* 6 (1997): 221–232.

———. "In a Constitutional Moment: Science and Social Order at the Millennium." In *Social Studies of Science and Technology: Looking Back, Ahead*, edited by Bernward Joerges and Helga Nowotny, pp. 155–180. Dordrecht: Kluwer, 2003.

———. "Cultural Aspects of Risk Assessment in Britain and the United States." In *The Social and Cultural Construction of Risk: Essays on Risk Selection and Perception*, edited by Branden B. Johnson and Vincent T. Covello, pp. 359–397. Dordrecht: Reidel, 1987.

————. "The Eye of Everyman: Witnessing DNA in the Simpson Trial." *Social Studies of Science* 28, 5–6 (1998): 713–740.

————. *The Fifth Branch: Science Advisers as Policymakers*. Cambridge: Harvard University Press, 1990.

————. "Image and Imagination: The Formation of Global Environmental Consciousness." In *Changing the Atmosphere: Expert Knowledge and Environmental Governance*, edited by Paul Edwards and Clark Miller, pp. 309–337. Cambridge: MIT Press, 2001.

————, ed. *Learning from Disaster: Risk Management after Bhopal*. Philadelphia: University of Pennsylvania Press, 1994.

————. "Ordering Life: Law and the Normalization of Biotechnology." *Politeia* 17, 62 (2001): 34–50.

————. "Product, Process, or Programme: Three Cultures and the Regulation of Biotechnology." In *Resistance to New Technology*, edited by Martin Bauer, pp. 311–331. Cambridge: Cambridge University Press, 1995.

————. *Risk Management and Political Culture*. New York: Russell Sage Foundation, 1986.

————. "Science and the Statistical Victim: Modernizing Knowledge in Breast Implant Litigation." *Social Studies of Science* 32, 1 (2002): 37–70.

————. *Science at the Bar: Law, Science, and Technology in America*. Cambridge: Harvard University Press, 1995.

————. "Science, Politics, and the Renegotiation of Expertise at EPA." *Osiris* 7 (1991): 195–217.

————. *States of Knowledge: The Co-Production of Science and Social Order*. London: Routledge, 2004.

————. "Technological Innovation in a Corporatist State: The Case of Biotechnology in the Federal Republic of Germany." *Research Policy* 14 (1985): 23–38.

————. "Technological Risk and Cultures of Rationality." In *Incorporating Science, Economics, and Sociology in Developing Sanitary and Phytosanitary Standards in International Trad*, by National Research Council, pp. 65–84. Washington, DC: National Academy Press, 2000.

Jasanoff, Sheila, and Brian Wynne. "Science and Decisionmaking." In *Human Choice and Climate Change*, edited by Steve Rayner and Elizabeth L. Malone, pp. 1–87. Washington, DC: Battelle Press, 1998.

Jasanoff, Sheila, and Marybeth Long Martello, eds. *Earthly Politics: Local and Global in Environmental Governance*. Cambridge: MIT Press, 2004.

Jasanoff, Sheila, et al., eds. *Handbook of Science and Technology Studies*. Thousand Oaks, CA: Sage Publications, 1995.

Jenkins, Simon. "Face it, the last thing we need is more scientists." *Times Higher Supplement*, September 11, 1998, pp. 19–20.

————. "This Constitutional Cloud-Cuckoo Land." *Times of London*, January 24, 2001, p. F16.

Joerges, Christian. "'Economic Order'—'Technical Realisation'—'the Hour of the Executive': Some Legal Historical Observations on the Commission White Paper on European Governance." In *Mountain or Molehill? A Critical Appraisal of the Commission White Paper on Governance*, edited by Christian Joerges, Yves Mény, and J.H.H. Weiler, pp. 128–129. Florence: European University Institute, 2001.

Joy, Bill. "Why the Future Doesn't Need Us." *Wired* 8.04 (April 2000): 1–11.

Judson, Horace F. *The Eighth Day of Creation*. New York: Simon and Schuster, 1979.

Kaczynski, Theodore. *The Unabomber Manifesto*. San Francisco: Jolly Roger Press, 1995.

Kagan, Robert. *Of Paradise and Power: America and Europe in the New World Order*. New York: Knopf, 2003.

Kahn, Patricia. "Germany's Gene Law Begins to Bite." *Science* 255 (1992): 524–526.

Kass, Leon. "The Wisdom of Repugnance." *New Republic* (June 2, 1997): 17–26.

Kaufman, Marc. "The Biotech Corn Debate Grows Hot in Mexico." *Washington Post*, March 25, 2002, p. A9.

Kay, Lily E. *The Molecular Vision of Life: Caltech, the Rockefeller Foundation, and the Rise of the New Biology*. New York: Oxford University Press, 1993.

———. *Who Wrote the Book of Life: A History of the Genetic Code*. Stanford: Stanford University Press, 2000.

Keller, Evelyn Fox . *The Century of the Gene*. Cambridge, MA: Harvard University Press, 2000.

———. *A Feeling for the Organism: The Life and Work of Barbara McClintock*, San Francisco: W. H. Freeman, 1983.

Keller, Rolf, Hans-Ludwig Günther, and Peter Kaiser, *Embryonenschutzgesetz* (Stuttgart: Kohlhammer, 1992).

Kelman, Stephen. *Regulating America, Regulating Sweden: A Comparative Study of Occupational Safety and Health Policy*. Cambridge: MIT Press, 1981.

Keohane, Robert O., and Joseph S. Nye, Jr. *Power and Interdependence*. New York: Longman, 2001.

Kessler, David A., et al. "The Safety of Foods Developed by Biotechnology." *Science* 256 (1992): 1747–1749, 1832.

Kevles, Daniel J. *The Baltimore Case: A Trial of Politics, Science, and Character*. New York: W. W. Norton, 1998.

———. *In the Name of Eugenics: Genetics and the Uses of Human Heredity*. Berkeley: University of California Press, 1985.

Kingdon, John W. *Agendas, Alternatives, and Public Policies*. 2d ed. New York: Longman's, 1995.

Kitcher, Philip. *Science, Truth, and Democracy*. Oxford: Oxford University Press, 2001.

Kleinman, Arthur. *Writing at the Margin: Discourse Between Anthropology and Medicine*. Berkeley: University of California Press, 1996.

Kleinman, Arthur, Renee C. Fox, and Allan M. Brandt, eds. *Bioethics and Beyond*. *Daedalus* 128, 4 (1999).

Klintman, Mikael. "Arguments Surrounding Organic and Genetically Modified Food Labelling: A Few Comparisons." *Journal of Environmental Policy and Planning* 4 (2002): 247–259.

Köhler, Georges, and César Milstein. "Continuous Cultures of Fused Cells Secreting Antibody of Predefined Specificity." *Nature* 256 (1975): 495–497.

Kohler-Koch, Beate. "The Commission White Paper and the Improvement of European Governance." In *Mountain or Molehill? A Critical Appraisal of the Commission White Paper on Governance*, edited by Christian Joerges, Yves Mény, and J.H.H. Weiler, pp. 177–184. Florence: European University Institute, 2001.

Kolata, Gina. "How Safe Are Engineered Organisms?" *Science* 229, 4708 (1985): 34–35.

REFERENCES

Kolinsky, Eva, ed. *The Greens in West Germany: Organisation and Policy Making*. Oxford: Berg, 1989.

Kopp, Vincent J. "Henry K. Beecher, M.D.: Contrarian (1904–1976)." *American Society of Anesthesiologists Newsletter* 63(9) (September 1999).

Krimsky, Sheldon. *Genetic Alchemy: The Social History of the Recombinant DNA Controversy*. Cambridge: MIT Press, 1982.

———. *Science in the Private Interest: How the Lure of Profits Has Corrupted the Virtue of Biomedical Research*. Lanham, MD: Rowman-Littlefield, 2003.

Krimsky, Sheldon, and Alonzo Plough. "The Release of Genetically Engineered Organisms into the Environment: The Case of Ice Minus." In *Environmental Hazards: Communicating Risks as a Social Process*, pp. 75–110. Dover, MA: Auburn House Publishing Company, 1988.

Kuhn, Thomas. *The Structure of Scientific Revolutions*. Chicago: University of Chicago Press, 1962.

Lake, Gordon. "Scientific Uncertainty and Political Regulation: European Legislation on the Contained Use and Deliberate Release of Genetically Modified (Micro)organisms." *Project Appraisal* 6, 1 (March 1991): 7–15.

Landy, Marc K., Marc J. Roberts, and Stephen R. Thomas. *The Environmental Protection Agency: Asking the Wrong Questions from Nixon to Clinton*. New York: Oxford University Press, 1994.

Latour, Bruno. "Drawing Things Together." In *Representation in Scientific Practice*, edited by Michael Lynch and Steve Woolgar. Cambridge: MIT Press, 1990.

———. *Science in Action: How to Follow Scientists and Engineers through Society*. Cambridge: Harvard University Press, 1987.

———. *We Have Never Been Modern*. Cambridge: Harvard University Press, 1993.

Laubicher, Manfred D. "Frankenstein in the Land of *Dichter* and *Denker*." *Science* 286 (1999): 1859–1860.

Law, John, and John Hassard. *Actor Network Theory and After*. Sociological Review Monographs. Oxford: Blackwell, 1999.

Leicht, Robert. "Ein Rat der Anständigen," *Die Zeit*, May 2, 2001.

Leopold, Aldo. *Game Management*. New York: Scribner's, 1933.

Lepkowski, Wil. "Biotech's OK Corral." *Science and Policy Perspectives* 13, July 9, 2002. http://www.cspo.org/s&pp/060902printer.html.

Levidow, Les. "The Oxford Baculovirus Controversy—Safely Testing Safety?" *Bioscience* 8, 45 (1995): 545–551.

Levidow, Les, and Joyce Tait. "The Greening of Biotechnology: GMOs as Environment-Friendly Products." *Science and Public Policy* 18, 5 (1991): 271–80.

Levidow, Les, et al. "Bounding the Risk Assessment of a Herbicide-Tolerant Crop." In *Coping with Deliberate Release: The Limits of Risk Assessment*, edited by Ad van Dommelen, pp. 81–102. Tilburg, NL: International Centre for Human Rights, 1996.

Liberatore, Angela. *The Management of Uncertainty: Learning from Chernobyl*. Amsterdam: Gordon and Breach, 1999.

Lilliston, Ben, and Ronnie Cummins. "Organic vs 'Organic': The Corruption of a Label." *The Ecologist* 28, 4 (July/August 1998): 195–199.

Lin, William, Gregory K. Price, and Edward Allen. "StarLink™: Where No Cry9C Corn Should Have Gone Before." *Choices* (Winter 2001–2002): 31–34.

Lippmann, Walter. *The Phantom Public*. New Brunswick: Transaction Publishers, 1993 [1925].

Litfin, Karen . *Ozone Discourses: Science and Politics in Global Environmental Cooperation*. New York: Columbia University Press, 1994.

Losey, John E., Linda S. Rayor, and Maureen E. Carter. "Transgenic Pollen Harms Monarch Larvae." *Nature* 399 (1999): 214.

McAdam, Doug , Sidney Tarrow and Charles Tilly. *Dynamics of Contention*. Cambridge: Cambridge University Press, 2001.

Mackenzie, Donald. *Inventing Accuracy: A Historical Sociology of Nuclear Missile Guidance*. Cambridge: MIT Press, 1990.

McLaren, Anne. "IVF: Regulation or Prohibition?" *Nature* 342 (1989): 469–470.

MacLeod, Roy. "Science and Democracy: Historical Reflections on Present Discontents," *Minerva* 35 (1997): 369–384.

McNaghten, Phil. "Animals in their Nature," *Sociology* 38,3 (2004): 533–551.

Magnette, Paul. "European Governance and Civic Participation: Can the European Union Be Politicised?" In *Mountain or Molehill? A Critical Appraisal of the Commission White Paper on Governance*, edited by Christian Joerges, Yves Mény, and J.H.H. Weiler, pp. 24–25. Florence: European University Institute, 2001.

Maier, Charles S. *The Unmasterable Past: History, Holocaust, and German National Identity*. Cambridge: Harvard University Press, 1988.

Marcus, George E. *Ethnography Through Thick and Thin*. Princeton: Princeton University Press, 1998.

Marks, Kathy. "Widow Wins Fight to Bear Child of Dead Husband." *The Daily Telegraph*, February 7, 1997, p. 6.

Marris, Claire, et al. *Public Perceptions of Agricultural Biotechnologies in Europe (PABE)*, Final Report of the PABE Research Project, Contract number: FAIR CT98-3844 (DG12—SSMI), Lancaster University, December 2001.

Marshall, Eliot. "The Genome Program's Conscience." *Science* 274 (1996): 488–490.

———. "NIMH to Screen Studies for Science and Human Risks." *Science* 283 (1999): 464–465.

Marx, Leo. "The Idea of 'Technology' and Postmodern Pessimism." In *Does Technology Drive History: The Dilemma of Technological Determinism*, edited by Merritt Roe Smith and Leo Marx, pp. 237–257. Cambridge: MIT Press, 1994.

Masood, Ehsan. "Gag on Food Scientists Lifted as Gene Modification Row Hots Up." *Nature* 397 (1999): 547.

May, Robert M. "The Scientific Wealth of Nations." *Science* 275 (1997): 793–796.

Medrano, Juan Díez. *Framing Europe: Attitudes toward European Integration in Germany, Spain, and the United Kingdom*. Princeton: Princeton University Press, 2003.

Merton, Robert K. "The Normative Structure of Science." In *The Sociology of Science: Theoretical and Empirical Investigations*, edited by Norman W. Storer, pp. 267–278. Chicago: University of Chicago Press, 1973.

Meyer, Peter. "Regulations for the Release of Transgenic Plants According to the German Gene Act and Their Consequences for Basic Research." *AgBiotech News and Information* 3, 6 (1991): 999–1001.

Meyer, Peter, et al. "Endogenous and Environmental Factors Influence 35S Promoter Methylation of a Maize A1 Gene Construct in Transgenic Petunia and Its Color Phenotype." *Molecular and General Genetics* 231 (1991): 345–352.

Meyer, Peter, et al. "A New Petunia Flower Colour Generated by Transformation of a Mutant with a Maize Gene." *Nature* 330 (1987): 677–678.

Miller, Henry I. "The Big Fed Freeze." *National Review On Line*, April 4, 2002. http://www.nationalreview.com/comment/comment-miller040402.asp.

Miller, Henry I., et al. "Risk-Based Oversight of Experiments in the Environment." *Science* 250 (1990): 490–491.

Monbiot, George. "The Fake Persuaders: Corporations are inventing people to rubbish their opponents on the internet." *The Guardian*, May 14, 2002, p. 15.

Monsanto, "Bt Corn and the Monarch Butterfly," Biotech Knowledge Center. http://www.biotechknowledge.monsanto.com/biotech/knowcenter.nsf/ f055f4dc645999ad86256ac4000e6b68/0231086dd38f9a3d86256af6005433ae?Open Document.

Moravcsik, Andrew. "If It Ain't Broke, Don't Fix It." *Newsweek*, March 4, 2002, p. 15.

Morris, James. *Pax Britannica: The Climax of an Empire*. New York: Harcourt, Brace and World, 1968.

Mukerji, Chandra. *A Fragile Power: Scientists and the State*. Princeton: Princeton University Press, 1989.

Mulkay, Michael. *The Embryo Research Debate: Science and the Politics of Reproduction*. Cambridge: Cambridge University Press, 1997.

———. "The Triumph of the Pre-Embryo: Interpretations of the Human Embryo in Parliamentary Debates over Embryo Research." *Social Studies of Science* 24 (1994): 611–639.

Müller-Hill, Benno. *Murderous Science: Elimination by Scientific Selection of Jews, Gypsies, and Others in Germany, 1933–1945*. Cold Spring Harbor: Cold Spring Harbor Press, 1998.

———. *Tödliche Wissenschaft: die Aussonderung von Juden, Zigeunern und Geisteskranken 1933–1945*. Reinbek bei Hamburg: Rowohlt, 1984.

Murphy, Walter F., and Joseph Tanenhaus. *Comparative Constitutional Law Cases and Commentaries*. New York: St. Martin's Press, 1977.

Murray, Thomas. *Journal of the American Medical Association* 286 (2001): 2331–2332.

National Organic Program, Regulatory Impact Assessment for Proposed Rules Implementing the Organic Foods Production Act of 1990. http://www.ams.usda.gov/nop/ archive/ProposedRule/RegImpAssess.html.

National Research Council. *Field Testing Genetically Modified Organisms—Framework for Decisions*. Washington, DC: National Academy Press, 1989.

———. *Mapping and Sequencing the Human Genome*. Washington, DC: National Academy Press, 1988.

National Science Foundation, *Science and Engineering Indicators 2002*. http://www.nsf.gov/sbe/srs/seind02.

NBAC. *Ethical Issues in Stem Cell Research*. Washington DC, September 1999.

Nelkin, Dorothy, and Susan M. Lindee. *The DNA Mystique: The Gene as a Cultural Icon*. New York: Freeman, 1995.

Nelkin, Dorothy, and Michael Pollak. *The Atom Besieged: Extraparliamentary Dissent in France and Germany*. Cambridge: MIT Press, 1981.

Nowotny, Helga. "Knowledge for Certainty: Poverty, Welfare Institutions and the Institutionalization of Social Science." *Discourses on Society: The Shaping of the Social Science Disciplines* 15 (1990): 23–41.

Nowotny, Helga, Peter Scott, and Michael Gibbons. *Re-Thinking Science: Knowledge and the Public in an Age of Uncertainty*. Cambridge: Polity, 2001.

Nuffield Council on Bioethics. *Annual Report 1991–1992*. London: Nuffield Foundation, 1992.

———. *The Ethics of Patenting DNA*. London: Nuffield Foundation, 2002.

———. *Genetically Modified Crops: The Ethical and Social Issues*. London: Nuffield Foundation, 1999.

Nye, Joseph S., Jr., and John D. Donahue, eds. *Governance in a Globalizing World*. Washington, DC: Brookings Institution Press, 2000.

Obeyesekere, Gananath. *The Apotheosis of Captain Cook: European Mythmaking in the Pacific*. Princeton: Princeton University Press, 1992.

Palevitz, Barry A., and Ricki Lewis. "Perspective: Fears or Facts? A Viewpoint on GM Crops." *The Scientist* 13, 20 (October 11, 1999): 10.

Parens, Erik. "Respect for the RAC." Letter, *Science* 272 (1996): 1569–1570.

PCBE. Executive Summary, *Human Cloning and Human Dignity: An Ethical Inquiry*, July 2002.

Perrow, Charles. *Normal Accidents: Living with High-Risk Technologies*. New York: Basic Books, 1984.

Petersen, Melody. "Uncoupling Campus and Company." *New York Times*, September 23, 2003, p. F2.

Philip, Kavita. "Imperial Science Rescues a Tree: Global Botanic Networks, Local Knowledge, and the Transcontinental Transplantation of Cinchona." *Environment and History* 1 (1995): 173–200.

*Phillips Inquiry Report*. http://www.bseinquiry.gov.uk/report/volume6/chapt413.htm.

Pinch, Trevor. "'Testing—One, Two, Three . . . Testing!': Toward a Sociology of Testing." *Science, Technology, and Human Values* 18, 1 (1993): 25–41.

Polanyi, Michael. "The Republic of Science." *Minerva* 1 (1962): 54–73.

Porritt, Jonathon. "Down-to-Earth Agenda; Suggestions to Mrs. Thatcher." *The Times (London)*, September 27, 1988.

Porter, Theodore M. *The Rise of Statistical Thinking 1820–1990*. Princeton: Princeton University Press, 1986.

———. *Trust in Numbers: The Pursuit of Objectivity in Science and Public Life*. Princeton: Princeton University Press, 1995.

President's Commission for the Study of Ethical Problems in Medicine and Biomedical and Behavioral Research. *Splicing Life: A Report on the Social and Ethical Issues of Genetic Engineering with Human Beings*. Washington, DC: The Commission, 1982.

Press, Eyal, and Jennifer Washburn. "The Kept University." *Atlantic Monthly* (March 2000): 39–54.

Price, Frances. "Now You See It, Now You Don't: Mediating Science and Managing Uncertainty in Reproductive Medicine." In *Misunderstanding Science? The Public Reconstruction of Science and Technology*, edited by Alan Irwin and Brian Wynne, pp. 84–106. Cambridge: Cambridge University Press, 1996.

Prince Charles, "My 10 Fears for GM Food." *The Daily Mail*, June 1, 1999, pp. 10–11.

Proctor, Robert N. *The Nazi War on Cancer*. Princeton: Princeton University Press, 1999.

———. *Racial Hygiene: Medicine under the Nazis*. Cambridge: Harvard University Press, 1988.

Quist, David, and Ignacio H. Chapela. "Transgenic DNA Introgressed into Traditional Maize Landraces in Oaxaca, Mexico." *Nature* 414 (2001): 541–543.

Rabinow, Paul. *Essays on the Anthropology of Reason*. Princeton: Princeton University Press, 1996.

———, ed. *The Foucault Reader*. New York: Random House, 1984.

———. *French DNA: Trouble in Purgatory*. Chicago: University of Chicago Press, 1999.

———. *Making PCR*. Chicago: University of Chicago Press, 1996.

Rai, Arti K., and Rebecca Eisenberg. "Bayh-Dole Reform and the Progress of Biomedicine." *American Scientist* 91 (2003): 52–59.

Reardon, Jennifer. "The Human Genome Diversity Project: A Case Study in Coproduction." *Social Studies of Science* 31 (2001): 357–388.

Reich, Michael. *Toxic Politics: Responding to Chemical Disaster*. Ithaca: Cornell University Press, 1991.

Rensberger, Boyce. "Making a Pink Petunia Turn Red." *Washington Post*, December 21, 1987, p. A3.

*Report of the Parliamentary Commission of Enquiry on Prospects and Risks of Genetic Engineering, German Bundestag*. Bonn, January 1987.

Rhodes, Richard. *Deadly Feasts*. New York: Simon and Schuster, 1997.

Rifkin, Jeremy. *Algeny*. New York: Viking Press, 1983.

Rissler, Jane, and Margaret Mellon. "A Real Hot Tomato." *Washington Post*, August 14, 1993, p. A19.

Roberts, Leslie. "NIH Gene Patents, Round Two." *Science* 255 (1992): 912–913.

Rowland, Shelley, and Jared Scarlett, "The World's Most Litigated Mouse." *NZ Bio Science* 13 (February 2003). http://www.bsw.co.nz/articles/xfactor13.html.

Royal Commission on Environmental Pollution. *The Release of Genetically Engineered Organisms to the Environment*, Thirteenth Report. London: HMSO, 1989.

Rueschemeyer, Dietrich, and Theda Skocpol, eds. *States, Social Knowledge, and the Origins of Modern Social Policies*. Princeton: Princeton University Press, 1996.

Sachs, Wolfgang. *Planet Dialectics: Explorations in Environment and Development*. Halifax, Nova Scotia: Fernwood Publishing, 1999.

Sahlins, Marshall D. *How "Natives" Think: About Captain Cook, for Example*. Chicago: University of Chicago Press, 1995.

Sandel, Michael J. "The Case against Perfection." *Atlantic* (April 2004): 51–62.

Sbragia, Alberta, ed. *Euro-Politics: Institutions and Policymaking in the "New" European Community*. Washington, DC: Brookings Institution, 1992.

Schaffer, Simon. "Late Victorian Metrology and Its Instrumentation: A Manufactory of Ohms." In *Invisible Connections: Instruments, Institutions and Science*, edited by Robert Bud and Susan E. Cozzens, pp. 23–56. Bellingham, WA: SPIE Optical Engineering Press, 1992.

Schiermeier, Quirin. "German Transgenic Crop Trials Face Attack." *Nature* 394 (1998): 819.

Schmemann, Serge. "The Coalition of the Unbelieving." *New York Times Book Review*, January 25, 2004, p. 12.

Schmitter, Philippe C. "What Is There to Be Legitimized in the European Union . . . and How Might This Be Accomplished?" Political Science Series, Institute for Advanced Studies, Vienna, May 2001.

Schön, Donald A., and Martin Rein. *Frame/Reflection: Toward the Resolution of Intractable Policy Controversies*. New York: Basic Books, 1994.

Schuller, Konrad. "Digging in the Dirt." *Frankfurter Allgemeine Zeitung* (English edition), April 6, 2002.

Schwartz, John. "FDA Clears Tomato with Altered Genes." *Washington Post*, May 19, 1994, p. A1.

Scientific and Technological Options Assessment (STOA). *Bioethics in Europe*, PE 158.453, Luxembourg, September 8, 1992.

"Scorpion has Sting in Tale," *The Splice of Life*, Bulletin of The Genetics Forum 1 (8/9) (May 1995).

Scott, James C. *Seeing Like a State: How Certain Schemes to Improve the Human Condition Have Failed*. New Haven: Yale University Press, 1998.

"Secondary Bioengineering Protein Product Permitted as Food Additive." *Food Drug Cosmetic Law Reports*, para. 40301 (1994).

SET Forum, "Shaping the Future: A Policy for Science, Engineering and Technology." Discussion Document for the Labour Party (1995).

Shapin, Steven. "Pump and Circumstance: Robert Boyle's Literary Technology." *Social Studies of Science* 14 (1984): 481–520

———. *A Social History of Truth*. Chicago: University of Chicago Press, 1994.

Shapin, Steven, and Simon Schaffer. *Leviathan and the Air-Pump: Hobbes, Boyle, and the Experimental Life*. Princeton: Princeton University Press, 1985.

Shapiro, Harold T. "Ethical Dilemmas and Stem Cell Research," *Science* 285 (1999): 2065.

Shiva, Vandana. *Biopiracy: The Plunder of Nature and Knowledge*. Toronto: Between The Lines, 1997.

———. *Monocultures of the Mind: Perspectives on Biodiversity and Biotechnology*. London: Third World Network, 1993.

———. *Yoked to Death: Globalisation and Corporate Control of Agriculture*. New Delhi: Research Foundation for Science, Technology and Ecology, 2001.

Silver, Lee. *Remaking Eden*. New York: Perennial, 2002.

Simm, Michael. "Violence Study Hits a Nerve in Germany." *Science* 264 (1994): 653.

Simon, Stephanie. "The Food Industry Loves Engineered Crops, but Not When Plants Altered to 'Grow' Drugs and Chemicals Can Slip into Its Products." *Los Angeles Times*, December 23, 2002, p. 1.

Singer, Maxine. "Genetics and the Law: A Scientist's View." *Yale Law and Policy Review* 3 (1985): 315–335.

———. "Hot Tomato." *Washington Post*, August 10, 1993, p. A15.

Singer, Peter. "On Being Silenced in Germany." *New York Review of Books* 38, 14 (August 15, 1991) pp. 36–42.

———. *Practical Ethics*. Cambridge: Cambridge University Press, 1979.

Skocpol, Theda. *Protecting Soldiers and Mothers: The Political Origins of Social Policy in the United States*. Cambridge: Harvard University Press, 1992.

Skolnikoff, Eugene. *The Elusive Transformation: Science, Technology, and the Evolution of International Politics*. Princeton: Princeton University Press, 1993.

355

REFERENCES

Slovic, Paul. "Beyond Numbers: A Broader Perspective on Risk Perception and Risk Communication." In *Acceptable Evidence: Science and Values in Risk Management*, edited by Deborah G. Mayo and Rachelle D. Hollander, pp. 48–65. New York: Oxford University Press, 1991.

Slovic, Paul, et al. "Characterizing Perceived Risks." In *Perilous Progress: Managing the Hazards of Technology*, edited by Robert W. Kates, Christoph Hohenemser, and Jeanne X. Kasperson, pp. 99–125. Boulder: Westview, 1985.

———. "Facts and Fears: Understanding Perceived Risk." In *Societal Risk Assessment: How Safe is Safe Enough?* edited by R. Schwing and W. A. Albers, Jr., pp. 181–214. New York: Plenum, 1980.

Smith, Crosbie, and M. Norton Wise. *Energy and Empire: A Biographical Study of Lord Kelvin.* Cambridge: Cambridge University Press, 1989.

Smith, Merritt Roe, and Leo Marx, eds. *Does Technology Drive History?: The Dilemma of Technological Determinism.* Cambridge: MIT Press, 1994.

The Social Learning Group. *Learning to Manage Global Environmental Risks: A Comparative History of Social Responses to Climate Change, Ozone Depletion, and Acid Rain.* Cambridge: MIT Press, 2001.

Solingen, Etel. "Between Markets and the State: Scientists in Comparative Perspective." *Comparative Politics* 26 (1993): 31–51.

Specter, Michael. "The Dangerous Philosopher." *New Yorker*, September 6, 1999, pp. 46–55.

Sperling, Stefan. "Managing Potential Selves: Stem Cells, Immigrants, and German Identity." *Science and Public Policy* 39, 2 (2004): 139–149.

Stafford, Ned. "GM Crop Sites Stay Secret." *The Scientist.* May 28, 2004. www.biomedcentral.com/news/20040528/02.

Star, Susan Leigh and James R. Griesemer. "Institutional Ecology, 'Translations' and Boundary Objects: Amateurs and Professionals in Berkeley's Museum of Vertebrate Zoology, 1907–39." *Social Studies of Science* 19 (1989): 387–420.

Starkey, David. *Elizabeth.* London: Vintage, 2001.

Stein, George. "Biological Science and the Roots of Nazism." *American Scientist* 76 (1988): 50–58.

Stern, Fritz. *Einstein's German World.* Princeton: Princeton University Press, 2001.

Stern, Paul, and Harvey Fineberg, eds. *Understanding Risk.* Washington, DC: National Academy Press, 1996.

Stokes, Donald E. *Pasteur's Quadrant: Basic Science and Technological Innovation.* Washington, DC: Brookings Institution, 1997.

Stolberg, Sheryl Gay. "The Biotech Death of Jesse Gelsinger." *New York Times*, Sunday Magazine, November 28, 1999. p. 137.

Stone, Richard. "Religious Leaders Oppose Patenting Genes and Animals." *Science* 268 (1995): 1126.

Storey, William K. *Science and Power in Colonial Mauritius.* Rochester: University of Rochester Press, 1997.

Straw, Jack. "By Invitation." *Economist* July 10, 2004, p. 40.

Sugarman, Jeremy. "Ethical Considerations in Leaping from Bench to Bedside." *Science* 285: 2071–2072.

Sugawara, Sandra. "For the Next Course, 'Engineered' Entrees? 'Genetic' Tomato May Launch an Industry." *Washington Post*, June 10, 1992, p. F1.

Sunder Rajan, Kaushik. "Genomic Capital: Public Cultures and Market Logics of Corporate Biotechnology." *Science as Culture* 12, 1 (2003): 87–121.

Sunstein, Cass. *Risk and Reason*. Cambridge: Cambridge University Press, 2002.

Swazey, Judith, et al. "Risks and Benefits, Rights and Responsibilities: A History of the Recombinant DNA Research Controversy." *Southern California Law Review* 51 (1978): 1019–1078.

Tarrow, Sidney. *Power in Movement: Social Movements and Contentious Politics*. 2d ed. Cambridge: Cambridge University Press, 1998.

Taylor, Michael R., and Jody S. Tick. "The StarLink Case: Issues for the Future." Pew Initiative on Food and Biotechnology and Resources for the Future (October 2001).

Thomas, Jo. "British Debate Embryo Research." *New York Times*. October 16, 1984, p. 6.

Tickell, Oliver. "Scorpion Gene Virus Experiment Abandoned." *Pesticides News* 25 (September 1994): 21.

Tickner, Joel A. ed., *Precaution: Environmental Science and Preventive Public Policy*. Washington, DC: Island Press, 2003.

Tokar, Brian. "Resisting the Engineering of Life." In *Redesigning Life? The Worldwide Challenge to Genetic Engineering*, edited by Brian Tokar. London: Zed Books, 2001, pp. 320–336.

Tribe, Laurence. "Clone as Outlaw? Reasons *Not* to Ban 'Unnatural' Ways of Making Babies." In *Clones and Clones*, edited by Martha C. Nussbaum and Cass R. Sunstein, pp. 223–234. New York: Norton, 1998.

Turney, Jon. "Public Understanding of Science." *Lancet* 347 (1996): 1087–1090.

Tyler, Christian. "Private View: Professor with Killer Gene Blues." *Financial Times*, April 8, 1995, p. 18.

U.K. Government. *The BSE Inquiry: The Report*. http://www.bseinquiry.gov.uk/report/volume4/chapterb.htm#886837.

U.S. Congress, Office of Technology Assessment, *Commercial Biotechnology: An International Analysis* (Washington, DC: U.S. GPO, 1984).

U.S. House of Representatives, Committee on Government Reform (Minority Report), *Politics and Science in the Bush Administration*, Washington, DC, August 7, 2003. http://www.house.gov/reform/min/politicsandscience/index.htm.

Vogel, David. "Consumer Protection and Protectionism in Japan." *The Journal of Japanese Studies* 18 (1992): 119–154.

———. "The Hare and the Tortoise Revisited: The New Politics of Consumer and Environmental Regulation in Europe." *British Journal of Political Science* 33 (2003): 557–580.

———. *National Styles of Regulation: Environmental Policy in Great Britain and the United States*. Ithaca: Cornell University Press, 1986.

Vogel, Gretchen. "Capturing the Promise of Youth." *Science* 286 (1999): 2238–2240.

———. "NIH Sets Rules for Funding Embryonic Stem Cell Research." *Science* 286 (1999): 2050–2051.

———. "Study of HIV Transmission Sparks Ethics Debate." *Science* 288 (2000): 22–23.

———. "U.K. Backs Use of Embryos, Sets Vote." *Science* 289 (2000): 1269–1273.

von Beuzekom, Brigitte. *Biotechnology Statistics in OECD Member Countries: Compendium of Existing National Statistics*. STI Working Papers 2001/6 (OECD, 2001).

Wagner, Peter, Björn Wittrock, and Richard Whitley, eds., *Discourses on Society: The Shaping of the Social Science Disciplines*. Dordrecht: Kluwer, 1991.

Walker, Neil. "The White Paper in Constitutional Context." In *Mountain or Molehill? A Critical Appraisal of the Commission White Paper on Governance*, edited by Christian Joerges, Yves Mény, and J.H.H. Weiler, pp. 33–53. Florence: European University Institute, 2001.

Warnock, Mary. *A Question of Life: The Warnock Report on Human Fertilisation and Embryology*. Oxford: Blackwell, 1985.

Waterton, Claire, and Brian Wynne. "Knowledge and Political Order in the European Environment Agency." In *States of Knowledge: The Co-Production of Science and Social Order*, edited by Sheila Jasanoff, pp. 87–108. London: Routledge, 2004.

Watson, James D. "In Defense of DNA," *New Republic* 170 (June 25, 1977): 11.

———. *The Double Helix: A Personal Account of the Discovery of the Structure of DNA*. New York: Atheneum, 1968.

———. "Trying to Bury Asilomar." *Clinical Research* 26 (1978): 113.

Watson, James D., and Francis H. C. Crick. "Genetical Implications of the Structure of Deoxyribonucleic Acid," *Nature* 171 (1953): 964–967.

———. "Molecular Structure of Nucleic Acids: A Structure for Deoxyribonucleic Acid." *Nature* 171 (1953): 737–738.

Watson, James D., and John Tooze. *The DNA Story: A Documentary of Gene Cloning*. San Francisco: W. H. Freeman, 1981.

Watts, Susan. "Controversy in the Cabbage Patch," *Independent*, May 17, 1994, p. 15.

———. "Genetic Riddle of 'Scorpion' Pesticide Virus." *Independent*, September 4, 1994, p. 2.

———. "Genetics Row Fueled by Scorpion's Venom." *Independent*, May 17, 1994, p. 3.

———. "Legal Fight Planned to Halt Scorpion Toxin Test." *Independent*, May 18, 1994, p. 3.

———. "Safety Scare on Eve of Mutant Virus Test." *Independent*, June 26, 1994, p. 1.

———. "Warning: This Thing Isn't Natural." *Independent*, May 26, 1994, p. 20.

Weart, Spencer. *Nuclear Fear: A History of Images*, Cambridge: Harvard University Press, 1988.

Weijer, Charles, and Ezekiel J. Emanuel. "Protecting Communities in Biomedical Research." *Science* 289 (2000): 1142–1144.

Weingart, Peter, Jürgen Kroll and Kurt Bayertz. *Rasse, Blut und Gene: Geschichte der Eugenik und Rassenhygiene in Deutschland*. Frankfurt: Suhrkamp, 1992.

Weiss, Rick. "'Organic' Label Ruled Out For Biotech, Irradiated Food." *Washington Post*, May 1, 1998, p. A2.

Wilkie, Tom. "Whose Genes Are They Anyway?" *Independent*, May 6, 1991, p. 19.

Wilmut, Ian, et al. "Viable Offspring Derived from Foetal and Adult Mammalian Cells." *Nature* 385 (1997): 810–813.

Wilson, Graham. *The Politics of Safety and Health*. Oxford: Clarendon Press, 1985.

Winner, Langdon. *The Whale and the Reactor: A Search for Limits in an Age of High Technology*. Chicago: University of Chicago Press, 1986.

Worster, Donald. *Nature's Economy: A History of Ecological Ideas*. Cambridge: Cambridge University Press, 1977.

Wortmann, Michael. "Multinationals and the Internationalization of R&D: New Developments in German Companies." *Research Policy* 19 (1990): 175–183.

Wright, Shirley J. "Human Embryonic Stem-Cell Research: Science and Ethics." *American Scientist* 87 (1999): 352–361.

Wright, Susan. *Molecular Politics: Developing American and British Regulatory Policy for Genetic Engineering, 1972–1982.* Chicago: University of Chicago Press, 1994.

Wuerth, Andrea. "National Politics/Local Identities: Abortion Rights Activism in Post-Wall Berlin." *Feminist Studies* 25, 3 (1999): 601–632.

Wynne, Brian. "Creating Public Alienation: Expert Cultures of Risk and Ethics on GMOs." *Science as Culture* 10, 4 (2001): 445–481.

———. "The Prince and the GM debate: Performing the Monarchy as Culture." http://domino.lancs.ac.uk/csec/bn.NSF/0/ c3bbc73ca660b14f802569df005d6cdd?OpenDocument.

———. "Public Understanding of Science," In *The Handbook of Science and Technology Studies*, edited by Sheila Jasanoff et al., pp. 361–388. Thousand Oaks, CA: Sage Publications, 1995.

———. "Public Uptake of Science: A Case for Institutional Reflexivity." *Public Understanding of Science* 2 (1992): 321–337.

———. "Unruly Technology." *Social Studies of Science* 18 (1988): 147–167.

Wynne, Brian, et al. "Institutional Cultures and the Management of Global Environmental Risks in the United Kingdom." In *Learning to Manage Global Environmental Risks*, by The Social Learning Group, pp. 93–113. Cambridge: MIT Press, 2001.

Yeats, William Butler. "Among School Children." *The Collected Poems of W.B. Yeats.* New York: Macmillan, 1956.

Yoxen, Edward. *The Gene Business: Who Should Control Biotechnology?* London: Pan Books, 1983.

Zalewski, Daniel. "Ties That Bind." *Lingua Franca* (June/July 1997): 51–59.

Zielke, Anne. "Im Disneyland der Kindermacher." *Frankfurter Allgemeine Zeitung*, Feuilleton no. 67 (March 20, 2002): 49.

Ziman, John. *Public Knowledge.* Cambridge: Cambridge University Press, 1968.

ZKBS, *Bericht über die zurückliegende Amtsperiode der Zentralen Kommission für die Biologische Sicherheit, (29.01.81 bis 30.06.88).* Bonn, 1989.

# Index